HEALTH EF[illegible]
EXPOSU[illegible]
LOW LEVELS OF
IONIZING RADIATION

BEIR V

1 National Research Council

Committee on the Biological Effects
of Ionizing Radiations
Board on Radiation Effects Research
Commission on Life Sciences
National Research Council

NATIONAL ACADEMY PRESS
Washington, D.C. 1990

National Academy Press • 2101 Constitution Avenue, N.W. • Washington, D.C. 20418

NOTICE: The project that is the subject of this report was approved by the Governing Board of the National Research Council, whose members are drawn from the councils of the National Academy of Sciences, the National Academy of Engineering, and the Institute of Medicine. The members of the committee responsible for the report were chosen for their special competences and with regard for appropriate balance.

This report has been reviewed by a group other than the authors according to procedures approved by a Report Review Committee consisting of members of the National Academy of Sciences, the National Academy of Engineering, and the Institute of Medicine.

The study summarized in this report was supported by the Oak Ridge Associated Universities, acting for the Office of Science and Technology Policy's Committee on Interagency Radiation Research and Policy Coordination (CIRRPC) under Purchase Order No. C-43892.

International Standard Book Number 0-309-03995-9 (paper) 0-309-03997-5 (cloth)

Library of Congress Catalog Card Number 89-64118

Printed in the United States of America
First Printing, December 1989
Second Printing, June 1990

COMMITTEE ON THE BIOLOGICAL EFFECTS OF IONIZING RADIATIONS

ARTHUR C. UPTON, *Chairman*, Department of Environmental Medicine, New York University Medical Center, New York, New York
DANIEL L. HARTL, *Vice Chairman*, Department of Genetics, Washington University School of Medicine, St Louis, Missouri
BRUCE B. BOECKER, Inhalation Toxicology Research Institute, Albuquerque, New Mexico
KELLY H. CLIFTON, University of Wisconsin Clinical Cancer Center, Madison, Wisconsin
CARTER DENNISTON, Department of Genetics, University of Wisconsin, Madison, Wisconsin
EDWARD R. EPP, Division of Radiation Biophysics, Massachusetts General Hospital, Boston, Massachusetts
JACOB I. FABRIKANT, Donner Laboratory, Univerisity of California, Berkeley, California
DOUGLAS GRAHN, Argonne National Laboratory, Argonne, Illinois
ERIC J. HALL, Radiological Research Laboratory, Columbia University, New York, New York
DONALD E. HERBERT, Department of Radiology, University of South Alabama, Mobile, Alabama
DAVID G. HOEL, National Institute of Environmental Health Science, Research Triangle Park, North Carolina
GEOFFREY R. HOWE, National Cancer Institute of Canada, University of Toronto, Toronto, Ontario, Canada
SEYMOUR JABLON, National Cancer Institute, Bethesda, Maryland
ANN R. KENNEDY, Department of Radiation Oncology, Hospital of the University of Pennsylvania, Philadelphia, Pennsylvania
ALFRED G. KNUDSON, JR., Fox Chase Cancer Center, Philadelphia, Pennsylvania
DUNCAN C. THOMAS, Department of Preventive Medicine, University of Southern California, Los Angeles, California
DALE PRESTON, Scientific Advisor to the Committee, National Cancer Institute, Bethesda, Maryland

National Research Council Staff

WILLIAM H ELLETT, Study Director, Board on Radiation Effects Research, Commission on Life Sciences
RAYMOND D. COOPER, Senior Program Officer, Board on Radiation Effects Research, Commission on Life Sciences
RICHARD E. MORRIS, Editor, National Academy Press

Sponsor's Project Officer

WILLIAM A. MILLS, Oak Ridge Associated Universities

Preface

BACKGROUND

The National Research Council's committees on the Biological Effects of Ionizing Radiations (BEIR) have prepared a series of reports to advise the U.S. government on the health consequences of radiation exposures. The most recent of these reports "Health Risks of Radon and Other Internally Deposited Alpha-Emitters—BEIR IV" was published in 1988. The last BEIR report to address health effects from external sources of penetrating electromagnetic radiation such as x rays and gamma rays was the report by the BEIR III Committee, "The Effects on Populations of Exposure to Low Levels of Ionizing Radiation: 1980." That report relied heavily on the mortality experience of the Japanese A-bomb survivors from 1950 through 1974 as a basis for the risk estimates it contains. The need for replacement of the BEIR III report became obvious when it was determined that the long standing estimates of the radiation exposures received by the A-bomb survivors, that had been utilized by the BEIR III Committee, required extensive revision. Following a binational research program by U.S. and Japanese scientists, a reassessment of A-bomb dosimetry was largely completed in 1986 and a new program of survivor dose estimation was initiated by the Radiation Effects Research Foundation (RERF) at Hiroshima and Nagasaki. In addition, RERF scientists extended their follow-up of A-bomb survivor mortality through the year 1985.

In April of 1986, the Office of Science and Technology Policy's Committee on Interagency Radiation Research and Policy Coordination (CIRRPC) asked the National Research Council to form a new BEIR committee to report on the effect of ionizing radiations on the basis of the new information that was becoming available. A purchase agreement between the Oak Ridge Associated Universities, acting for CIRRPC, and the National Research Council to fund the BEIR V Committee was concluded in June of 1986.

CHARGE TO THE COMMITTEE

The new BEIR V Committee was asked by CIRRPC to conduct a comprehensive review of the biological effects of ionizing radiations focusing on information that had been reported since the conclusion of the BEIR III study, and to the extent that available information permitted, provide new estimates of the risks of genetic and somatic effects in humans due to low-level exposures of ionizing radiation. These risk estimates were to address both internal and external sources of radiation, and the procedure by which these risk estimates are derived was to be documented.

The Committee was also asked to discuss the uncertainty in their risk estimates and, where possible, quantitate these uncertainties including the consequences of any necessary assumptions. Finally, the Committee was asked to prepare a detailed final report of their findings in a form suitable for making health risk assessments and calculating the probability that an observed cancer may be due to radiation. The conclusions of the BEIR IV Committee concerning alpha particle emitters were to be summarized in this final report to an extent consistent with the BEIR V Committee's presentation, but additional review of the scientific literature on the effects on alpha particle radiation was not required. While the BEIR V Committee was asked to summarize radiation risk information in a way that is useful for formulating radiation control decisions, recommendations on standards or guidelines for radiation protection were specifically excluded under the terms of this study.

ORGANIZATION OF THE STUDY

To carry out the charge, the NRC appointed a committee of scientists experienced in radiation carcinogenesis, epidemiology, radiobiology, genetics, biostatistics, pathology, radiation dosimetry, radiology, mathematical modeling, and risk assessment. The study was conducted under the general guidance of the Board on Radiation Effects Research of the Commission on Life Sciences.

To facilitate its work and to augment its expertise so as to encompass a wider spectrum of scientific subjects, the Committee solicited specific contributions from a number of scientific experts other than its own members.

These experts participated in the Committee's deliberations throughout the course of its work.

The Committee held eight meetings over a period of 30 months—seven in Washington, D.C., and one in Woods Hole, Massachusetts. The second meeting, on March 2, and March 3, 1987 included a public meeting, at which open discussion and contributions from interested scientists and the public at large were invited. In addition, over a dozen meetings of subgroups of the Committee were held to plan and carry out specific work assignments.

The Committee organized its work according to the main objectives of the charge and divided the study into the following categories:

- Heritable genetic effects.
- Cellular radiobiology and carcinogenic mechanisms.
- Radiation carcinogenesis.
- Radiation effects on the fetus.
- Radiation epidemiology and risk modeling.

The expertise of the Committee, including its invited participants, permitted considerable overlapping of assignments among the different categories, ensuring interaction between scientific specialists in different disciplines.

Acknowledgments

In order to respond to the broad charge to the Committee, the work of the Committee was assisted by a number of experts in selected scientific disciplines. The Committee wishes to acknowledge with thanks the valuable contributions of the Directors and Staff of the Radiation Effects Research Foundation, Hiroshima, Japan, for providing the most current Life Span Study data on the Japanese atomic-bomb survivors, and for new organ dose estimates based on the revised atomic-bomb dosimetry. These records have emerged as the most complete data base on the health effects of low-LET radiation exposure in human populations, and continue to be the most comprehensive that have been analyzed for purposes of risk estimation. The analyses presented in the Committee's report were made possible by computational programs developed at the Radiation Effects Research Foundation; Dale Preston, who was responsible for much of their development, served as Scientific Advisor to the Committee, and provided invaluable aid during the course of its deliberations.

The preparation of the report required broad scientific experience in several interrelated disciplines. In this regard, the Committee acknowledges the special help, effort and time of a number of invited participants, and especially Sarah C. Darby, James V. Neel, Susan Preston-Martin, Elaine Ron, William J. Schull, Oddvar Nygaard, and Roy Shore. All of these scientists provided scientific data, advice, and help in the preparation or review of scientific sections of the report, and gave freely of their time and scientific expertise. The Committee would also like to thank Dr. Alice

Stewart for meeting with the Committee's cancer risk group and providing advance copies of the paper on the A-bomb survivors she presented at the 14th Gray Conference, Oxford, 1988.

Very special thanks are extended to Lea Arnold, Doris E. Taylor, and Collette A. Carmi, for their administrative support and for preparation of the many drafts of the report. Their tasks were done with speed and good humor; they were invaluable in assisting the members of the Committee in the completion of their work.

ARTHUR UPTON, *Chairman*

Contents

Executive Summary

INTRODUCTION

This report, prepared by the National Research Council's Committee on the Biological Effects of Ionizing Radiations (BEIR), is the fifth in a series that addresses the health effects of exposure of human populations to low-dose radiation. Ionizing radiations arise from both natural and man-made sources and can affect the various organs and tissues of the body. Late health effects depend on the physical characteristics of the radiation as well as biological factors. Well demonstrated late effects include the induction of cancer, genetically determined ill-health, developmental abnormalities, and some degenerative diseases (e.g., cataracts). Recent concern has centered on the risks of these effects following low-dose exposure, in part because of the presence of elevated levels of radon progeny at certain geographical sites and fallout from the nuclear reactor accidents at Three Mile Island in Pennsylvania in 1979 and Chernobyl in the USSR in 1986. In addition, there is concern about radioactivity in the environment around nuclear facilities and a need to set standards for cleanup and disposal of nuclear waste materials.

Since the completion of the 1980 BEIR III report, there have been significant developments in our knowledge of the extent of radiation exposures from natural sources and medical uses as well as new data on the late health effects of radiation in humans, primarily the induction of cancer and developmental abnormalities. Furthermore, advanced computational techniques and models for analysis have become available for radiation risk assessment. The largest part of the committee's report deals with radiation carcinogenesis in humans, primarily because: (1) there is extended follow-up in major epidemiological studies, particularly those of the Japanese A-bomb survivors and radiotherapy patients treated for benign and malignant conditions, and (2) the revision by a binational group of experts of the dosimetric system for A-bomb survivors in Hiroshima and Nagasaki allows improved analyses of the Japanese data. The report also addresses radiation-induced genetic injury and health effects associated with prenatal irradiation. While only limited application of the advances in our understanding of the molecular mechanisms of cancer induction and genetic disease is possible, these have been examined with the aim of narrowing the range of uncertainties and assumptions inherent in the risk estimation process.

RISK ASSESSMENT

The 1988 BEIR IV report addressed the health effects of exposure to internally-deposited, alpha-emitting radionuclides: radon and its progeny, polonium, radium, thorium, uranium and the transuranic elements. The current BEIR V Committee report includes information and analyses from the BEIR IV report that are appropriate for cancer and genetic risk assessment. In addition, this report addresses the delayed health effects that are induced by low linear energy transfer (LET) radiations such as x rays and gamma radiation and, where possible, makes quantitative risk estimates based on statistical analyses of the results of human epidemiological studies and laboratory animal experiments.

The human data on cancer induction by radiation are extensive; the most comprehensive studies are of the survivors of the atomic bombings of Hiroshima and Nagasaki, x-rayed tuberculosis patients, and persons exposed during treatment for ankylosing spondylitis, cervical cancer, and tinea capitis. Radiation associated cancer risk estimates have been calculated for a number of different organs and tissues, including bone marrow (leukemia), breast, thyroid, lung, and the gastrointestinal organs. To the extent possible, the biological differences among human beings that may modify susceptibility to radiation-induced cancer have been taken into account.

Considerable progress has been made in our understanding of the mutation process on genes and chromosomes and its expression as genetic disorders. Due to a lack of direct evidence of any increase in human heritable effects resulting from radiation exposure, the estimates of genetic risks in humans are based, primarily, on experimental data obtained with laboratory animals. As in all experimental animal studies, the extent to which the results can be extrapolated to humans and the confidence that can be placed on such extrapolation remain uncertain. At present, no data are available to provide reliable estimates of the risks of most complex, multifactorial hereditary disorders. Such risks were not evaluated by the committee.

During the past decade, extensive data have become available on the developmental anatomy of the mammalian brain, and this information has aided the interpretation of effects observed among Japanese survivors irradiated in utero during the atomic bombings. New analyses of the data on A-bomb survivors exposed in utero, together with the reassessment of the A-bomb dosimetry, have permitted delineation of the time-specific susceptibility to radiation-induced mental retardation, the most prevalent developmental abnormality to appear in humans exposed prenatally, and has allowed the risk of these effects to be estimated.

In preparing risk estimates, the committee has relied chiefly on its own evaluations, using recently developed methods for the analysis of population cohort data, rather than relying solely on information in the scientific literature. The Committee recognizes that the application of more sophisticated statistical methods for estimating risks reduces, but does not eliminate, the uncertainties inherent in risk estimation. Throughout the Committee's deliberations consideration was given to both the sources of uncertainty in the data and the potential effect of the assumptions on which the risk estimates are based. The degree of uncertainty in the Committee's risk estimates is presented as an integral part of the risk estimates in this report.

STRUCTURE OF THE REPORT

The report consists of seven chapters. The first chapter reviews the scientific principles, epidemiological methods and the experimental evidence for the biological and health effects in populations exposed to low levels of ionizing radiation. Chapter 2 summarizes the scientific evidence for heritable effects. Chapter 3 includes a discussion of mechanisms involved in the initiation, promotion and progression of cancer induction. Chapter 4 describes the Committee's radiation risk models and the total risk of cancer following whole body exposure. Chapter 5 addresses site-specific cancer risks in the various organs and tissues of the body. Chapter 6 reviews the evidence for fetal and other radiation-induced somatic effects, and the concluding chapter reviews low dose epidemiological studies.

As in previous reports, the Committee on the Biological Effects of Ionizing Radiation cautions that the risk estimates derived from epidemiological and animal data should not be considered precise. Information on the lifetime cancer experience is not available for any of the human studies.

Therefore, the overall risk of cancer can only be estimated by means of models which extrapolate over time. Likewise, estimates on the induction of human genetic disorders by radiation are based on limited data from studies of human populations and therefore rely largely on studies with laboratory animals. It is expected that the risk estimates derived by the Committee will be modified as new scientific data and improved methods for analysis become available.

SUMMARY AND CONCLUSIONS

Of the various types of biomedical effects that may result from irradiation at low doses and low dose rates, alterations of genes and chromosomes remain the best documented. Recent studies of these alterations in cells of various types, including human lymphocytes, have extended our knowledge of the relevant mechanisms and dose-response relationships. In spite of evidence that the molecular lesions which give rise to somatic and genetic damage can be repaired to a considerable degree, the new data do not contradict the hypothesis, at least with respect to cancer induction and hereditary genetic effects, that the frequency of such effects increases with low-level radiation as a linear, nonthreshold function of the dose.

Heritable Effects

The effects of radiation on the genes and chromosomes of reproductive cells are well characterized in the mouse. By extrapolation from mouse to man, it is estimated that at least 1 Gray (100 rad) of low dose-rate, low LET radiation is required to double the mutation rate in man. Heritable effects of radiation have yet to be clearly demonstrated in man, but the absence of a statistically significant increase in genetically related disease in the children of atomic bomb survivors, the largest group of irradiated humans followed in a systematic way, is not inconsistent with the animal data, given the low mean dose level, < 0.5 gray (Gy), and the limited sample size. The Committee's estimates of total genetic damage are highly uncertain, however, as they include no allowance for diseases of complex genetic origin, which are thought to comprise the largest category of genetically-related diseases. To enable estimates to be made for the latter category, further research on the genetic contribution to such diseases is required.

Carcinogenic Effects

Knowledge of the carcinogenic effects of radiation has been significantly enhanced by further study of such effects in atomic bomb survivors.

Reassessment of A-bomb dosimetry at Hiroshima and Nagasaki has disclosed the average dose equivalent in each city to be smaller than estimated heretofore; furthermore, the neutron component of the dose no longer appears to be of major importance in either city. As a result, lifetime risk of cancer attributable to a given dose of gamma radiation now appears somewhat larger than formerly estimated.

Continued follow-up of the A-bomb survivors also has disclosed that the number of excess cancers per unit dose induced by radiation is increased with attained age, while the risk of radiogenic cancer relative to the spontaneous incidence remains comparatively constant. As a result, the dose-dependent excess of cancers is now more compatible with previous "relative" risk estimates than with previous "absolute" risk estimates; the Committee believes that the constant absolute or additive risk model is no longer tenable.

A-bomb survivors who were irradiated early in life are just now reaching the age at which cancer begins to become prevalent in the general population. It remains to be determined whether cancer rates in this group of survivors will continue to be comparable to the increased cancer risk that has been observed among survivors who were adults at the time of exposure. For this reason, estimation of the ultimate magnitude of the risk for the total population is uncertain and calls for further study.

The quantitative relationship between cancer incidence and dose in A-bomb survivors, as in other irradiated populations, appears to vary, depending on the type of cancer in question. The dose-dependent excess of mortality from all cancer other than leukemia, shows no departure from linearity in the range below 4 sievert (Sv), whereas the mortality data for leukemia are compatible with a linear-quadratic dose response relationship.

In general, the dose-response relationship for carcinogenesis in laboratory animals also appears to vary with the quality (LET) and dose rate of radiation, as well as sex, age at exposure and other variables. The influence of age at exposure and sex on the carcinogenic response to radiation by humans has been characterized to a limited degree, but changes in response due to dose rate and LET have not been quantified.

Carcinogenic effects of radiation on the bone marrow, breast, thyroid gland, lung, stomach, colon, ovary, and other organs reported for A-bomb survivors are similar to findings reported for other irradiated human populations. With few exceptions, however, the effects have been observed only at relatively high doses and high dose rates. Studies of populations chronically exposed to low-level radiation, such as those residing in regions of elevated natural background radiation, have not shown consistent or conclusive evidence of an associated increase in the risk of cancer.

For the purposes of risk assessment, the Committee summarized the epidemiological data for each tissue and organ of interest in the form of an exposure-time-response model for relative risk. These models were fitted to the data on numbers of cases and person-years in relation to dose equivalent, sex, age at exposure, time after exposure, and attained age. Standard lifetable techniques were used to estimate the lifetime risk for each type of cancer based on these fitted models.

On the basis of the available evidence, the population-weighted average lifetime excess risk of death from cancer following an acute dose equivalent to all body organs of 0.1 Sv (0.1 Gy of low-LET radiation) is estimated to be 0.8%, although the lifetime risk varies considerably with age at the time of exposure. For low LET radiation, accumulation of the same dose over weeks or months, however, is expected to reduce the lifetime risk appreciably, possibly by a factor of 2 or more. The Committee's estimated risks for males and females are similar. The risk from exposure during childhood is estimated to be about twice as large as the risk for adults, but such estimates of lifetime risk are still highly uncertain due to the limited follow-up of this age group.

The cancer risk estimates derived with the preferred models used in this report are about 3 times larger for solid cancers (relative risk projection) and about 4 times larger for leukemia than the risk estimates presented in the BEIR III report. These differences result from a number of factors, including new risk models, revised A-bomb dosimetry, and more extended follow-up of A-bomb survivors. The BEIR III Committee's linear-quadratic dose-response model for solid cancers, unlike this Committee's linear model, contained an implicit dose rate factor of nearly 2.5; if this factor is taken into account, the relative risk projections for cancers other than leukemia by the two committees differ only by a factor of about 2.

The Committee examined in some detail the sources of uncertainty in its risk estimates and concluded that uncertainties due to chance sampling variation in the available epidemiological data are large and more important than potential biases such as those due to differences between various exposed ethnic groups. Due to sampling variation alone, the 90% confidence limits for the Committee's preferred risk models, of increased cancer mortality due to an acute whole body dose of 0.1 Sv to 100,000 males of all ages range from about 500 to 1,200 (mean 760); for 100,000 females of all ages, from about 600 to 1,200 (mean 810). This increase in lifetime risk is about 4% of the current baseline risk of death due to cancer in the United States. The Committee also estimated lifetime risks with a number of other plausible linear models which were consistent with the mortality data. The estimated lifetime risks projected by these models were within the range of uncertainty given above. The committee recognizes that its risk estimates become more uncertain when applied to very low doses. Departures from a linear model at low doses, however, could either increase or decrease the risk per unit dose.

Mental Retardation

The frequency of severe mental retardation in Japanese A-bomb survivors exposed at 8-15 weeks of gestational age has been found to increase more steeply with dose than was expected at the time of the BEIR III report. The data now reveal the magnitude of this risk to be approximately a 4% chance of occurrence per 0.1 Sv, but with less risk occurring for exposures at other gestational ages. Although the data do not suffice to define precisely the shape of the dose-effect curve, they imply that there may be little, if any, threshold for the effect when the brain is in its most sensitive stage of development. Pending further information, the risk of this type of injury to the developing embryo must not be overlooked in assessing the health implications of low-level exposure for women of childbearing age.

RECOMMENDATIONS

There are a number of important radiobiological problems that must be addressed if radiation risk estimates are to become more useful in meeting societal needs. Assessment of the carcinogenic risks that may be associated with low doses of radiation entails extrapolation from effects observed at doses larger than 0.1 Gy and is based on assumptions about the relevant dose-effect relationships and the underlying mechanisms of carcinogenesis. To reduce the uncertainty in present risk estimation, better understanding of the mechanisms of carcinogenesis is needed. This can be obtained only through appropriate experimental research with laboratory animals and cultured cells.

While experiments with laboratory animals indicate that the carcinogenic effectiveness per Gy of low-LET radiation is generally reduced at low doses and low dose rates, epidemiological data on the carcinogenic effects of low-LET radiation are restricted largely to the effects of exposures at high dose rates. Continued research is needed, therefore, to quantify the extent to which the carcinogenic effectiveness of low-LET radiation may be reduced by fractionation or protraction of exposure.

The carcinogenic and mutagenic effectiveness per Gy of neutrons and other high-LET radiations remains constant or may even increase with decreasing dose and dose rate. For reasons which remain to be determined, the relative biological effectiveness (RBE) for cancer induction by neutrons and other high-LET radiations has been observed to vary with the type of cancer in question. Since data on the carcinogenicity of neutrons in human populations are lacking, further research is needed before confident estimates can be made of the carcinogenic risks of low-level neutron irradiation for humans. Similarly, the relative mutagenic effectiveness of neutron and other high LET radiation varies with the specific genetic end point. Therefore, additional data are also needed on the mutagenicity of low neutron doses to permit more confident projection of genetic risks from animal data to man.

The extrapolation of animal data to the human is necessary for genetic risk assessment. No population appears to exist, other than the A-bomb survivors, that could provide a substantial basis for genetic epidemiological study. The scientific basis of the extrapolation must therefore rely upon cellular and molecular homologies. Research needs in this area are clear.

As noted previously, the Committee's genetic risk assessment did not attempt to project risk for the category of diseases with complex genetic etiologies. Because genetically related disorders comparable to those in this heterogeneous category of human disorders may have no clearly definable counterparts in laboratory and domestic animals, the required research should be directed towards human diseases whenever feasible.

The dose-dependent increase in the frequency of mental retardation in prenatally irradiated A-bomb survivors implies the possibility of higher risks to the embryo from low-level irradiation than have been suspected heretofore. It is important that appropriate epidemiological and experimental research be conducted to advance our understanding of these effects and their dose-effect relationships.

Finally, further epidemiological studies are needed to measure the cancer excess following low doses as well as large doses of high and low LET radiation. Most of the A-bomb survivors are still alive, and their mortality experience must be followed if reliable estimates of lifetime risk are to be made. This is particularly important for those survivors irradiated as children or in utero who are now entering the years of maximum cancer risk. Studies on populations exposed to internally deposited radionuclides should be continued to assess the risks of nuclear technologies and the effects of radon progeny. Low-dose epidemiological studies may be able to supply information on the extent to which effects observed at high doses and high dose rates can be relied on to estimate the effects due to chronic exposures such as occur in occupational environments. The reported follow-up of A-bomb survivors has been essential to the preparation of this report. Nevertheless, it is only one study with specific characteristics, and other large studies are needed to verify current risk estimates.

1

Background Information and Scientific Principles

PHYSICS AND DOSIMETRY OF IONIZING RADIATION

All living matter is composed of atoms joined into molecules by electron bonds. Ionizing radiation is energetic enough to displace atomic electrons and thus break the bonds that hold a molecule together. As described below, this produces a number of chemical changes that, in the case of living cells, can lead to cell death or other harmful effects. Ionizing radiations fall into two broad groups: 1) particulate radiations, such as high energy electrons, neutrons, and protons which ionize matter by direct atomic collisions, and 2) electromagnetic radiations or photons such as x rays and gamma rays which ionize matter by other types of atomic interactions, as described below.

Absorption and Scattering of Photons

Photons ionize atoms through three important energy transfer processes: the photoelectric process, Compton scattering, and pair production. For photons with low energies (<0.05 megaelectron volt [MeV]) the photoelectric process dominates in tissue. The photoelectric process occurs when an incoming photon interacts with a tightly bound electron from one of the inner shells of the atom, and causes the electron to be ejected with sufficient energy to escape the atom. Characteristic x rays and Auger electrons follow from this process, but the biological effects are due mainly to excitations and ionizations in molecules of tissue caused by the ejected electron. The probability of the photoelectric process occurring is strongly dependent on the average atomic number of the tissue with an equally strong inverse dependence on the photon energy.

At higher photon energies (0.1-10 MeV), Compton scattering is the most probable process that takes place in irradiated tissue. It occurs when the photon energy greatly exceeds the electron binding energy, so that an orbital electron appears to the photon as a free electron. The photon scatters off the electron, giving up part of its energy to the electron, which proceeds to ionize and excite tissue molecules. The scattered photon with reduced energy continues to interact with other electrons and repeats the above process many times until the photon either escapes the absorbing material or its energy is sufficiently degraded for the photoelectric process to occur. Within the energy range of 0.1-10 MeV, the Compton process has a modest dependence on energy and is almost independent of atomic number.

Above a threshold energy of 1.02 MeV, the pair-production process is possible. Here a photon converts its energy in the presence of an atomic nucleus to a positron-electron pair, which, in turn, proceeds to interact with tissue atoms and molecules, leading to eventual biological effects. When the positron slows down it is almost always annihilated with an electron, producing two 0.511 MeV photons. The probability of pair-production in tissue increases slowly with photon energy but does not outweigh that of the Compton process until the photon energy reaches 20 MeV. The process depends upon the average atomic number of the tissue.

Photon Spectral Distributions

As seen from the description presented above, the absorption and scattering of photons depend critically on photon energy. The initial photon energy depends on the source of the radiation. Gamma rays resulting from radioactive decay consist of monoenergetic photons with energies that do not exceed several MeV in energy. Because of scattering and absorption within the radioactive source itself and in the encapsulating material, the photons that are emitted do have a spectrum of energies but it is fairly narrow.

Relatively broad energy distributions are the rule for x-ray photons produced from electrical devices. X rays are effectively produced by the rapid deceleration of charged particles (usually electrons) by a material of high atomic number. This results in a continuous distribution of energies with a maximum at an energy about one third that of the most energetic electron. As photons interact with matter, their spectral distribution is further altered in a complex manner as the photons transfer energy to the absorbing medium by the processes described above.

Electron Spectral Distributions and LET

When monoenergetic photons interact with a tissue medium, the electrons that are set in motion, particularly from the Compton process, proceed to interact with the atoms and molecules of the medium, losing energy through collisions and excitations, and are scattered in the process. The result is a complex shower of electrons, the energy distribution of which is continuously degraded as the electrons give up their energy to the medium at a rate defined by the electron stopping power of the medium. As the electron proceeds through tissue, it creates a track of excited and ionized molecules that, for energetic electrons, are relatively far apart. For example, the dimension of this spacing is such that there is a finite probability that the energetic electron can pass through a DNA molecule, with about 3 nm separating the two strands, without releasing any of its energy and therefore without causing damage. The spatial energy distribution, stated in terms of the amount of energy deposited per unit length of particle track, is defined as the *linear energy transfer* (LET) of the radiation. X rays and gamma rays set in motion electrons with a relatively low spatial rate of energy loss and thus are considered *low LET* radiations. The photon and electron energy degradation processes described above result in a broad distribution of LET values occurring in irradiated tissue. A typical value of LET for the electrons set in motion by cobalt-60 gamma rays (average energy 1.25 MeV) would be about 0.25 keV/μm. This can be contrasted with a densely ionizing 2 MeV alpha particle which produces about 1000 times more ionization per unit distance, 250 keV/μm. Such particles are characterized as *high LET* radiation. Knowledge of LET is important when considering the relative biological effectiveness (RBE) of a given radiation; LET is commonly used as a measure of radiation quality, as discussed below.

Microdosimetry

Various limitations in the concept of LET and absorbed dose in subcellular tissue volumes led to the introduction of microdosimetry. Microdosimetry takes account of the fact that energy deposition by ionizing radiations is a stochastic (random) process. Identical particles of the same energy interacting in a small volume of material deposit differing amounts of energy due to chance alone. The specific energy, z, is defined as the ratio ϵ/m where ϵ is the energy imparted by a single ionizing particle in a volume element of mass m. The mean value of z for a large number of particles is equal to the absorbed dose. The microdosimetric analogue to LET is the quantity lineal energy, defined as ϵ/d, where d is the mean chord length in the volume occupied by mass m. Distributions of absorbed dose in terms of lineal energy can be measured by proportional counters filled with tissue-equivalent gas at pressure levels appropriate for simulating spheres of tissue with diameters on the order of 1 μm. The principles of microdosimetry are extensively discussed in the BEIR IV report (NRC88) and ICRU report 36 (ICRU83).

Energy Transfer—Kerma and Absorbed Dose

The transfer of energy from photons to tissue takes place in two stages: (1) the interaction of the photon with an atom, causing an electron to be set in motion, and then (2) the subsequent absorption by the medium of kinetic energy from the high energy electron through excitation and ionization.

The first stage can be identified with the quantity called *kerma*, *K*, which stands for *k*inetic *e*nergy *r*eleased in the *ma*terial.

$K = dE_{tr}/dm$, where dE_{tr} is the kinetic energy transferred from photons to electrons in a volume element of mass *dm*.

The second stage, energy absorption, is more important for understanding radiobiological effects. The absorbed *dose*, the energy absorbed per unit mass, differs from kerma in that the dose may be smaller due to lack of charged particle equilibrium, *bremsstrahlung* escaping from the medium, etc. Another difference is that the kerma refers to energy transfer at a point, whereas the energy is absorbed over a distance equal to the electron range. Of the two quantities, absorbed dose is the easier one to approach experimentally and can be determined by a number of well-defined techniques, including gas ionization methods, calorimetry, and thermoluminescent techniques. On the other hand, kerma is often more easily calculated.

Radiation Chemical Effects Following Energy Absorption

After the electron produced by a photon interaction passes through tissue, exciting and ionizing atoms and molecules, a number of important chemical events that precede the biological effects take place. Most of the energy absorption takes place in water, since cells are made up of more then 70% water. When an ionizing particle passes through a water molecule, it may ionize it to yield an ionized water molecule, H_2O^+, and an electron by the reaction:

$$H_2O \xrightarrow{\text{radiation}} H_2O^+ + e- .$$

The electron can be trapped, polarizing water molecules to produce the so-called hydrated electron, e_{aq}. On the other hand, the ionized water molecule, H_2O^+, reacts at the first collision with another water molecule to produce an hydroxyl radical, $OH^\bullet$ according to the reaction:

$$H_2O^+ + H_2O \rightarrow OH^\bullet + H_3O^+.$$

The free radical $OH^\bullet$ has an unpaired electron and is therefore highly reactive as it seeks to pair its electron to reach stability. At the high initial concentrations, certain back reactions occur producing hydrogen molecules, hydrogen peroxide and water. The initial species produced in water radiolysis can then be written as:

$$H_2O \xrightarrow{\text{radiation}} e_{aq}, H^\bullet, H_2O_2, H_2.$$

Instead of being ionized, the water molecule may simply be excited according to the reaction:

$$H_2O \xrightarrow{\text{radiation}} H_2O^*.$$

where H_2O^* is the excited molecule. But H_2O^* soon breaks up into the $H^\bullet$ radical and the $OH^\bullet$ radical according to:

$$H_2O^* \rightarrow H^\bullet + OH^\bullet.$$

As a result of the above processes, three important reactive species are produced: the aqueous electron, $OH^\bullet$, and $H^\bullet$, with initial relative yields of about 45%, 45%, and 10%, respectively. These reactive species attack molecules in the cell leading to the production of biological damage. The $OH^\bullet$ radical is believed to be the most effective of the three species in causing damage. Because it is an oxidizing agent, it can abstract a hydrogen atom from the deoxyribose moiety of DNA, for example, yielding a highly reactive site on DNA in the form of a DNA radical. Since this process arises from the irradiation of a water molecule rather than the DNA itself, the process is known as the *indirect effect.* Electrons set in motion by photons can, of course, directly excite or ionize cell macromolecules by direct interaction with the critical molecule. This is called the *direct effect.* Both mechanisms can produce cellular damage. There is strong evidence that the DNA is the most critical site for lethal damage, but other sites such as the nuclear membrane or the DNA-membrane complex may also be important.

Ward (Wa88) has derived an approximation of the damage yields expected in various moieties of DNA within an irradiated cell, in which consideration is given to the direct deposition of energy in DNA and other molecules. Table 1-1 shows the amount of energy deposited per Gray in each moiety of DNA within a cell that is assumed to contain 6 pg of DNA.

TABLE 1-1 Amount of Energy Deposited in DNA per Cell per Gray

Constituent	Mass per Cell (pg)	eV Deposited	Number of 60-eV Events
Deoxyribose	2.3	14,000	235
Bases	2.4	14,700	245
Phosphate	1.2	7,300	120
Bound water	3.1	19,000	315
Inner hydration	4.2	25,000	415

SOURCE: J. F. Ward, C. L. Limoli, P. Calabro-Jones, and J. W. Evans (Wa88).

Calculated from this is the number of events since 60 eV is the average amount of energy deposited per event.

The yields of DNA damage necessary to kill 63% of mammalian cells (63% of cells killed means that, on average, each cell has sustained one lethal event) can be assessed for various lethal agents (Wa88), as shown in Table 1-2. The high efficiency with which ionizing radiation (and bleomycin) kill cells is not simply due to individual OH radical-induced lesions, as witnessed by the large-scale production of single-strand breaks with hydrogen peroxide. Ward et al. (WA87) suggest that the efficiency of cell killing by ionizing radiation at relatively low levels of DNA damage is due to the production of damage in more than one moiety in a localized region, i.e., lesions resulting from multiply damaged sites in a single location or locally multiply damaged sites (LMDS).

Recent studies (Wi85, Gr85, Ei81), as analyzed by Ward (Wa88), support the importance of indirect effects of ionizing radiation in producing damage to intracellular DNA. This is of particular significance in view of the suggestion that most intracellular DNA damage is caused by direct ionization and that radicals produced in water cannot access the macromolecule. It appears from the above analysis (Wa88) that the volume of water in the DNA-histone complex (nucleosome) is at least equal to the DNA volume and that radiation-produced OH radicals in the water volume have ready access to the DNA molecule.

Some of the current assessments of DNA damage caused by ionizing radiation in mammalian cells (Wa88) are as follows: (1) direct and indirect effects are both important; (2) the quantity of damage produced by ionizing radiation is orders of magnitude lower than for most other agents for equal cell-killing efficiency; (3) individual damage moieties are not biologically significant since they can be repaired readily by using the undamaged DNA strand as a template; (4) LMDS are more likely the lethal lesion in cellular

TABLE 1-2 Yields of DNA Damage Necessary to Kill 63% of the Cells Exposed

Agent	DNA Lesion	Number of Lesions per Cell per D_{37}[a]
Ionizing radiation	ssB	1,000
	dsB	40
	Total LMDS[b]	440
	DPC[c]	150
Bleomycin A2	ssB	150
	dsB	30
UV light	T<>T dimer	400,000
	ssB	100
Hydrogen peroxide		
0°	ssB	<2,600,000
37°C	?	
Benzo[a]pyrene 4,5-oxide	Adduct	100,000
Aflatoxin	Adduct	10,000
1-Nitropyrene	Adduct	400,000
Methylnitrosourea	7-Methylguanine	800,000[d]
	O^6-Methylguanine	130,000[d]
	3-Methyladenine	30,000[d]
2-(*N*-Acetoxy-*N*-acetyl)amino-fluorene	Adduct	700,000
Other similar aromatic amides produce about the same number of adducts per lethal event		

[a] D_{37} = dose of agent required to reduce survival of cells to 37% of the number exposed.
[b] Calculated, LMDS = locally multiply damaged sites.
[c] DPC = DNA-protein cross-links.
[d] D_{37} calculated from individual exposures; no survival curves available.

SOURCE: J. F. Ward, C. L. Limoli, P. Calabro-Jones, and J. W. Evans (Wa88).

DNA; these result from a high local energy deposition in the DNA (in such a volume, multiple radicals cause multiple lesions locally); (5) the individual lesions making up an LMDS can be widely separated on the opposite strands of the DNA; if they are separated too much, they could be repaired as individual lesions.

Physics and Dosimetry of High-LET Radiation (Neutrons)

Interactions of Neutrons with Tissue Elements

When neutrons impinge on a tissue medium, they will either penetrate it without interacting with its constituent atoms or they will interact with its atoms in one or more of the following ways: (1) elastically, (2) inelastically, (3) nonelastically, (4) by capture reactions, or (5) through spallation processes.

Elastic scattering is the most important interaction in tissue irradiated with neutrons at energies below 20 MeV. This would include the energy range for fission neutrons (<10 MeV), neutrons produced with 16 MeV deuterons bombarding a beryllium target (<20 MeV), and neutrons produced with 150 keV deuterons on tritium (<20 MeV). The neutron, an uncharged particle, interacts primarily by collisions with nuclei in the absorbing medium. If the total kinetic energy of the neutron and the nucleus remains unchanged by the collision, the collision is termed elastic. During an elastic collision, the maximum energy is transferred from the neutron to the nucleus if the two masses are equal. In soft tissue, the most important neutron interaction is with hydrogen. There are three reasons for this: (1) Nearly two-thirds of the nuclei in tissue are protons, (2) the energy transfer with protons is maximal (about one-half), and (3) the interaction probability (cross-section) for hydrogen is larger than that for any other element. The result is that about 90% of the energy absorbed in tissue from neutrons with energy of less than 20 MeV comes from protons that are recoiling from elastic collisions. The remaining energy is absorbed by other recoiling tissue nuclei in the following decreasing order of importance: oxygen, carbon, and nitrogen.

Inelastic scattering refers to reactions in which the neutron interacts with the nucleus but is promptly reemitted with reduced energy and usually with a changed direction. The scattering nucleus, which is left in an excited state, then emits a nuclear deexcitation gamma ray. For neutrons with kinetic energies of greater than 10 MeV, inelastic scattering contributes to energy loss in tissue; about 30% of the energy deposited in tissue by 14-MeV neutrons, for example, comes from inelastic interactions. The important inelastic interactions of neutrons in soft tissue are not with hydrogen but with carbon, nitrogen, and oxygen.

Nonelastic scattering defines reactions in which the neutron-nucleus interaction results in the emission of particles other than a single neutron such as alpha particles and protons [e.g., $^{16}O(n,\alpha)^{13}C$, $^{14}N(n,p)^{14}C$]. The cross-sections for nonelastic scattering in tissue become significant at energies greater than 5 MeV and increase as the neutron energy approaches 15 MeV. These reactions are usually accompanied by deexcitation gamma rays, but their importance is due to the high LET of the charged particles emitted, especially alpha particles. At neutron energies greater than 20 MeV, even though nonelastic cross-sections do not increase appreciably, nonelastic processes become increasingly important contributors to the total dose because of the increased average energy of the charged particles resulting from the interaction.

The capture of low-energy neutrons in the thermal and near-thermal regions provides a significant contribution to tissue dose. The reactions of importance are $^{14}N(n,p)^{14}C$ and $^{1}H(n,\gamma)^{2}H$. The former reaction produces locally absorbed energy of 0.62 MeV from the proton and the recoil nucleus. The latter reaction yields a 2.2-MeV gamma ray that, in general, deposits energy at a distance from the capture site and that has a reasonable probability of escaping altogether from a mass as large as a rodent. For thermal neutrons the $^{14}N(n,p)^{14}C$ reaction is the major contributor of absorbed energy in tissue samples with a dimension of less than 1 cm because of the short range (<10 μm) of the 0.58-MeV proton. However, for larger masses of tissue (e.g., the human body), the 2.2-MeV gamma rays from the $^{1}H(n,\gamma)^{2}H$ reaction are a significant dose contributor.

In the spallation process the neutron-nucleus interaction results in the fragmentation of the nucleus with the emission of several particles and nuclear fragments. The latter are heavily ionizing, so the local energy deposition can be high. Several neutrons and deexcitation gamma rays also can be emitted, yielding energy carriers that escape local energy deposition. The spallation process does not become significant until neutron energies are much greater than 20 MeV.

In summary, elastic and nonelastic scattering and the capture process are by far the most important reactions in tissue for neutrons in the fission energy range. Inelastic and nonelastic scattering begin at about 2.5 and 5 MeV, respectively, and become important at an energy of about 10 MeV. As the neutron energy goes higher, nonelastic scattering and spallation reactions increase in importance, and elastic scattering becomes of less importance for energies greater than 20 MeV.

POPULATION EXPOSURE TO IONIZING RADIATION IN THE UNITED STATES

A new assessment of the average exposure of the U.S. population to ionizing radiation has recently been made by the National Council on Radiation Protection and Measurements (NCRP87b). Six main radiation sources were considered: natural radiation and radiation from the following five man-made sources: occupational activities (radiation workers), nuclear fuel production (power), consumer products, miscellaneous environmental sources, and medical uses.

For each source category, the collective effective dose equivalent was obtained from the product of the average per capita effective dose equivalent received from that source and the estimated number of people so exposed. The average effective dose equivalent for a member of the U.S. population was then calculated by dividing the collective effective dose equivalent value by the number of the U.S. population (230 million in 1980). As discussed below, the dose equivalent is defined as the product of the absorbed dose, D, and the quality factor Q, which accounts for

TABLE 1-3 Average Annual Effective Dose Equivalent of Ionizing Radiations to a Member of the U.S. Population

	Dose Equivalent[a]		Effective Dose Equivalent	
Source	mSv	mrem	mSv	%
Natural				
Radon[b]	24	2,400	2.0	55
Cosmic	0.27	27	0.27	8.0
Terrestrial	0.28	28	0.28	8.0
Internal	0.39	39	0.39	11
Total natural	—	—	3.0	82
Artificial				
Medical				
x-ray diagnosis	0.39	39	0.39	11
Nuclear medicine	0.14	14	0.14	4.0
Consumer products	0.10	10	0.10	3.0
Other				
Occupational	0.009	0.9	<0.01	<0.3
Nuclear fuel cycle	<0.01	<1.0	<0.01	<0.03
Fallout	<0.01	<1.0	<0.01	<0.03
Miscellaneous[c]	<0.01	<1.0	<0.01	<0.03
Total artificial	—	—	0.63	18
Total natural and artificial	—	—	3.6	100

[a] To soft tissues.
[b] Dose equivalent to bronchi from radon daughter products. The assumed weighting factor for the effective dose equivalent relative to whole-body exposure is 0.08.
[c] Department of Energy facilities, smelters, transportation, etc.

SOURCE: National Council on Radiation Protection and Measurements (NCRP87b).

differences in the relative biological effectiveness of different types of radiation. The effective dose equivalent relates the dose-equivalent to risk. For the case of partial body irradiation, the effective dose equivalent is the risk-weighted sum of the dose equivalents to the individually irradiated tissues.

As seen in Table 1-3 and Figure 1-1, three of the six radiation sources, namely radiation from occupational activities, nuclear power production (the fuel cycle), and miscellaneous environmental sources (including nuclear weapons testing fallout), contribute negligibly to the average effective dose equivalent, i.e., less than 0.01 millisievert (mSv)/year (1 mrem/year).

A total average annual effective dose equivalent of 3.6 mSv (360 mrem)/year to members of the U.S. population is contributed by the other three sources: naturally occurring radiation, medical uses of radiation, and radiation from consumer products. By far the largest contribution (82%) is made by natural sources, two-thirds of which is caused by radon and its

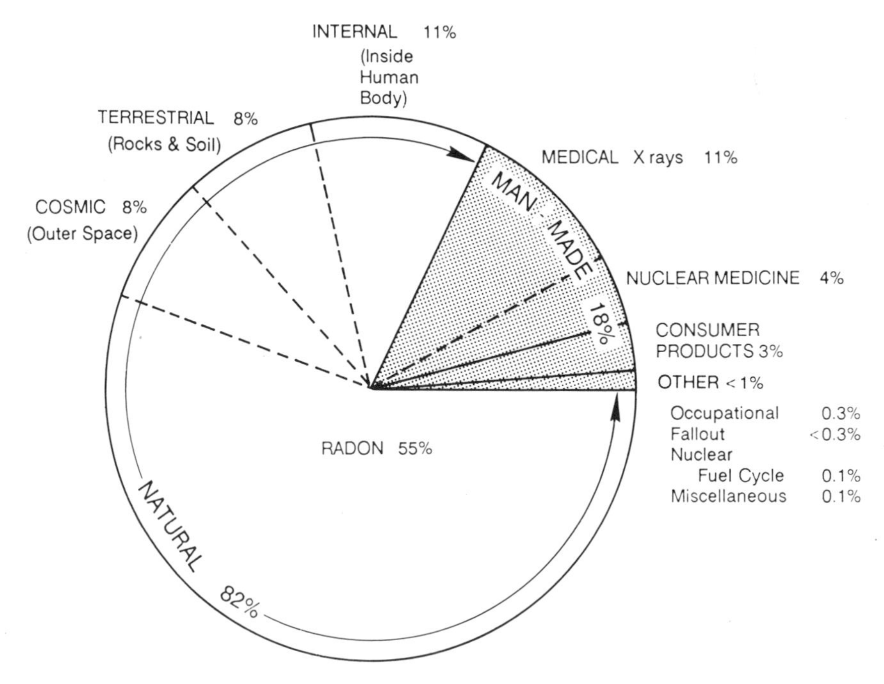

FIGURE 1-1 Sources of radiation exposure to the U.S. population (NCRP87b).

decay products. Approximately equal contributions to the other one-third come from cosmic radiation, terrestrial radiation, and internally deposited radionuclides. The importance of environmental radon as the largest source of human exposure has only recently been recognized.

The remaining 18% of the average annual effective dose equivalent consists of radiation from medical procedures (x-ray diagnosis, 11% and nuclear medicine, 4%) and from consumer products (3%). The contribution by medical procedures is smaller than previously estimated. For consumer products, the chief contributor is, again, radon in domestic water supplies, although building materials, mining, and agricultural products as well as coal burning also contribute. Smokers are additionally exposed to the natural radionuclide polonium-210 in tobacco, resulting in the irradiation of a small region of the bronchial epithelium to a relatively high dose (up to 0.2 Sv per year) that may cause an increased risk of lung cancer (NCRP84).

Uncertainties exist in the data shown in Table 1-3. Uncertainties for exposures from some consumer products are greater than those for exposures from cosmic and terrestrial radiation sources. The estimates for the most important exposure, that of lung tissue to radon and its decay products, have many associated uncertainties. Current knowledge of the average radon concentration, the distribution of radon indoors in the United States, and alpha-particle dosimetry in lung tissue is limited. In addition, knowledge of the actual effective dose equivalent is poorly quantified. Further uncertainties are caused by difficulties in combining data for exposure from different sources that actually are from different years, mainly from 1980 to 1983.

RADIOBIOLOGICAL CONCEPTS

Experiments on radiation-induced cell killing have given rise to a number of radiobiological principles and concepts. Many of these principles and concepts are inferred to apply to mutagenesis and carcinogenesis, as well as to cell killing, although this is often not known for certain since it is not possible to perform comparable experiments with all of these endpoints. Some of the major concepts are discussed below.

The first concept is that the principal target for radiation-induced cell killing is DNA. Although it is not the exclusive target, it is generally the most consequential. While the evidence for this conclusion is circumstantial, it is also convincing (Le56). As noted above, the consequences of the absorption of radiant energy arise from excitations and ionizations along the tracks of the charged particles that are set in motion when radiant energy is absorbed. Biological damage may be a consequence of a *direct* interaction between the charged particles and the DNA molecule, or the biological effects may be mediated by the production of free radicals (Mi78). In the latter case, which is the *indirect* action of radiation, the absorption of the radiation may occur in, for example, a water molecule, and the consequent free radical produced may diffuse to the DNA, where it gives up its energy to produce a biological lesion. In the case of sparsely ionizing radiations, such as x rays and gamma rays, about two-thirds of the biological effects are produced by this indirect action, and this component of the radiation damage is amenable to modification by a variety of physical and chemical factors. As the quality of the radiation changes from low to high LET, the balance shifts from the indirect action to the direct action.

The second major concept concerns the shape of the dose-response relationship. With cell lethality, R, as the endpoint, the dose-response relationship for low-LET radiations often approximates a linear-quadratic function of the dose, D.

$$R = \alpha D + \beta D^2.$$

The relative importance of the linear and quadratic terms varies widely for different cells and tissues. The ratio α/β, which is the dose at which the linear and quadratic contributions to the biological effect are equal, may vary from about 1 Gray (Gy) to more than 10 Gy. As the LET of the radiation is increased, the ratio α/β also increases for a given cell or tissue, and for very high LET radiations, survival (1-R) approximates an exponential function of dose at doses of interest. For carcinogenesis in laboratory animals, dose-response relationships with a wide variety of shapes have been reported. At higher doses there is the complication of a balance between increased cell transformation and increased cell killing.

The linear-quadratic formulation had its origins in the 1930s, when it was used to fit data for radiation induced chromosome aberrations (Sa40). Many chromosome aberrations appear to be the consequence of the interaction between breaks in two separate chromatids. This applies to aberrations, such as dicentrics, that lead to cell lethality, as well as to aberrations such as translocations that, in some cases, lead to cancer through the activation of an oncogene.

Thus, the interpretation of the linear-quadratic formulation is that the characteristic shape of the dose-response curve reflects a predominance of single-track events, which are proportional to the dose at low doses and low dose rates, and of two-track events which are proportional to the square of the dose and result in the upward bending of the cancer induction curve at high doses received at high dose rates.

This biophysical model has been challenged in recent years, largely on the basis of data with soft x rays, which are highly effective biologically even though the length of the secondary tracks they produce is too short to enable a single track to break two independent chromosomes (Th86). Hence, although the data have been interpreted in terms of the more conventional linear-quadratic formulation (Br88), an alternative model has been proposed in which all biological damage is presumed to result from single track effects, with the additional factor of a repair process that saturates at higher doses. Biological experiments that allow an unequivocal choice to be made between the models have not yet been performed.

The third concept is that the biological consequence of a given dose of radiation varies with the quality of the radiation. With cell killing as the endpoint, the relative biological effectiveness (RBE) of many types of radiation has been studied in detail (Ba63). Although the RBE varies with the LET of the radiation, it also varies with the dose, dose rate, type of cell or tissue used to score the biological effect, and the endpoint in question (Br73, Ba68). The pattern of variation of the RBE with LET appears to be similar for mutagenesis as for cell killing, but it has not been established to be the same for carcinogenesis as an endpoint. The quality factor (Q) rather than RBE is widely used in radiation protection. The International Commission on Radiological Protection (ICRP) has suggested, however, that the quality factor should be based on a microdosimetric quantity such as lineal energy (ICRU86).

For cell lethality as an endpoint, cell sensitivity to radiation varies as a function of its stage in the cell cycle. This is the fourth major radiobiological concept. In general, cells are most sensitive in the G_2 phase or in mitosis, and they are most resistant during the phase of DNA synthesis (Si66, Ter63). In the case of mutagenesis, it appears, in some instances at least, that the most sensitive phase of the cycle is G_1. There is little or no information concerning the variation of cellular sensitivity with the phase of the cell cycle for oncogenic transformation in vitro.

The fifth concept is that the effect of a given dose may be influenced greatly by the dose rate. The influence of the dose-rate effect has been widely studied and is well established for cell lethality as an endpoint (Ha64, Ha72). In general, the effectiveness of a given dose tends to decrease with decreasing dose rate. In the case of low-LET radiations, the reduced effectiveness of a dose delivered at low dose rates is a consequence of the interaction of a number of factors, most notably the repair of sublethal damage, the redistribution of the cells within the mitotic cycle, and the compensatory cellular proliferation during a protracted exposure. In the case of high-LET radiations, the dose-rate effect is much reduced, at least those components of it that are a consequence of repair and redistribution. These general considerations appear to be equally valid for mutagenesis and carcinogenesis, although there is some evidence that for high-LET radiations, protracting an exposure may lead to an increase in the induction of cancer and mutations (Ha79, 80, He88, Hi84, Vo81, Ul84, and Fr77) in some situations.

There is the important practical problem of allowing for dose-rate effects in the analysis of site-specific cancer risks (see Chapter 4.) The cumulative knowledge of dose rate factors in experimental radiobiology was summarized in NCRP Report 64 (NCRP80) and has been discussed in several reports of the United Nations Scientific Committee on the Effects of Atomic Radiation (UNSCEAR) concerned with risk estimates for the carcinogenic effects of radiation (UN77, UN86). These reports noted that any value from 2 to 10 for the extent to which a given dose of low-LET radiation may be assumed to decrease in effectiveness at low dose rates, could be rationalized on the basis of experiments with laboratory animals, but suggested a factor of 2.5 for use in risk assessment for human leukemia at low doses and dose rates. They further suggested that this risk be multiplied by 5 to get the risk for all cancers. There are scant human data that allow an estimate of the dose-rate effectiveness factor (DREF). If the apparently nonlinear dose-incidence curve for leukemia in atomic bomb survivors (see Chapter 5) is assumed to reflect a linear quadratic relationship between the incidence and the dose, the contribution of the quadratic dose term can be expected to be reduced at low doses and low dose rates. According to this interpretation, fitting linear and

TABLE 1-4 Summary of Dose-Rate Effectiveness Factors for Low-LET Radiation

Source of Data	Observed Full Range of Values	Limited for Narrow Range of Values	Single Best Estimate
Human leukemia (present report)	—	—	2.1
BEIR III	—	—	2.0 to 2.5
Laboratory animal studies			
Specific locus mutation	3–10	3–7	5
Reciprocal transloc.	5–10	5–7	5
Life shortening	3–10	3–5	4
Tumorigenesis	2–10	2–5	4

linear-quadratic models to these data, the ratio of the linear coefficients for the two fits yields an estimate of the DREF. This Committee's analysis in Chapter 5 yields a DREF of 2. This compares with the estimate of 2.25 made by the BEIR III Committee, based on essentially the same data set but with the obsolete T65D dose estimates (see Annex 4B).

The much more extensive animal data include four basic sets from which DREF values can be derived. These include (1) the induction of specific locus mutations, (2) the induction of reciprocal chromosome translocations, (3) life shortening induced by whole-body external irradiation, and (4) tumor induction in small mammals. All of these studies are relevant for the selection of DREF values for estimating human risks for neoplastic disease. Table 1-4 provides a summary of the experimental findings for these categories of radiation injury.

The observed full range of values in Table 1-4 closely reflects findings from many individual studies. The upper limit of 10 for all four endpoints is a repeatedly observed value; there are some higher values, but these are not recurring findings. The lower limit depends on exact experimental conditions regarding instantaneous dose rate, protraction period, fractionation pattern, and, for tumorigenesis, the specific type of tumor involved. The narrow range recognizes that the upper limit may include some experimental conditions that are not entirely relevant. For example, the highest values come from studies of the effects of continuous daily irradiation until death, which may be an unlikely circumstance for humans except as a result of natural background radiation. The single best estimate values are appropriate for all low-dose-rate, low-LET radiation exposures delivered intermittently, or even continuously, over periods of months to years.

The sixth radiobiological concept is that a variety of chemicals can modify the cell killing effects of radiation. Oxygen and other agents that mimic oxygen by being electron affinic tend to sensitize cells to the effects of a given dose of radiation, while radical scavengers, such as sulfhydryl compounds, tend to protect cells (Mo36, Pal84, Pat49, and Yu80). In general, the redox status of the cell affects its response to radiation. There is little available evidence suggesting that the same considerations apply to mutagenesis and carcinogenesis.

The seventh radiobiological concept is that modifiers exist which have little influence on cell killing but may greatly modify the multistep process of carcinogenesis and its in vitro counterpart, oncogenic cell transformation (Ha87). These modifiers include: (1) hormones (Gu80); (2) tumor promoters, that is, agents that do not affect initiation but that dramatically affect the later stages of carcinogenesis in vivo or transformation in vitro (Ke80); (3) protease inhibitors, such as antipain (Bo79, Ke81).

These factors, which have little influence on cell lethality, can exert a profound effect on the response to radiation when carcinogenesis, transformation or both, are the endpoints being studied. Indeed, such biological factors can dwarf in magnitude the effect of such physical factors as radiation quality and dose rate. Promoters, for example, can alter the shape of the dose-response relationship and can modify the absolute frequency of transformation produced by a given dose of radiation. This is discussed in more detail in Chapter 3.

Differences in Relative Biological Effectiveness (RBE) Among Radiations

Absorbed dose (which is most often referred to simply as dose) is a physical quantity that, all other things being equal, correlates well with biological effect. However, when the quality of radiation changes, absorbed dose alone no longer specifies biological effect. In other words, a given absorbed dose of x rays, does not necessarily result in the same biological effect as the identical dose of neutrons or alpha particles.

To characterize this difference, the concept of RBE was introduced; that is, the RBE of radiation 1 relative to that of radiation 2 is the inverse ratio of the doses of each, (D_2/D_1) required to produce the same biological effect. When the dose-response relationships for the two types of radiation differ in shape, RBE is necessarily dependent on the level of the effect that is considered and should be specified as such.

In the 1963 "Report of the RBE Committee to the International Commission on Radiological Protection and the International Commission on Radiological Units and Measurements" (ICRP63), the comparison of low-LET or standard radiation was designated as x rays, gamma rays, electrons, or positrons of any specific ionization; and an RBE of unity was assigned to any radiation with an average LET in water of 3.5 keV/μm or less. RBE values relative to this standard were then tabulated for a variety of LET values and biological endpoints as a basis for deriving the risk per unit dose of any high-LET radiation relative to the risk per unit dose of the standard low-LET radiation at low doses and dose rates.

In the ICRP-ICRU, 1963 report, it was pointed out that knowledge of the RBE of different types of radiation is used in two ways in radiological protection: first, to provide a basis for setting occupational dose limits for high-LET radiation in relation to accepted limits for low-LET radiation (and to allow the reverse procedure for certain bone-seeking isotopes) and, second, to provide a basis for summing the doses of radiations of different qualities to which a person may have been exposed. This latter use of RBE generally has only limited validity, however, since the prediction of biological effects on the basis of doses of different radiations weighted by their RBEs is a correct procedure only if (1) radiations act independently (a condition rarely met), and (2) their dose-response curves are linear. An example illustrating this point is the fact that the biological effect of neutron fields contaminated by various amounts of photons *cannot* be predicted from knowing the neutron RBE only, except, perhaps, at very low doses.

The ICRP-ICRU Report clearly differentiated the radiobiological concept of RBE from that of the quality factor (now designated Q). Conceptually, Q has a meaning similar to that of RBE; however, it was recognized that Q may not necessarily be identical to RBE. Q is defined as the ratio of occupational exposure dose limits, while RBE values are determined experimentally from radiobiological data. Thus, the concept of Q cannot be considered independently of the general philosophy that is to be applied to the derivation of dose limits for different radiations in the context of radiation protection.

In dealing with the limited data on RBE then available, particularly on the more relevant endpoints of mutagenesis and carcinogenesis, it was assumed that the dose-response curve for high-LET radiation generally tended to be linear, at least at low doses.

For the low-LET standard radiation, discussion oriented largely around the linear quadratic dose-response curve, with an initial linear component dominating at low doses and dose rates. The linear component of the low-LET radiation curve, interpreted as resulting from a single-track mechanism, was thought to be due almost entirely to the high-LET radiation regions at the end of particle tracks. The slope of this linear component of the total dose-response curve was expected to be largely independent of dose rate and dose fractionation. Dose-rate effects were expected only at higher doses, where the dose squared or multitrack mechanisms were associated with the nonlinear component of the overall dose-effect curve. A similar formulation has been used repeatedly in the literature, including a report by the National Council on Radiological Protection and Measurements, NCRP Report 64 (NCRP80).

With higher-LET radiations, the initial linear term generally extends to higher doses than those seen with low-LET radiations. Frequently, it is as difficult to demonstrate a quadratic term with high-LET radiations as it is to demonstrate the initial linear term with low-LET radiations.

On the basis of the linear-quadratic model, the RBE derived from data obtained at high dose rates would be expected to be highly dependent on dose, with a sharp increase in RBE as the dose decreases (Figure 1-2). With decreasing dose rate, the slope of the high-LET curve would be expected to change only minimally. With low-LET radiation, however, at very low doses or with higher doses at low dose rates (or with a very high degree of fractionation), the curve would ultimately be expected to become linear with a slope equal to that of the linear component of the linear-quadratic dose-response curve. Thus, with the limiting conditions of very low dose, any dose at very low dose rates, or both, the limiting RBE should be equal to the slope of the high-LET dose-response relationship, divided by the slope of the linear term of the linear-quadratic dose-response relationship. This ratio was designated in ICRP-ICRU 63 as RBE_m, which is the maximum RBE which is obtained at minimal doses. Thus, emphasis was put on RBE values that were obtained at very low doses, very low dose rates, or both, which were considered to be most relevant to radiation protection. It was made clear by ICRP-ICRU, 1963 that essentially all of the increase in RBE at low doses is caused by a decrease in the slope of the low-LET curve as the dose decreases. This is a basic problem with the current definition of RBE in which low-LET radiation is the "standard" relative to which RBE is evaluated.

Currently, the biological effectiveness of all photon and electron radiations are assumed to be the same, although there is experimental evidence that medium energy (200-250 kVp) x rays are twice as effective as Cobalt-60 gamma rays for low doses on the order of 1 rad, at least for some endpoints such as oncogenic transformation and chromosome aberrations (Bo83, Un76, Sc74). Microdosimetric measurements lead to similar conclusions (El72).

Factors Affecting RBE

Radiation Quality (LET)

The current use of LET as a measure of radiation quality is based essentially on (1) its simplicity (easy to calculate, easy to understand), and (2) the recognition that there exists an association between the spatial patterns of energy deposition and biological effectiveness. As such, LET is a reasonable *qualitative* index for ranking radiations on an ordinal scale of biological effect. For *quantitative* predictions, however, LET has severe limitations (ICRU83, ICRU86).

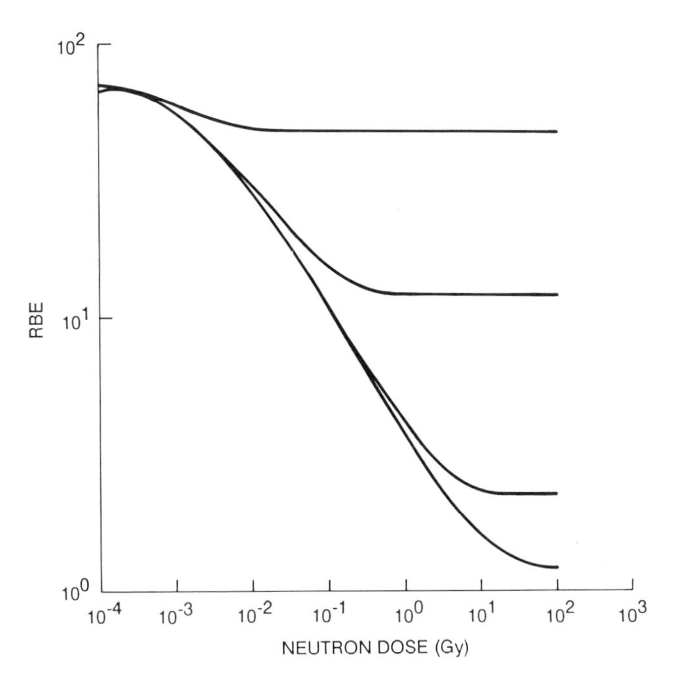

FIGURE 1-2 Dependence of RBE on dose and dose rate for situations in which a linear-quadratic dose-effect relationship applies. The four curves correspond (from top to bottom) to increasing values of the dose rate. The RBE shown here is representative of such endpoints as chromosomal damage or cell killing.

To provide a more adequate description of energy deposition and, implicitly, radiation quality, a number of microdosimetric-based concepts have been developed in the past 20 years. These range from lineal energy (the stochastic counterpart of LET) to distributions of distances between elementary deposits of energy (proximity functions) and radial dose distributions. These quantities are often used in making more successful predictions of RBE as a function of both radiation type and dose. In practical applications the fact remains, however, that they are used only by a restricted group of specialists, so that LET continues to dominate common perceptions of radiation quality (see Glossary).

Variation of RBE with LET

For charged particles of defined LET in the track segment mode, RBE has been determined as a function of LET, by using monolayers of mammalian cells and scoring cell lethality, mutation, and oncogenic transformation as biological endpoints. In all cases, RBE increases with LET, reaching a maximum at about 100 keV/μm, and subsequently falling

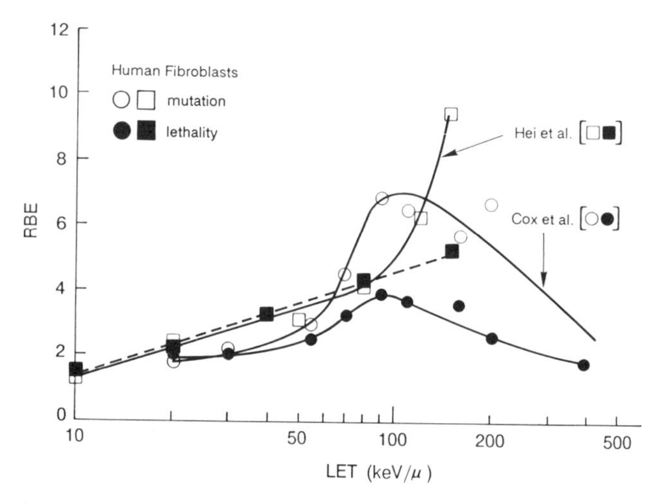

FIGURE 1-3 Radiobiological effectiveness, RBE, as a function of linear energy transfer, LET, in cells of human origin, with cell lethality or mutation at the HGPRT locus as endpoints (Co77, He88).

for higher-LET values. In general, a given LET predicts the same biological effect for a given dose if it is produced by particles with different masses and charges, such as protons, deuterons, or helium ions (Figure 1-3). However, the concept of LET breaks down, and in the case of very heavy particles having an atomic number close to that of uranium, anomalous results have been reported, together with a complex relationship between RBE and LET (Kr82). There is some evidence that, in the same cell system, higher RBE values are found for mutation than for cell lethality, even at the same radiation dose.

Variation of RBE with Dose Rate and Fractionation

For low-LET radiations, the consensus is that decreasing the dose rate or dividing a given dose into a number of fractions spread over a period of time reduces the biological effectiveness. In most cases, for high-LET radiations such as neutrons, the effect of a given dose is relatively unchanged when the dose rate is lowered or when fractionation is used. In a few important instances, including neoplastic transformation in vitro, carcinogenesis in experimental animals, and mutagenesis, dose protraction by use of a low dose rate or by fractionation actually *enhances* the biological effectiveness of a given dose (Figures 1-4, 1-5). The overall conclusion is that the RBE of high-LET radiations compared with that of low-LET radiations may be larger for a low dose rate than for a single acute exposure at a high dose rate.

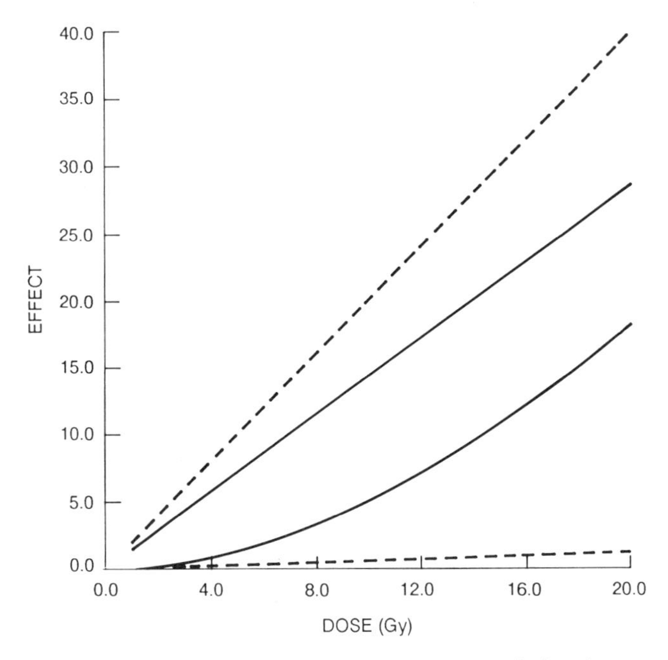

FIGURE 1-4 Hypothetical dose-effect curves for high-LET radiation (upper two curves) and low-LET radiation. It is assumed that lowering the dose rate, (dashed line) results in enhancement of the effect for the high-LET field and a decrease in the yield for the low-LET radiation. This situation has been observed in certain transformation experiments.

Variation of RBE with the Biological System or Endpoint Used

Even for a given dose or dose per fraction, the RBE of a given type of radiation can vary greatly according to the cell or tissue exposed and according to the endpoint scored. At higher doses and with cell lethality as an endpoint, there is a strong tendency for RBE values to be higher for cells and tissues in which the x-ray dose-response relationship has a large initial shoulder and for RBE values to be lower for cells and tissues for which the cell survival curve more closely approximates a simple exponential function of dose. For lower doses and dose rates and with mutation, neoplastic transformation, or carcinogenesis in vivo as an endpoint, a wide range of RBE_m values has been reported. Values have ranged from less than 10 to greater than 100.

The Need for the Concept of RBE

It would be desirable to have human dose-response information, and therefore risk estimates, for somatic and genetic effects for all types of

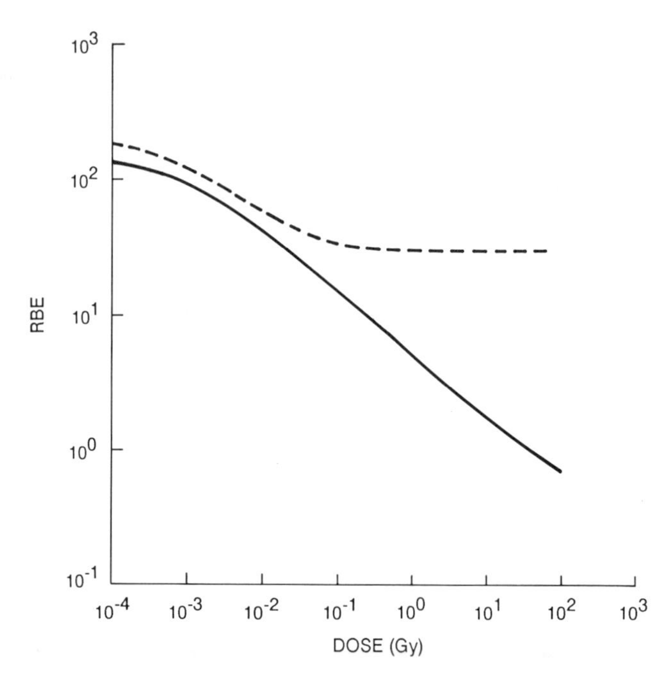

FIGURE 1-5 RBE versus dose for the curves of Figure 1-4. Dashed line, low dose rate; solid line, high dose rate.

radiations, including x rays, neutrons, and alpha particles. Human risk estimates for low-LET radiations are available for many effects from various populations, including the Japanese atomic-bomb survivors; however, the recent revision of the dosimetry from Hiroshima and Nagasaki essentially negates previous RBE estimates for neutrons obtained from the Japanese data (see Annex 4-2). For neutrons, therefore, human risk estimates must result from a two-step process, namely, low-LET effects data from human studies and RBE estimates from animal experiments.

The body of radiobiological data available indicate that, in principle, RBE increases with decreasing dose, with limiting higher values generally reached at low doses or at low dose rates. This relationship results from the fact that the dose-response for low-LET radiation is often a linear-quadratic function of dose, whereas for neutrons it approximates a linear function of dose.

In general, the biological effects of x rays or gamma rays decrease with fractionation or reduction in the dose rate, whereas with neutrons the effectiveness per rad remains the same or even increases as the dose rate is reduced or the time over which the dose is delivered is protracted. For this reason, the RBE is usually quite different for a protracted exposure from that for a single acute exposure.

The limiting value of the RBE at low doses or low dose rates varies with the tissue or cell irradiated (Br73, Fi69, Fi71). This has been documented extensively with cell lethality as an endpoint; but there appears to be at least as much variation between systems when carcinogenesis, mutation, or transformation in vitro is the endpoint. The limiting value of the RBE also varies by a factor of about 2, depending on whether x rays or gamma rays are used as the low-LET radiation (Bo83). This is consistent with the difference in microdosimetric spectra that are characteristic of 250-keV x rays, as compared with those which are characteristic of high-energy gamma-rays (El72). There is some evidence, at least in C3H10T1/2 cells, that the RBE of neutrons relative to x rays may depend on the level of tumor promoting agent present, since TPA has a larger influence on the incidence of oncogenic transformation induced by x rays than neutrons (Ha82).

RBE was a relatively simple concept when it was first introduced, during an era in which radiobiological experimentation was characterized by measurements of the dose which was lethal to 50% of the laboratory animals (LD_{50}) (Bo78). It has now become a complex quantity as a result of the sophistication of the biological systems that are available. While the RBE is complicated by its dependence on dose and dose rate, there is no prospect, at present, that this useful concept can be dropped. A vast body of additional human data will be needed before the concept of RBE can be replaced. However, selection of an appropriate RBE in a specific situation is often difficult. An intensive review of RBE values from experimental systems, including in vitro studies and studies of carcinogenesis in laboratory animals, leads to the conclusion that, for fission spectrum neutrons, RBE values range from about 2 to greater than 100 (ICRU86).

In the analysis of a-bomb survivor data in Chapter 4 of this report, the committee elected to assume a value of 20 for the RBE of bomb neutrons relative to gamma rays for radiocarcinogenesis. This is consistent with the value of Q recommended by national and international groups concerned with radiation protection (NCRP87a, ICRP85). It is also consistent with many experimentally determined RBE values obtained for a variety of tumors in experimental animals, although it was recognized that lower, as well as higher, values have been reported for some neoplasms.

EFFECTS OF RADIATION ON GENES AND CHROMOSOMES

The Genome

The human genome is composed of DNA that is contained principally in the chromosomes and, to a much lesser extent, in the mitochondria. The chromosomes, of which there are 23 pairs, contain about 6×10^9

pairs of DNA bases (3×10^9 per haploid set of chromosomes) and each chromosome includes a single supercoiled molecule of DNA associated with chromosomal proteins. The organization of this material can be visualized microscopically only to a limited degree. With contemporary cytogenetic techniques, fixed chromosomal metaphase spreads reveal 500 or so bands, although refined techniques can reveal about 2,000 bands per haploid set of chromosomes. The total number of genes is unknown but has been estimated to be in the range of 50,000 to 100,000 per haploid set of chromosomes. This genetic material comprises approximately 3,000-4,000 units of recombination (centimorgans). Thus, a visible chromosomal band at a resolution of 500 bands per haploid set, may include 6×10^3 kilobase pairs (kb) of DNA, 100-200 genes, and 6-8 centimorgans of recombining genome. The range in gene size is extreme, with some of the order of magnitude of 10 kb, the retinoblastoma gene about 200 kb, and the muscular dystrophy gene almost 2000 kb of DNA. The parts of genes translated into proteins constitute a minority of total DNA, with many proteins being coded for one kb or so of DNA. Some of the untranslated DNA is important in the regulation of gene expression, while much DNA seems to be extragenic and of unknown function.

Not only does the genome recombine in each generation but it can also undergo mutation, a term applied here to denote all changes in chromosomes, their genes, and their DNA. Thus, alterations in chromosome number and structure are included, as are changes that are not visible microscopically. These latter, submicroscopic changes include deletions, rearrangements, breaks in the sugar-phosphate backbone, errors in DNA replication, and base alterations. Most mutations occur during cellular replication. Mutation occurs in both germ cells and somatic cells, although it is much less apparent in somatic cells unless the mutation occurs during tissue proliferation, as happens with some congenital defects and with cancer. On the other hand, many mutations in the germ line are lethal during embryonic development. Thus, the same mutation might be more common in somatic cells than in germ cells because of the lack of tissue-specific selection against it.

Chromosomal Abnormalities

Three classes of chromosomal abnormalities are known to occur in both germ cells and somatic cells. The best known changes in the germ line are those that affect chromosomal number. Thus, Down syndrome is the result of a mutation in which a parental (usually maternal) germ cell acquires two copies of chromosome 21 as a result of chromosomal nondisjunction during gametogenesis. Fertilization by a normal sperm then yields a zygote with 47 chromosomes. Such trisomy is common at conception, although trisomy (and monosomy) for most chromosomes is invariably lethal to the embryo. The cause of the increase in trisomy with advancing maternal age has focused on differences between male and female gametogenesis. In the female, oogonial mitoses occur during fetal life, and maturation of eggs proceeds to the dictyotene stage, where it is arrested until the time of ovulation. Eggs in a 40-year-old woman have been at this stage for twice as long as in a 20-year-old woman. In contrast, male gametogenesis continues without interruption from puberty to death. Changes in chromosome number can also occur in somatic cells, although the frequency is difficult to estimate because of selection against monosomic and trisomic cells. However, in cancer cells such changes are common.

A second class of chromosomal abnormality is the chromosomal break. When a chromosome break occurs in the cell cycle before DNA replication (G1 or early S phase), it will be observed at the following mitosis as a chromosome break (both chromatids are broken). If the break occurs later in the S phase or in the G2 phase, it will be observed as a chromatid break. For each such break that is observed, there may be many others that rejoin and are not observed. Single breaks, both chromosomal and chromatid, are readily induced by ionizing radiation, and their number increases linearly with dose.

A third class of visible chromosomal abnormality is the structural rearrangement, which embraces unstable forms, such as rings and dicentrics, and stable forms, including interstitial deletions, inversions, and translocations. These result from the inappropriate joining of two breaks at different sites. The number of these aberrations is generally proportional to the square of the x-ray dose, since two events are necessary. However, there is also a linear component, because a single densely ionizing tail of a particle track can produce both events, so that a linear-quadratic equation more properly describes the dose-response relationship (see Figure 1-6). At low doses only the linear term dominates. Neutrons, on the other hand, because they are more densely ionizing particles, often produce two breaks as the result of a single event, so the dose-response relationship is more nearly linear. At low doses, neutrons are much more biologically effective; i.e., the RBE of neutrons relative to that of x rays is significantly greater than unity.

The frequency of two-break aberrations in human lymphocytes irradiated in culture approximates 0.1 aberration per cell per Sv in the low-to-intermediate dose range (L181). The frequency of such aberrations is increased correspondingly in radiation workers, as well as in accidentally or therapeutically irradiated persons, in whom it may serve as a biological dosimeter (L181; IAEA86). Since chromosome aberrations are preponderantly deleterious to the cells in which they occur, the affected cells tend to be gradually eliminated with time.

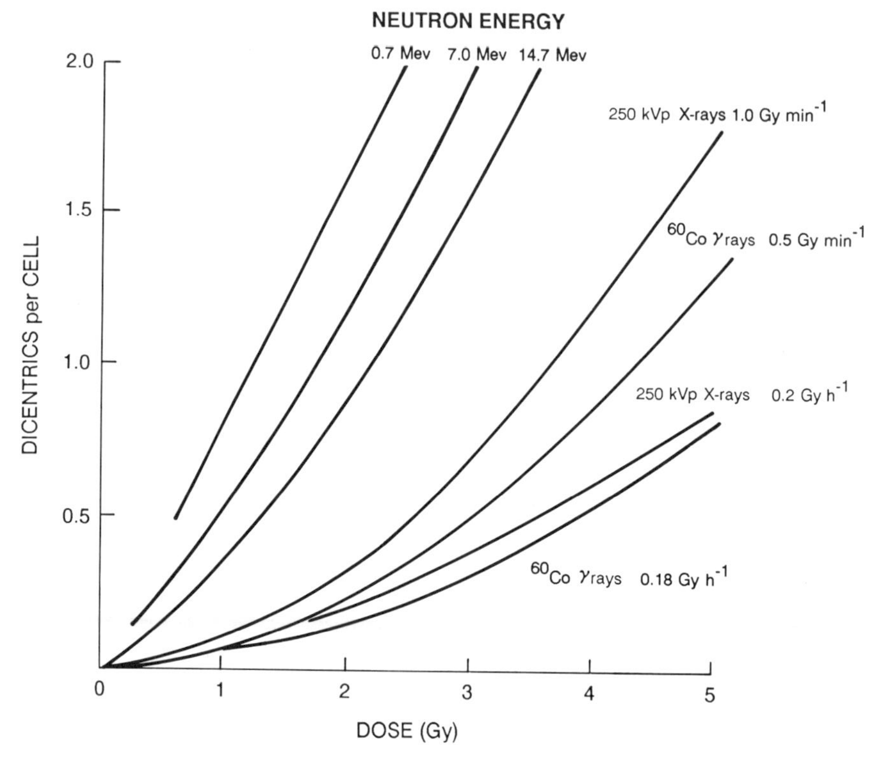

FIGURE 1-6 Frequency of dicentric chromosome aberrations in human lymphocytes irradiated in vitro in relation to dose, dose rate, and quality of radiation (LI81).

Although chromosome aberrations can be induced by relatively low doses of radiation, only a small percentage of them is attributable to natural background radiation. The majority result from other causes, including certain viruses, chemicals, and drugs. The health implications, if any, of an increase in the frequency of such aberrations in circulating lymphocytes is uncertain.

All of these classes of chromosomal abnormalities (non-diploid number, breaks, and structural rearrangements) occur as either germ line (constitutional) mutants or somatic mutants. The Down, Turner, and Klinefelter syndromes are all examples of abnormalities in chromosome number. Many examples of disease-specific constitutional deletions and rearrangements are known. There are no examples of constitutional breaks in all cells examined, but there are about 18 known heritable fragile sites, in which breakage at a specific site can be elicited under certain in vitro conditions, such as folate deficiency (He84). In addition, there are three recessively inherited conditions in which chromosomal breakage and rearrangement occur, namely, ataxia telangiectasia (AT), Fanconi's anemia (FA), and Bloom's syndrome (BS) (He87, Sc74). All three predispose a person to cancer. Patients with AT are unusually sensitive to ionizing radiation, as are their cells in vitro. Cells from patients with BS show a high rate of quadriradial figures, which are caused by homologous chromosomal exchanges, and a high rate of sister chromatid exchanges. A fourth recessive disorder, xeroderma pigmentosum (XP), is not associated with spontaneous chromosomal breakage, but it does predispose a person to chromosomal aberrations induced by ultraviolet light. XP predisposes a person to ultraviolet radiation-induced skin cancers.

Somatic chromosome abnormalities can be found at a low rate in the general population, but they are found almost universally in cancer cells. Abnormalities of both number and form are typical. Cancer cells generate abnormalities at an increased rate, but some of them are so specific that they are regarded as being important in the origin of cancer (Ro84). Thus, about 90% of patients with chronic myelocytic leukemia have an aberration known as the Philadelphia chromosome in their leukemia cells. The Philadelphia chromosome is a reciprocal translocation between chromosomes 9 and 22. Every person with Burkitt lymphoma shows a translocation between chromosome 8 and chromosomes 14, 2, or 22; again, this is confined to the tumor cells. Several other tumor-specific translocations are known. Monosomy for chromosome 22 is common in people with meningiomas. Deletions of various chromosomes are found to be associated at a high frequency with certain cancers; e.g., deletion of the short arm of chromosome 3 (3p-) in persons with small-cell carcinoma of the lung and renal carcinoma; 1p- in persons with neuroblastoma; 11p- in persons with Wilms' tumor; and deletion of the long arm of chromosome 13 (13q-) in persons with retinoblastoma and osteosarcoma. There are also two other kinds of aberrations: homogeneous staining regions and double minute chromosomes; these are found in certain cancers, especially neuroblastoma and small-cell carcinoma of the lungs, and do not occur constitutionally.

The most compelling evidence that a specific aberration may be causal for cancer can be seen in retinoblastoma and Wilms' tumor; that is, persons are predisposed to these tumors if they inherit the same type of constitutional deletion (at chromosome band 13q14 or 11p13, respectively) as is found confined to the tumor cells in other cases. This finding suggests that both the hereditary and the nonhereditary forms of these tumors are initiated by an abnormality at the same chromosomal site, with the abnormality being a visible chromosomal deletion in some cases and a submicroscopic mutation in others (Kn85).

DNA Abnormalities

Abnormalities in chromosome number are not necessarily associated with structural changes in DNA, but chromosomal breaks and aberrations involve such changes, as do the many mutations that are not visible microscopically. The mechanisms by which mutations are caused have of course, been of considerable interest. Some in vitro studies with DNA provide an example of the changes that can occur even at 37° C. For example, one of the most frequently noted changes is deamination of cytosine, in which cytosine is converted to uracil (Li74). Uracil then pairs with adenine instead of guanine, so the coding sequence is changed following replication. Deamination of adenine, although less frequent, also leads to mutation, because the product, hypoxanthine, pairs with cytosine instead of thymine (Li72). Another important change concerns the methylation of guanine, which may be caused by the presence of the active methyl donor S-adenosylmethionine (Ry82). This change alters both the geometry and the base pairing of guanine. Two products of thymine, the cyclobutane pyrimidine dimer and 6,4-pyrimidine-pyrimidone, which are produced by ultraviolet irradiation, distort the DNA helix (Mi85).

Mutations would occur at much higher rates than are actually observed, if it were not for the existence of repair mechanisms. In the case of the thymine photoproducts noted above and in the case of bulky adducts of DNA with certain chemicals, repair proceeds via sequential steps, the first being a cutting of the abnormal strand of DNA on each side of the site of the abnormal nucleotide by an endonuclease. This leads to deletion of a DNA segment that includes the dimer or adduct. The gap, which may be enlarged by an exonuclease, is then filled by a polymerase-catalyzed DNA strand that is complementary to the intact strand of DNA. The final reaction is closure at the growing end by a ligase. This is the classical excision repair pathway first described for bacteria. It is a relatively slow process, but it is very accurate. Recognition of the DNA repair pathway came in humans with the discovery that the disease xeroderma pigmentosum involves a defect in excision repair (Cl68). This was the first known example of a DNA repair defect in humans. It is thought to account for the propensity of individuals with this disease to develop cancer of the skin, because ultraviolet light induces thymine photoproducts in the exposed skin cells. If not excised, these thymine photoproducts in turn impair faithful DNA replication, causing induced mutations and chromosome aberrations at an increased rate, as has been observed in vitro. Presumably, these mutations may occur in one or more "cancer genes" that are involved in carcinogenesis in skin cells and melanocytes.

Many spontaneous and induced mutations do not affect the gross configuration of DNA. Such mutations include those resulting from the removal, destruction, or mutation of bases; destruction of deoxyribose residues; and breakage of DNA chains. Such damage, which is common with exposure to ionizing radiation, is also corrected by excision repair, but an array of specific enzymes different from those employed in the classical mechanism is used (Li82). These mechanisms are much faster, but less accurate, so residual mutation is more likely. DNA chain breaks are, of course, associated with all chromatid and chromosome breaks. The dose-response curves for single-strand and double-strand breaks may both be linear with x rays, apparently because the former are caused by single ionizations and the latter are caused by the dense tails of ionization tracks. Most chain breaks are repaired following modification of the break termini, filling the defect with polymerase activity and ligation. It may be that the same ligase can function in both slow and fast repair processes. It has recently been reported that ligase deficiency is a feature of Bloom's syndrome (Ch87, Wi87). This would explain the propensity for chromosome breakage and aberration found in patients with that syndrome. It would also explain the increased mutation rate that has been reported in vitro (Vi83) and recently in vivo (La89).

An important kind of damage to DNA, and one frequently produced by ionizing radiation, is removal of a base, with the formation of an apurinic or an apyrimidinic (AP) site (Li82). This damage can be repaired by an AP endonuclease that excises the remaining deoxyribose phosphate. There are reports that some cases of xeroderma pigmentosum and ataxia telangiectasia may have reduced AP endonuclease activity. After the creation of AP sites, the AP site itself can be mutagenic if the sites are not removed by AP endonuclease. During the next round of cell division, DNA polymerase may copy past the AP site by inserting a purine, usually adenine, without regard to what is present opposite the site in the other strand. This kind of repair is obviously prone to error.

Alterations in DNA caused by deamination of cytosine or adenine and by disruption of purine or pyrimidine rings can also be repaired. The mismatched or degraded base is removed by one of several specific glycosylases, enzymes that are relatively abundant and rapidly acting, leaving an AP site, which is then handled by AP endonuclease as noted above (Li82). While genetic defects in these enzymes are not known in humans, bacterial mutants lacking uracil glycosylase show considerably altered mutation rates.

One other alteration in DNA is processed in a unique way. As noted earlier, methylation of guanine (of an oxygen atom at position 6) may occur under physiological conditions, but it is also produced by certain alkylating agents. An unusual enzyme has been discovered that removes this methyl group and transfers it to a cysteine residue of the enzyme itself, restoring the DNA to its normal configuration, but inactivating the enzyme in the process (Ha83). This methyl transferase is literally a suicide enzyme; in fact, it is not strictly an enzyme because it is not regenerated. No inherited defect has been reported for this enzyme in humans, but it has been found that in some cancer cells the function of this enzyme is defective. It may be that such cancer cells undergo further mutations relevant to tumor progression more readily. If so, they should be more susceptible to killing by alkylating agents.

Conclusions

Although the human genome is highly stable in both germ-line and somatic cells, errors do occur in its transmission from one generation of individuals or cells to the next. These errors occur at a spontaneous rate that can be increased by environmental agents, including radiation. These errors can be so macroscopic that they are detectable cytogenetically, as in the case of abnormalities of number or structure of chromosomes. Other errors cannot be detected cytogenetically, but can be detected as changes in the nucleotide sequence of a gene. Many such errors (mutations) are repaired. The importance of the existence of repair mechanisms is underscored by the predisposition to cancer that is associated with some rare hereditary disorders in which one of these repair mechanisms is defective.

INTERNALLY DEPOSITED RADIONUCLIDES: SPECIAL CONSIDERATIONS

Exposure to ionizing radiation occurs from radionuclides deposited within the body as well as from sources outside the body. Differences in the characteristics of these two types of exposure must be considered when interpreting studies of irradiated populations and estimating the possible health effects of different patterns of irradiation.

With an internally deposited radionuclide, the radionuclide enters the body at the time of exposure but the doses it delivers to various organs and tissues of the body continue to accumulate until the radionuclide is removed by physical or biological processes. Thus, the radiation is delivered to various organs gradually, at changing dose rates, over what may be an extended range of ages. An internally deposited radionuclide also frequently produces nonuniform irradiation to the organs and tissues in which or near which it is incorporated, depending on its radioactive emissions and metabolic characteristics. In this respect, the spatial and temporal patterns of the doses delivered by internally deposited radionuclides differ from those typically delivered by external irradiation (Figure 1-7).

These and other differences in both dosimetry and biological response have a direct impact on the characteristics of the resulting dose-response relationships. Accordingly, any quantification of human health risks from

External Irradiation

Exposure and Dose Occur at Same Time
Dose - Relatively Uniform
Majority of Data on Exposed People

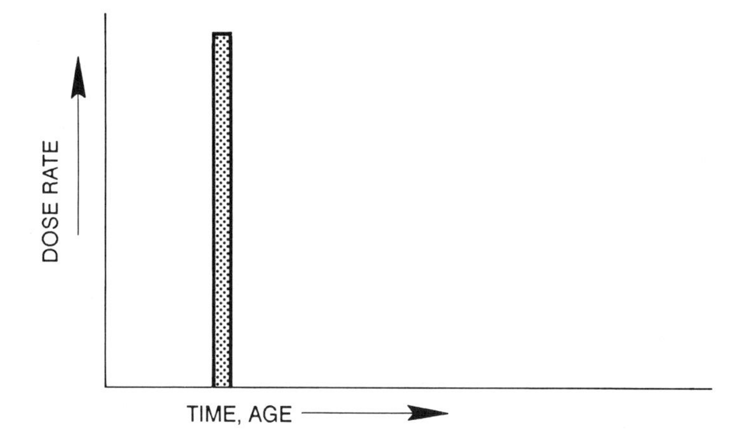

Radionuclides

Exposure - May Be Chronic
Dose - Very Likely to Be Chronic and Nonuniform

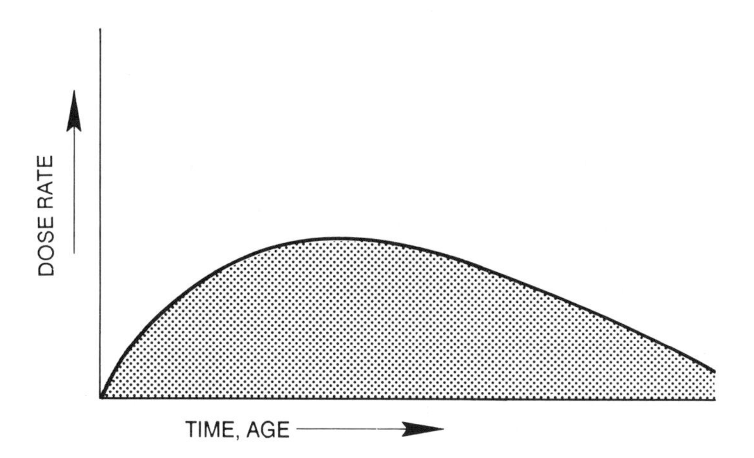

FIGURE 1-7 Temporal patterns of dose distribution.

exposure to ionizing radiation must consider, first, the determination of risk factors for exposure situations in which adequate data on dose and response are available, and second, the relative importance of various dosimetric or response factors that can alter the resulting risk estimates. This applies to both external and internal irradiation conditions. In this section, the general characteristics of exposure and dose-response relationships are discussed for internally deposited radionuclides as they apply to estimation of site-specific radiation-induced cancer risks in exposed human populations (see Chapter 4). The health effects of radon progeny and other internally deposited alpha-emitting radionuclides were examined in depth in the BEIR IV report (NRC88).

Radionuclide Dose-Modifying Factors

The intake of radionuclides can occur by inhalation, ingestion, injection, and absorption through the skin and mucous membranes or through cuts and abrasions (ICRP79). The relative importance of these different routes of intake depends on the particular exposure situation considered, for example, environmental or occupational exposure, accidental exposure, or medical administration of radionuclides. Each of these exposure routes has its own characteristic pattern of initial deposition on or in various parts of the body such as the lungs, the gastrointestinal tract, or skin. As long as a radionuclide is present at one of these sites of intake, the surrounding tissue will be irradiated, and the extent of this irradiation will be determined by dosimetric factors (see the section on physics and dosimetry earlier in this chapter and see below).

A portion of the radionuclide present at these sites of intake may dissolve and be absorbed into the blood. Once uptake to body fluids has occurred, the radionuclide will be deposited in other organs and tissues, depending on its physical and chemical properties. Chemical, physical, and biological processes can also influence the effective retention time for a given radionuclide, thereby influencing the period of time during which the irradiation of the surrounding tissues occurs.

The description and quantification of the deposition, retention, and excretion of internally deposited radionuclides are generally well understood. The most extensive reviews of metabolic and dosimetric data for the different radionuclides currently available are those given by the International Commission on Radiological Protection (ICRP) Publication 30 (ICRP79). Additional information on the dosimetric approaches incorporated in the current ICRP system is available in reports by Johnson (Jo85) and the National Council on Radiation Protection and Measurements (NCRP85). The methodology and values given by ICRP were assembled for radiological protection planning purposes. Thus, the values chosen for the various parameters are conservative; that is, they can lead to overestimates of risk factors. These values may not be appropriate for estimation of risk when the organ and tissue doses received by exposed individuals are considered.

Some of the relevant data have been derived from human studies, in particular studies on the deposition of inhaled particles and gases (e.g., radon progeny in uranium miners), whole-body retention of radionuclides with emissions that are detectable outside the body (e.g., radiocesium from worldwide fallout), and excretion (feces and urine) samples collected primarily in occupational exposure situations (e.g., transuranic radionuclides). Concentrations of radionuclides in some tissues have also been measured at autopsy. The remainder of the data have been and continue to be obtained from studies of various species of laboratory animals conducted under controlled laboratory conditions. The study of laboratory animals makes it possible to examine radionuclide biokinetics and metabolism, for which human data are sparse or nonexistent, and to determine the effects of various modifying factors on the resulting dosimetry (NRC88).

Each laboratory animal species has its own anatomic and physiological characteristics that need to be considered when the resulting dosimetric parameters are extrapolated to human exposure situations. For instance, the mechanical clearance of insoluble particles from the pulmonary region is strongly species-dependent; mice and rats clear these particles by mucociliary activity much more rapidly than do guinea pigs, dogs, or humans (Sn83, Sn84). Knowledge of these differences is necessary for appropriate dose calculations in studying dose-response relationships in different species as surrogates for humans. Similarly, the hepatic turnover of actinide and lanthanide radionuclides in mice and rats is considerably faster than that in dogs and nonhuman primates (ICRP86, Bo74).

Other factors that need to be considered when determining the dose received by critical cells include an identification of the target cells of concern and how the patterns of cellular irradiation are influenced by nonuniform radionuclide deposition or clearance, age, and health status (Sm84, Fi83).

Radionuclide Response-Modifying Factors

There are only a few groups of human subjects with radionuclide burdens of sufficient magnitude to produce long-term biological effects. Major groups in this category include patients treated with ^{226}Ra, ^{224}Ra, Thorotrast (^{232}Th and progeny), or ^{131}I; uranium miners exposed to ^{222}Rn and its progeny; and uranium workers exposed to ^{238}U, ^{235}U, and ^{234}U. All of these study populations, except those exposed to ^{131}I, involve people exposed to high-LET (alpha) radiations and were discussed in detail in the BEIR IV report (NRC88).

With the exception of the special case of exposure of the human thyroid to ^{131}I, discussed in Chapter 5, long-term biological effects of internally deposited low-LET-emitting radionuclides have not been observed in human populations. Estimations of the potential health risks of such radionuclides must be sought by other means. To provide such data, a large number of life-span studies on the effects of radionuclides have been conducted in laboratory animals. Major studies currently in progress include those examining the effects of inhaled $^{239}PuO_2$ in baboons (Me88), inhaled $^{238}PuO_2$ or $^{239}PuO_2$ in dogs (Pa86, Mc86), inhaled $^{239}Pu(NO_3)_4$ in dogs (Pa86), inhaled fission products (^{90}Sr, ^{144}Ce, ^{137}Cs, ^{91}Y, and ^{90}Y) in relatively soluble or insoluble forms in dogs (Mc86), injected ^{226}Ra in dogs (Go86), and intravenously injected ^{239}Pu, ^{226}Ra, ^{228}Th, ^{228}Ra, and ^{90}Sr in dogs (Wr86). Comparative life-span studies involving large numbers of rats exposed to low doses of high- or low-LET radiation include studies of inhaled $^{239}PuO_2$ (Sa88), inhaled $^{144}CeO_2$ (Lu87a) and thoracic or whole-body x-irradiation (Lu87b). These studies should provide a critical link between observations on laboratory animals and existing human data.

It is expected that such studies, many of which are currently nearing completion (Th86), will contribute to our understanding of the relative importance of possible risk modifiers such as dose, dose rate, nonuniformity of dose distribution, species, age, health status, and exposure to other carcinogenic agents in combination with radiation.

USE OF ANIMAL STUDIES

Observations on the biological effects of ionizing radiation began to be made soon after the discovery of x rays in 1895. Already in 1896, there were reports of dermatitis and alopecia in those experimenting with x-ray generators (Fu54). By 1902-1903, the first reports had appeared describing skin carcinomas on the hands of radiologists, and less than a decade later, sarcomas had been induced in rats by repeated exposure to irradiation (Fu54). In 1906, from studies of radiation effects on the testes of goats, J. Bergonie and L. Tribondeau formulated their well-known generalization that:

> X-rays are more effective on cells which have a greater reproductive activity; the effectiveness is greater on those cells which have a longer dividing future ahead, on those cells the morphology and function of which is least fixed (Translation by G.H. Fletcher, Be06).

Two decades later, H. J. Muller reported the mutagenic effects of radiation on the germ cells of *Drosophila melanogaster* (Mu28). During the 1920s congenital abnormalities were recognized in children whose mothers had been irradiated while they were pregnant (Go29), and in the following decades radiation teratogenesis was widely investigated in mice and rats

(Ru54). Research in these areas and on the systemic cellular and molecular effects of ionizing radiations have continued in a variety of animal species, in parallel with continued observation on radiation effects in humans.

In many respects the human data and the animal data are complementary. There are several important areas in which the human data are inadequate for risk estimation and must be interpreted in the light of concepts developed from experiments with animals. In particular, information from experimental animals is useful for human risk assessment in the following areas:

1. prediction of the effects of high-LET external radiations, including neutrons;
2. prediction of the effects of low or varying dose rates and of various patterns of fractionation of exposure to low-LET radiations, high-LET radiations, or both;
3. clarification of the mechanisms of radiogenic damage including mutagenesis, carcinogenesis, and developmental effects; this is crucial to the development of appropriate interpretations and mathematical models of radiation effects in humans; and
4. prediction on the uptake, distribution, retention, dose distribution, and biological effects of internally deposited radionuclides for which there are inadequate data in humans.

The validity of quantitative extrapolation from animals to humans is of great concern. Such a procedure may be defined better as the "transposition" of concepts and parameters derived from animal studies to humans in order to compensate for inadequate or unavailable information. Opinions vary about appropriate methods for extrapolating data and concepts between species, but there are times when it is essential. It is unlikely that humans are so physiologically unique among mammalian species as to invalidate selective use of animal data.

Consideration has been given recently in two areas to direct extrapolation of dose-incidence ratios for carcinogenesis from experimental animals to humans. The first of these includes the use of ratios of the relative effectiveness of two internally deposited radionuclides in animals in order to estimate the relative risks of the nuclides for man when human data are available on only one of the nuclides. This was examined in the BEIR IV report (NRC88). The second is the direct application of the relative risk (per Gy) of cancer in animals to prediction of the relative risk of cancer in irradiated humans (St88).

Much of the information on radiation-induced and spontaneous mutation rates in humans is based on chromosomal aberrations and specific locus mutations in somatic cells, the latter primarily in culture systems. Estimations of human genetic risk are thus made in the light of dose-response relationships and mechanistic considerations derived from experimental studies on inherited genetic effects, primarily in mice, and interpreted in the light of the large body of data from other biological systems. Radiation-induced mutation is a process which is completed within a relatively short interval after exposure. Interspecies extrapolation of experience with dose-mutation effects can therefore be done with somewhat greater confidence than can comparable extrapolations of effects on multistage prolonged processes such as carcinogenesis and lifeshortening.

There are extensive experimental data concerning radiation effects on embryogenesis with specific reference to the development of gross abnormalities of the central nervous system and disruption of neuroblast proliferation, migration, differentiation, and establishment of neural pathways. Measurements of effects of exposure during embryogenesis on neurological function, including learning capacity and cognition, are less common and more difficult to perform in experimental animals. Although interpretation and application of the experimental data to human risk estimation requires careful comparison of equivalent developmental stages, the data are valuable in complementing sparse human information.

EPIDEMIOLOGICAL STUDIES: SPECIAL CONSIDERATIONS

Epidemiologic studies are a critical tool in assessing radiation risks, since they alone provide data directly applicable to humans. However, epidemiologic studies of individuals exposed to radiation have methodologic limitations which should be kept in mind when assessing the results of such studies. This section briefly summarizes these concerns. Further discussion of these issues can be found in standard textbooks on epidemiologic methods (Ma70, Ro86). Most epidemiologic studies of low-LET radiation have focused on cancer as the outcome. This discussion of epidemiologic methods and their limitations also focuses on cancer, although most of the considerations also apply to studies of other outcomes.

High-Dose Studies

The use of high-dose studies to quantify risk estimates involves a two-stage process. First, risk parameters that apply to the particular high-dose group under observation must be estimated from the empirical data. Second, mathematical models must then be used to extrapolate from the experience of the specific high-dose population to that of the low-dose population of interest, taking into account differences both in exposure factors such as dose and dose rate and host factors such as age, sex and race. Both steps are, of course, subject to error, and the assumptions and limitations involved in the second step will be discussed in detail later in this chapter. The problems and limitations involved in the first step are discussed here.

Studies reported to date have essentially been of the retrospective cohort type. Populations receiving high doses of low-LET radiation are rare, and exposure to such doses is unlikely to occur in the future, apart from the therapeutic irradiation of patients. Such studies are subject to both sampling variability and bias. Sampling variability should generally be adequately expressed by the confidence intervals around the parameters estimated by the particular mathematical model, but bias represents a greater problem. Biases in epidemiology are generally classified as resulting from selection, information, or confounding.

Selection bias can be defined as arising from any design problem that tends to make the study subjects unrepresentative of their source population. Such a bias can prevent generalization of the results. For example, if the survivors of the atomic bombings at Hiroshima and Nagasaki were healthier than the general population, their susceptibilities to radiation carcinogenesis could be different from those of the general population. In addition, selection may lead to internally biased results when the follow-up is selective. This occurs when those individuals selected for follow-up are different for differing categories of exposure and when that difference is associated with a differing underlying cancer risk. For example, if only 50% of the atomic-bomb survivors had been followed, and there were more smokers in the high-dose group that were followed than in the low-dose group, there would be an excess of lung cancer in the high-dose group that was not caused by radiation. Such a selection bias is likely to occur only when there is substantial loss to follow-up. It is unlikely that this plays a role in the major high-dose epidemiologic studies on which risk estimates are currently based, since follow-up has been essentially complete for these studies.

Information bias, which refers to any process which distorts the true information on either exposure or disease status, is likely to be of more importance than selection bias. Misclassification of exposure is likely to be a major potential source of error in making risk estimates. Nondifferential misclassification with respect to exposure level leads to an underestimation of risk and tends to reduce any upward curvature in the dose-response relationship. This occurs, for example, when the distribution of errors in dose estimates is the same in the diseased and the nondiseased, as will generally be the case for most cohort studies. Other biases may be more subtle. Misclassification of disease status is particularly important when such status is determined from death certificates which are often unreliable for a number of cancer types. These errors are more likely to be differential, i.e., dependent upon a subject's exposure status, and could bias a dose-response curve away from the null.

Finally, confounding—i.e., distortion of risk estimates due to the association of both exposure and disease with some other covariate, such as smoking—is unlikely to be of substantial importance in affecting risk estimation based on comparison of groups of individuals with varying degrees of exposure, but it could be of importance when an unexposed control group is also used in the estimation procedure. For example, the characteristics of the "not in the city" group in the Japanese atomic-bomb survivor study may be somewhat different from the group exposed to the radiation, and if these characteristics are associated with differing cancer risks, such confounding would have an effect on the risk estimates. This may be a particular problem with studies of patients irradiated for medical conditions if risk estimation is carried out with an unexposed comparison group, such as the general population: the condition for which irradiation is used could well be associated with an altered cancer risk.

The three types of bias discussed above could all play roles in affecting the internal validity of risk estimates (i.e., the validity of the results for the particular population being studied). However, even in the absence of such biases, there remains a fundamental problem in extrapolating the risks from one population to another, for example, from the Japanese to North Americans. The method of such extrapolation depends on the mathematical model chosen; and, although empirical evidence may be available from studies carried out in both countries, there often is considerable uncertainty about the validity of the procedure that is used.

The quantitative risk estimates developed in Chapter 4 of this report are based primarily on extrapolation from studies of populations exposed to high doses of radiation over relatively short periods of time. The rationale for this approach is that only these studies provide sufficiently precise estimates of risk at any dose. Risk estimates for low doses and protracted exposure could therefore be in error because of (1) an inappropriate mathematical model, or (2) biases in the high-dose epidemiologic studies used to estimate the parameters of the chosen model, as discussed above.

The committee has attempted to mitigate the first problem by using sufficiently general model classes that include most of the widely accepted alternatives and by providing estimates of the range of uncertainty in the estimates. In general, the estimates of risks derived in this way for doses of less than 0.1 Gy are too small to be detectable by direct observation in epidemiologic studies. However, it is important to monitor the experience of populations exposed to such low levels of radiation, in order to assess whether the present estimates are in error by some substantial factor.

Low-Dose Studies

A number of low-dose studies have reported risks that are substantially in excess of those estimated in the present report. These include risks to populations exposed to high background levels of radiation, diagnostic x rays, and fallout from nuclear weapons testing or nuclear accidents, and to individuals with occupationally derived exposures. Some of these studies are discussed in more detail subsequently. Although such studies do not provide sufficient statistical precision to contribute to the risk estimation procedure per se, they do raise legitimate questions about the validity of the currently accepted estimates.

The discrepancies between estimates based on high-dose studies and observations made in some low-dose studies could, as indicated above, arise from problems of extrapolation. An alternative explanation could be inappropriate design, analysis, or interpretation of results of some low-dose studies. This section discusses the particular methodologic problems which can arise in such studies, and the section on low-dose studies in Chapter 7 summarizes a number of these studies and assesses their results, taking into account the methodological limitations discussed here.

The problem of random error caused by sampling variability is relatively more important for low-dose than for high-dose studies. (Sampling variation means the range of results to be expected by exact replication of the study, if this were possible; its major determinant is sample size and its distribution across exposure and disease categories.) To understand why this is so, suppose that two studies were conducted, one in a population exposed to 1 Gy and one in a population exposed to 0.01 Gy, in which similar sample sizes and designs were used, and suppose that the resulting standard errors on the log relative risk were the same. Thus, suppose the relative risk in the high-dose population was 11 with 95% confidence intervals of 5.5 and 22 and the relative risk in the low-dose population was 1.1 with confidence intervals of 0.55 and 2.2. The point estimates on the relative risk coefficient from the two studies would be identical at 10/Gy, but the confidence intervals on the high-dose estimate are 4.5 and 21 and on the low dose estimate are −4.5 and 12.0. This comparison emphasizes the importance of considering sampling variability in assessing the results of low-dose studies. In fact, the problem of sampling variation is even more serious than this simple example would indicate. The standard error of the relative risk in a simple 2 × 2 table of exposure by disease status is determined primarily by the size of the smallest cell in the table, which is usually the number of exposed cases. In most studies of low-dose effects, this cell may be quite small, so the resulting standard error is larger than that for high-dose studies, even if the overall sample sizes were the same.

In general, systematic biases are also relatively more important for the objectives of low-dose studies than they are for those of high-dose studies. Because of the existence of more and larger populations exposed to low doses, low-dose studies are often ecological (correlational) or case-control studies rather than cohort studies. The ecological and case-control studies are particularly prone to bias in their design.

Selection bias is a major potential problem in case-control studies: the major concern is over the appropriateness of the control group. This is a particular problem for those studies in a medical setting.

Information bias leading to misclassification of either exposure or disease status, if random, leads to underestimated risk, and several low-dose studies could well involve substantial systematic misclassification, for example, misclassification because of recall bias by cases in case-control studies. Similarly, tumors which can be induced radiogenically could be overestimated in radiation-exposed individuals.

Confounding may be more important for low-dose than for high-dose studies. An observed relative risk of 2 is much more likely to be produced solely by confounding than a relative risk of 10 (Br80). The possibility of confounding can only be judged on a study-by-study basis, but some generalizations are possible. Ecological correlation studies, such as the studies of areas with high levels of background radiation, are probably the most susceptible to confounding. Residents of areas with high levels of background radiation are likely to differ in many ways from those in areas with low levels of background radiation. This could affect cancer rates, but data on the relevant characteristics are unlikely to be available for analysis. As an example, exposure to radiation from terrestrial sources may vary with housing structure, which, in turn, may reflect a socioeconomic status that correlates with such factors as smoking and alcohol use. This possibility alone generally makes such studies uninterpretable, and when the ecological fallacy discussed below is also considered, these two problems alone are enough to make such studies essentially meaningless. Case-control studies, on the other hand, generally offer the greatest opportunity to control for confounding by matching or obtaining information on definable covariates for use in analysis. However, the extent to which this has been done varies from study to study. It is necessary, of course, to collect data on such confounders, and, if the confounders are not recognized in advance, the appropriate data may not be available.

Finally, three other potential biases of low-dose studies should be mentioned (Be88). The first is the ecological fallacy, that is, that in correlational studies, any excess risk occurring in a population with increased exposure may be occurring in individuals other than the individuals who are actually receiving the excess exposure. Second, is the possibility of selective reporting. Epidemiologists are more likely to report and journal editors are more likely to accept positive findings than null findings. Thus, information in the literature on populations exposed to low doses of radiation may be slanted in favor of those studies that show higher risks than the conventional estimates, since those that show estimates consistent with the accepted values would not be seen as significant. The magnitude of this potential effect is unquantifiable, but it almost certainly exists and plays a role in the plethora of low-dose studies with a reported positive risk. Third, there is the problem of multiple comparisons. This arises if a number of tests of significance are made with respect to elevated risks for a number of cancer sites. Such a process invalidates the conventional value quoted for the test of significance and leads to more significant results than nominally would be expected by chance. For example, in following a cohort of occupationally exposed individuals, if comparisons are made for 10 cancer sites with a p value of 0.05, which nominally would be expected 5% of the time by chance for a single comparison, significant excesses would arise 40% of the time by chance for at least one of those outcomes. *Interpretation of such results must be guided by prior hypotheses, and by consistency of results among studies, a major criterion for causality.*

RISK ASSESSMENT METHODOLOGY

Need for Models in Risk Assessment

One of the major aims of this report, as of previous BEIR reports, is to provide estimates of the risks of cancer resulting from various patterns of exposure to ionizing radiation. In principle, such estimates could be derived by identifying a group of individuals with similar exposures and similar backgrounds and following them to compare the proportion of the group who eventually developed cancer with the proportion who developed cancer in a comparable unexposed group or in the general population. For situations in which it is not possible to measure the risks directly, statistical models must be used to derive estimates.

Large sample sizes are needed in any such comparisons, to minimize random variation; the rarer the disease and the smaller the effect of exposure, the larger the sample needs to be. For example, the BEIR III report estimated that a single exposure to 0.1 Gy (10 rads) of low-LET radiation might cause, at most, about 6,000 excess cases of cancer (other than leukemia and bone cancer) per million persons, as opposed to a natural incidence of about 250,000. To identify this number as a statistically significant excess, a cohort of about 60,000 people with the same exposure would have to be followed for a lifetime, or an even larger number of people would have to be studied if follow-up were for a shorter period of time. Under ideal conditions, a case-control study to identify the same excess would have to consist of at least 120,000 cases and 120,000 controls. It is unlikely that such large groups with similar exposures could be identified, let alone feasibly studied. Furthermore, even if the random variation could be overcome by the large sample sizes needed, estimates of such small excess risks (2%) could easily be biased by confounding, misclassification, or selection effects. Epidemiologists generally agree that excess risks of less than 50% are difficult to interpret causally (Br80). In practice, therefore, it is necessary to obtain risk estimates by extrapolation from smaller and less homogeneous groups who have been exposed to larger doses by using statistical dose-response models.

The second problem is that there are many other factors that are known to contribute to cancer risks or to modify the effects of radiation on cancer risks, and these factors need to be taken into account. While it is theoretically possible to control for such factors by cross-classifying the data into subgroups that are homogeneous with respect to all relevant factors, it is again unlikely that sufficiently large subgroups will be available to allow for stable estimates, particularly if the number of factors is large. For investigating lung cancer, for example, it might be necessary to control for sex, age, time since exposure, and smoking habit; if four levels were used for grouping each factor other than sex, a total of 128 subgroups would be needed, each of which would need to be the minimum size if risk estimates specific to each group were to be observed directly. Since this is not generally feasible, it is necessary to rely on multivariate statistical models to identify the consistent patterns across the variables simultaneously and to predict the risks for subgroups in which the sample sizes are inadequate.

The third problem is that direct estimates of lifetime risk can only be obtained after an exposed population has been followed for a lifetime. Few populations have been followed so long, and even the atomic-bomb survivors, one of the populations followed for the longest period, has been followed only for just over 40 years. As the risks for many cancers in this population are still elevated, it is an open question whether the excess risk will continue for the remainder of the population's life and, if so, at what rate. It is not appropriate to wait until follow-up is complete, however, since interim estimates of risk are needed now for public health purposes. Again, to provide such estimates, one must fall back on statistical models that adequately describe the data available so far and the range of uncertainty around them.

Epidemiologic data have increasingly been called on to help resolve claims for compensation by exposed individuals. Because a radiation-induced cancer is clinically indistinguishable from cancers caused by other factors, such claims must be settled on the "balance of probabilities," in other words, by determining what was the most likely cause, given the individual's history of exposure to radiation, and taking into account confounding and modifying factors. The calculation of these probabilities of causation depends on the availability of suitable multivariate exposure-response models. A recent National Institutes of Health working group

(NIH85) has provided tables of such probabilities; these were based on data that were available at the time.

Approaches to Model Construction and Fitting

Exposure-Time-Response Models

The last 20 years have seen a rapid increase in the use of multivariate models in the analysis of epidemiologic studies. The incidence of cancer and other diseases that are characterized as binary endpoints (present or absent) has usually been analyzed in terms of either the logistic model for the probability of disease $P(z)$ as a function of exposure and other variables, where $z = (z,...,z_p)$,

$$P(z) = 1/[1 + \exp(-\alpha - z'\beta)] \tag{1-1}$$

or the proportional hazards model for the instantaneous rate of disease, $\lambda(t \mid z)$, at age t,

$$\lambda(t \mid z) = \lambda_0(t)\exp(z'\beta), \tag{1-2}$$

where α and β are unknown regression coefficients that must be estimated, and $\lambda_0(t)$ is the baseline rate in unexposed subjects ($z = 0$). These functions have a number of desirable mathematical properties that make them convenient to use under a wide range of circumstances, but they are not based on any particular biological theory. Thus, while they are useful for describing patterns and testing associations in which there is relatively little prior knowledge or biological theory, more reliable predictions can be made by using models that exploit such prior knowledge.

The Committee has chosen, instead, to base its reanalyses of original epidemiologic data and risk assessments on the radiobiological principles and theories of the carcinogenesis process that are described elsewhere in this report. From this discussion, several considerations have emerged that need to be considered in designing statistical models.

Dose-Response Relations Radiobiological theory indicates that at low doses, the risk of a biological lesion being formed should depend linearly on dose if a single event is required or on the square of dose if two events are required. It is commonly held that high-LET radiation can cause lesions by the traversal of a single particle, but that for low-LET radiation, either one or two photons might be required. At higher doses, radiation can cause cell sterilization or cell death, which competes with the process of malignant transformation. The probability of avoiding sterilization and death follows the usual laws of survival, which indicate that it should have a negative exponential dependence on either dose or the square of dose (again, depending on whether one or two events are needed). When these principles are combined, one obtains the general dose-response model used in the BEIR III report and extensively throughout the radiation literature:

$$F(D) = (\alpha_0 + \alpha_1 D + \alpha_2 D^2)\exp(-\beta_1 D - \beta_2 D^2), \tag{1-3}$$

where D is the radiation dose, and $F(D)$ is the incidence rate of cancer, a quantity that will be defined more rigorously below.

Dependence on Time Cancer rates vary over several orders of magnitude as a function of age, and the excess risk caused by radiation exposure also varies as a complex function of age and time since exposure. Numerous mathematical theories of carcinogenesis have been devised to predict the dependence of incidence rates on exposure, age, and other time-related factors, but so far none has won universal acceptance and there have been few attempts to fit these models to epidemiologic data. Although the committee felt that stronger inferences about lifetime risk might be possible by exploiting these biomathematical models, it was unable to arrive at a consensus as to the particular models to use. Thus, there remains a need for simpler methods of summarizing the basic patterns of excess risk over time that do not depend on unproven hypotheses. Because leukemia and bone cancer appear to differ in temporal distribution from other cancers, these have generally been treated separately.

Leukemia and Bone Cancer Following an instantaneous exposure to radiation, the rates of leukemia and bone cancer appear to follow a wave like pattern, rising within 5 years after exposure and then returning to near baseline rates within 30 years. For populations that have been followed for at least that long, no problems of projection arise. One simply models the risk of leukemia over the study period as a function of dose, $F(D)$, and treats that as a lifetime excess risk estimate. The only complication is that the parameter estimates in $F(D)$ may depend on sex s and age at exposure t. For populations with incomplete follow-up, the BEIR III Committee (NRC80) modelled the mortality rate, λ (s,t,D), and applied that estimate as a constant to the period from 2 to 27 years after exposure.

All Other Cancers In contrast to the rates for leukemia and bone cancer, the rates for most other cancers appear to have remained in excess for as long as most exposed populations have been followed. Whether they will continue to remain elevated for the rest of the population's life remains an important unanswered question, but most risk assessments have been based on the assumption that they will, although not necessarily at the same level. The BEIR III committee (NRC80) and much of the radioepidemiologic literature has relied on two simple models for projecting risks of these cancers: *absolute risk* and *relative risk* models. Letting $\lambda[T,D(t)]$ represent the incidence rate of cancer at age T resulting from an instantaneous exposure to dose D at age t, and letting $\lambda_0(t)$ represent the baseline rate in unexposed persons, the two models can be represented as follows:

$$\text{Absolute risk:}\lambda[T, D(t)] = \lambda_0(t) + F_A(D); \tag{1-4}$$

$$\text{Relative risk:}\lambda[T, D(t)] = \lambda_0(t)F_R(D), \tag{1-5}$$

where $F(D)$ is given by Equation (1-3) with α_0 constrained to 0 for the absolute risk model and 1 for the relative risk model. The BEIR III Committee adopted two minor modifications to these models: first, the excess risk was taken to be 0 for the first 10 years following exposure; second, the coefficients of $F(D)$ were allowed to depend on sex and age at exposure. These modifications were extended in the BEIR IV Committee's (NRC88) reanalyses of the data on radon and lung cancer by adopting a general relative risk model of the form:

$$\lambda[T, D(t)] = \lambda_0(t)\{1 + \alpha_1 D\exp[f(T) + g(t) + h(T-t)]\}, \tag{1-6}$$

where α_1 is the average slope of a linear dose-response relationship and $f(T)$, $g(t)$, and $h(T-t)$ represent modifying effects of age at risk, age at exposure, and time since exposure to be estimated, respectively. A general model of this type is also used in this report, except that the dose term $\alpha_1 D$ is replaced by $(\alpha_2 D + \alpha_3 D^2)$.

Incorporation of Other Risk Factors In addition to the time-related factors discussed above, there are numerous risk factors that have been identified as having a direct effect on cancer rates; some of these may also modify the effects of radiation exposure on cancer rates. Unfortunately, there are relatively few studies that have assessed these other risk factors in combination with radiation. For lung cancer, the most important risk factor is smoking. The BEIR IV Committee (NRC88) has reviewed the studies reporting on the joint effects of smoking and radiation exposures and concluded that there was evidence of a *synergistic* (greater than additive) effect, but that there was also some evidence that the effect was less than multiplicative. They did not, however, consider the three-way interaction of age, smoking, and radiation. For low-LET radiation, the only data available on this point came from the Japanese atomic-bomb survivors and appeared to be too sparse to merit further modeling by incorporating age. Good human data on the interaction between radiation and other exposures do not appear to exist. The present Committee has therefore decided not to pursue analysis of interaction effects further at this time.

Approaches to Model Fitting

The approach that is taken to fitting risk models to epidemiologic data depends on the form in which the data are available. Some of the more complex models require access to the raw data on individual subjects and their entire history of exposures. However, most models can be fitted with very little loss of information by placing the subjects into subgroups with similar values of the relevant characteristics, particularly dose and age at exposure, and then tabulating their person-years at risk and the numbers of cases of each type of cancer as a function of age and time since exposure. The study data can then be summarized by two arrays, one of person-years, Y_{ijkl}, for dose group i, age at exposure group j, attained age group k, and time since exposure interval l, and one of numbers of cancers, N_{ijklm}, in each subgroup $ijkl$ from each type of cancer m. Admittedly, the numbers of cases in most of the cells will be small, but this does not pose a problem for the method of analysis to be used. Next, one assumes that the numbers of cases in each cell follows a Poisson distribution, with the expected value given by the product of the rate predicted by the model and the person-years for that cell. The data can then be fitted by the technique of maximum likelihood. The likelihood is the probability of the observed data given a particular choice of model parameters, which, in this circumstance, is obtained from the product of the Poisson probabilities for each cell of the cross-tabulation. A Newton-Raphson search is used to find the parameter values which maximize this likelihood. Confidence limits and significance tests can be derived from large sample theory (Co74). The committee used a computer program known as AMFIT for fitting a general class of regression models for the Poisson data. Further details of the fitting program can be found in Annex 4C to Chapter 4.

In any model fitting analysis, it is important to know how well the model describes the data. There are several aspects to this question. First, one would like an overall assessment of whether the model fits; such an assessment is known as a goodness-of-fit test. A poor fit might be an indication either that the chosen model is incorrect or that there is some problem with the data; a good fit does not prove that the model is correct—it simply means that there is insufficient evidence that the model is wrong. Next, assuming that the model fits, one would like to know the range of parameter values that is also consistent with the data; this range is known as a confidence interval and is important in evaluating the uncertainty in the fitted model. Next, one would like to be assured that the model is not unduly influenced by a few observations at the expense of the bulk of the data or by the inclusion of variables that are too highly correlated to be separated. Techniques to identify these types of problems are known as *diagnostics* and were used by the Committee throughout these analyses, as discussed in Annex 4F.

Special Problems

Pooling Data From Multiple Studies For many cancer sites, information was available from more than one epidemiologic study, raising the issue of how these data should be combined for risk assessment purposes. Because the studies generally differed in the nature of the exposures, the populations, and numerous methodological details, it was considered inappropriate to simply combine all of the raw data into a single data set. Instead, each of the studies for which original data were available to the committee were analyzed separately to obtain an estimate of the relevant parameters and their uncertainties. Formal tests of homogeneity were carried out to assess whether any differences in results could reasonably be ascribed to chance. If the results appeared to be consistent, an overall estimate could be obtained by a matrix weighted average and an estimate of the uncertainty of the pooled estimate could easily be derived. On the other hand, if the results appeared to be discrepant, the committee had to make a subjective judgment as to the quality and relevance of each of the studies.

Use of Animal Data The committee felt strongly that its risk assessments should be based on human data to the extent that they were available and that animal data should be used only to address questions for which human data were unavailable or inadequate. Questions in the latter category included the RBE of neutrons and gamma rays and the effect of dose rate.

Treatment of the RBE One of the problems for which the human data are inadequate is that of estimating the RBE for neutrons. The BEIR III Committee (NRC80) attempted to estimate the RBE for leukemia from the data from Japanese atomic-bomb survivors and then applied their estimate to the data on solid tumors. Aside from the inappropriateness of treating this point estimate as if it were known with certainty, the approach is no longer valid because reassessment of the atomic-bomb dosimetry has largely eliminated the differences in responses between Hiroshima and Nagasaki on which the previous estimate of the RBE was based. It therefore became necessary for the present Committee to rely on animal data for this purpose. For all analyses of the Radiation Effects Research Foundation

(RERF) data, a value of 20 for the RBE for neutrons was assumed as a fixed constant. The justification for this choice is given in Chapter 4.

Projection of Lifetime Risk Estimates

Once the epidemiologic and animal data are summarized in the form of an exposure-time-response model, the final stage of risk assessment involves the calculation of lifetime risk for patterns of exposure of particular interest. This is done with standard life table (mortality table) techniques (Bu81). Consider the case of lifetime exposure at a constant annual rate. A life table analysis would proceed as follows. Starting with a hypothetical population of 1 million newborn infants, the first column in the life table gives the number of infants that are expected to survive to each age. The second column gives the cancer rate predicted by the exposure-time-response model, and the third column gives the number of cases of cancers that would result; this is given by the product of the first two columns. The fourth column gives the number of deaths from other causes, based on current mortality rates, which are not assumed to depend on radiation. The number of survivors at the beginning of the next age interval is therefore the number at the start of the interval minus the number of radiogenic and nonradiogenic deaths, and the process continues until the entire cohort is dead (although, in practice, the calculations are usually terminated at age 100). The total number of excess cases of cancer is estimated by subtracting the number of deaths obtained from a similar life table for persons with no radiation exposure.

For protracted exposures, these calculations assume that each increment of exposure contributed independently to the cancer rates. Thus, the risk at age T is given by the background rate plus the sum over the entire exposure history of the excess rate attributable to each exposure increment; that is, if $D(t)$ represents the history of radiation doses at each age t and $[T, D(t)]$ represents the postulated dependence of cancer rates on age and each increment of exposure then the risk from the the entire history of exposure is given by:

$$\lambda[T, D] = \lambda_0(T) + \int_0^T \{\lambda[T, D(t)] - \lambda_0(t)\}dt. \qquad (1\text{-}7)$$

This implies that the rate is a function of cumulative exposure (possibly weighted by a function of age at exposure or time since exposure). There is evidence, however, that the contributions of extended exposures are not simply additive: for low-LET radiation, protracted exposures appear to be less hazardous than instantaneous exposures of the same total dose, possibly because sublesions caused by the first event can be repaired before additional events occur; for high-LET radiation, the effect may simply be additive, or protracted exposures may even be more hazardous, possibly because subsequent radiation exposure can promote already initiated cells. The committee acknowledges this problem but, as explained earlier in this chapter, it does not believe that sufficient information is available to deal with this question in a definitive manner. The committee therefore chose to retain the assumption of independence for the calculations but to present the results in such a way that the reader can make adjustments for protracted exposure when warranted.

Uncertainty of the Risk Estimates

Unlike the BEIR III Committee (NRC80) which presented a range of lifetime risks based on relative and absolute risk models for several choices of dose-response functions, the present committee has chosen to assess the uncertainty of the projected lifetime risks by using a Monte Carlo simulation approach. The committee's preferred exposure-time-response model for a particular site of cancer or group of sites was characterized by a vector of parameter estimates and a covariance matrix which describes the uncertainty in each parameter. By repeated sampling from the set of possible parameter values, with sampling probabilities determined by their covariance matrices, 1,000 sets of possible parameters were obtained. Each combination was then applied to the life table calculation described above to obtain a set of predicted lifetime risks. The resulting distribution, presented in Chapter 4, gives a measure of the statistical uncertainty in the committee's risk estimates under the preferred model. Other sources of uncertainty, external to the preferred model and its statistical uncertainty, are discussed in Annex 4F.

A number of other models fit the data nearly as well. The Monte Carlo simulation could, in principle, have been extended to include sampling over alternative models. However the committee invoked a number of non-statistical criteria, e.g., biological plausibility, to chose between alternative models, and felt that using a simple goodness-of-fit criteria as weights in the Monte Carlo simulation would not adequately reflect this process. Life table results are presented in Annex 4D for a number of alternative models. It is of interest that the range of life table risks estimated under these alternative models is less than the uncertainty estimated by the Monte Carlo simulation.

REFERENCES

Ad87 Adams, L. M., S. P. Ethier, and R. L. Ullrich. 1987. Enhanced *in vitro* proliferation and *in vivo* tumorigenic potential of mammary epithelium from BALB/c mice exposed *in vivo* to gamma-radiation and/or 7,12-dimethylbenz(a)anthracene. Cancer Res. 47:4425-4431.

Ba63 Barendson, G. W., H. M. D. Walter, J. F. Fowler, and D. K. Bewley. 1963. Effects of different ionizing radiations on human cells in tissue culture. III. Experiments with cyclotron - accelerated alpha particle and deuterons. Radiat. Res. 18:106.

Ba68 Barendson, G. W. 1968. Responses of cultured cells, tumors, and normal tissues to radiation of different linear energy transfer. Curr. Top. Radiat. Res. Q. 4:293-356.

Be88 Begg, C. B., and J. A. Berlin. 1988. Publication bias: a problem in interpreting medical data (with discussion). J. Roy. Statist. Soc. Ser A. 151:419-463.

Be06 Bergonie, J., and L. Tribondeau. 1906. De quelques resultats de la radiotherapie et essai de fixation d'une technique rationelle. C. R. Sceances Acad. Sci. 143:983. (English translation by G. H. Fletcher, Radiat. Res. 11:587, 1959.)

Bo74 Boecker,B. B., and R. G. Cuddihy. 1974. Toxicity of ^{144}Ce Inhaled as $^{144}CeCl_3$ by the beagle: Metabolism and dosimetry. Radiat. Res. 60:133-154.

Bo78 Bond, V. P., C. B. Meinhold, and H. H. Rossi. 1978. Low dose RBE and Q for X ray compared to gamma-ray radiations. Health Phys. 34(5):433-438.

Bo79 Borek, C., R. Miller, C. Pain, and W. Trom. 1979. Conditions for inhibiting and enhancing effects of the protease inhibitor antipain on X-ray induced neoplastic transformation in hamster and mouse cells. Proc. Natl. Acad. Sci. USA 76:1800-1803.

Bo83 Borek, C, E. J. Hall, and M. Zaider. 1983. X rays may be twice as potent as X rays for malignant transformation at low dose. Nature 301:156-158.

Br88 Brenner, D.J. 1988. Concerning the nature of the initial damage required for the production of radiation induced exchange aberrations. Int. J. Radiat. Biol. 52: 805-809.

Br80 Breslow, N. E., and N. E. Day. 1980. Statistical Methods in Cancer Research, vol. 1: The Analyses of Case Control Studies. Publication No. 32. Lyon, France: IAR Scientific Publications.

Br73 Broerse, J. J., and G. W. Barendsen. 1973. Relative biological effectiveness of fast neutrons for effects on normal tissue. Curr. Top. Radiat. Res. Q. 8:305-350.

Bu81 Bunger B. M., J. R. Cook, and M. K. Barrick. 1981. Life Table Methodology for Evaluating Radiation Risk: An Application Based on Occupational Exposures. Health Phys. 40:439-455.

Ch87 Chan, J. Y. H, F. F. Becker, J. German, and J. H. Ray. 1987. Altered DNA ligase I activity in Bloom's syndrome cells. Nature 325:357-359.

Cl68 Cleaver, J. E. 1968. Defective repair replication of DNA in xeroderma pigmentosum. Nature 218:652-656.

Co74 Cox, D. R. and D. V. Hinkley. 1974. Theoretical Statistics. London: Chapman and Hall.

Co77 Cox, R., J. Thacker, and D. T. Goodhead. 1977. Inactivation and mutation of cultured mammalian cells by aluminum characteristics, ultrasoft X rays, and radiation of different LET. Int. J. Radiat. Biol. 31:561-576.

Ei81 Eisenberg, H., and G. Felsenfeld. 1981. Hydrodynamic studies of the interaction between nucleosome core particles and core histones. J. Mol. Biol. 150:537-555.

El72 Ellett, W. H. and L. A. Braby. 1972. The Microdosimetry of 250 kVp and 65 kVp X Rays, ^{60}Co Gamma Rays, and Tritium Beta Particles. Radiat. Res. 51:229-243.

Fi69 Field, S. B. 1969. The relative biological effectiveness of fast neutrons for mammalian tissues. Radiology 93:915-920.

Fi71 Field, S. B. and S. Hornsey. 1971. RBE values for cyclotron neutrons for effects on normal tissues and tumors as a function of dose and dose fractionation. Eur. J. Cancer 7:151-169.

Fi83 Fisher, D. R. 1983. Current concepts in lung dosimetry. Report PNL-SA-11049. Springfield, Va.: National Technical Information Service.

Fr77 Fry, R. J. M. 1977. Radiation carcinogenesis. Int. J. Radiat. Oncol. Biol. Phys. 3:219-226.

Fu54 Furth, J., and E. Lorenz. 1954. Carcinogenesis by ionizing radiations. Pp. 1145-1201 in Radiation Biology, vol. I, part II, A. Hollaender, ed. New York: McGraw-Hill.

Go86 Goldman, M., L. S. Rosenblatt, and S. A. Book. 1986. Lifetime radiation effects research in animals: An overview of the status and philosophy of studies at University of California-Davis Laboratory for Energy Related Health Research. Pp. 53-65 in Life-Span Radiation Effects Studies in Animals: What Can They Tell Us?, R. C Thompson and J. A. Mahaffey, eds. U.S. Department of Energy Report CONF-830951. Springfield, Va.: National Technical Information Service.

Go29 Goldstein, L., and D.P. Murphy. 1929. Etiology of the ill-health in children born after maternal pelvic irradiation. II. Defective children born after post-conception pelvic irradiation. Am. J. Roentgenol. 22:322-331.

Gr53 Gray, L. H., A. D.Conger, M. Ebert, S. Hornsey, and O. C. A. Scott. 1953. The concentration of oxygen dissolved in tissues at the time of irradiation as a factor in radiotherapy. Br. J. Radiol 26:638-648.

Gr85 Greulich, K. O., J. Ausio, and H. Eisenberg. 1985. Neucleosome core particle structure and structural changes in solution. J. Mol. Biol. 186:167-173.

Gu80 Guernsey, D. L., A. Ong, and C. Borek. 1980. Thyroid hormone modulation of x ray induced in vitro neoplastic transformation. Nature 288:591-592.

Ha64 Hall, E. J., J. S. Bedford. 1964. Dose rate: Its effect on the survival of HeLa cells irradiated with gamma rays. Radiat. Res. 22:305-315.

Ha72 Hall, E. J. 1972. Radiation dose-rate: A factor of importance in radiobiology and radiotherapy. Br. J. Radiol. 45:81-97.

Ha87 Hall, E. J., and T. K. Hei. 1987. Oncongenic transformation by radiation and chemicals. Pp. 507-512 in Proceedings of the 8th International Congress of Radiation Research, E. M. Fielden, J. F. Fowler, J. H. Hendry, and D. Scott, eds. Taylor and Francis.

Ha79 Han, A., and M. M. Elkind. 1979. Transformation of mouse C3H/101/2 cells by single and fractionated doses of X rays and fission-spectrum neutrons. Cancer Res. 39:123-130.

Ha80 Han, A., C. K. Hill, and M. M. Elkind. 1980 Neoplastic transformation of lOT 1/2 cells by ^{60}Co gamma-rays: Evidence of repair of damage at reduced dose rate. Int. J. Radiat. Biol. 37:585-589.

Ha82 Han, A. and M. M. Elkind. 1982. Enhanced transformation of mouse 10T 1/2 cells by 12-O-tetradecanoylphorbol-13- acetate following exposure to x-rays or to fission-spectrum neutrons. Cancer Res. 42:477-483.

Ha83 Harris, A. L., P. Karran, and T. Lindahl. 1983. O6-Methylguanine-DNA methyltransferase of human lymphoid cells: Structural and kinetic properties and absence in repair-deficient cells. Cancer Res. 43:3247-3252.

He84 Hecht, F., and G. R. Sutherland. 1984. Fragile sites and cancer breakpoints. Cancer Genet. Cytogenet. 12:179-181.

He87 Hecht, F., and B. K. Hecht. 1987. Chromosome changes connect immunodeficiency and cancer in ataxia telangiectasia. Am. J. Pediatr. Hematol. Oncol. 9:185-188.

He88 Hei, T. K., D. J. Chen, D. J. Brenner, and E. J. Hall. 1988. Mutation induction by charged particles of defined linear energy transfer. Carcinogenesis 9:1233-1236.

Hi84 Hill, C. K., A. Han, and M. M. Elkind. 1984. Fission-spectrum neutrons at a low dose rate enhance neoplastic transformation in the linear, low dose region (0-10 cGy). Int. J. Radiat. Biol. 46:11-15.

IAEA86 International Atomic Energy Agency. 1986. Biological dosimetry: Chromosomal Aberration Analysis for Dose Assessment. Technical Reports Series No. 260. Vienna: International Atomic Energy Agency.

ICRU83 International Commission on Radiation Units and Measurements. 1983. Microdosimetry. ICRU Report 36. Bethesda, Md.: International Commission on Radiation Units and Measurements.

ICRU86 International Commission on Radiation Units and Measurements. 1986. The Quality Factor in Radiation Protection. ICRU Report 40. Report of Joint Task Group of the ICRP and the ICRU.

ICRP63 International Commissions on Radiological Protection and on Radiological Units and Measurements. 1963. Report of the RBE Committee to the International Commissions on Radiological Protection and on Radiological Units and Measurements. Health Physics 9:357-386.

ICRP79 International Commission on Radiological Protection. 1979-1988. Limits for intakes of radionuclides by workers. ICRP Publication 30. Vols. 2(3/4), 3, 4(3/4), 5, 6(2/3), 7, 8, and 19(4). Oxford: Pergamon.

ICRP86 International Commission on Radiological Protection. 1986. The metabolism of plutonium and related elements. ICRP Publication 48. Oxford: Pergamon.

Jo85 Johnson, J. R. 1985. Internal dosimetry for radiation protection, Chapter 6 in The Dosimetry of Ionizing Radiation, (K. R. Kase, B. E. Bjarngard, and F. H. Attix, eds. Orlando: Academic Press, Inc.

Ke78b Kennedy, A. R., and J. B. Little. 1978. Protease inhibitors suppress radiation-induced malignant transformation in vitro. Nature 276:825-826.

Ke81 Kennedy, A. R. and J. B. Little. 1981. Effects of protease inhibitors on radiation transformation in vitro. Cancer Res. 41:2103-2108.

Ke78a Kennedy, A. R., S. Monjal, C. Heidelberger, and J. B. Little. 1978. Enhancement of X ray transformation by 12-O-tetradecanoyl phorbol 13 acetate in a cloned line C3H mouse embryo cells. Cancer Res. 38:439-443.

Ke80 Kennedy, A. R., G. Murphy, and J. B. Little. 1980. Effect of time and duration of exposure to 12-O-tetradecanoyl-phorbol-13-acetate aon x-ray transformation of C_3H 10T1/2 cells. Cancer Res. 40:1915-1920.

Kn85 Knudson, A. G. 1985. Hereditary cancer, oncogenes, and antioncogenes. Cancer Res. 45:1437-1443.

Kr82 Kraft, G., W. Kraft-Weyrather, H. Meister, H. G. Miltenburger, R. Roots, and H. Wulf. 1982. The influence of radiation quality on the biological effectiveness of heavy charged particles. Pp. 743-73 in Radiation Protection, J. Booz and H. G. Ebert eds. Commission of European Communitics.

La87 Laird, N. M. 1987. Thyroid cancer risk from exposure to ionizing radiation: A case study in the comparitive potency model. Risk Analysis 7:299-309.

La89 Langlois, R. G., W. L. Bigbee, R. H. Jensen, and J. German. 1989. Evidence for increased in vivo mutation and somatic recombination in Bloom's syndrome. Proc. Natl. Acad. Sci. 86:670-674.

Le56 Lea, D. E. 1956. DEA: Actions of radiations on living cells, 2nd ed. Cambridge, England: Cambridge University Press.

Le42 Lea and Catcheside. 1942. The mechanism of the induction by radiation of chromosome aberrations in Tradescantia. J. Genet. 44:216-249.

Li82 Lindahl, T. 1982. DNA repair enzymes. Annu. Rev. Biochem. 1:61-87.

Li72 Lindahl, T. and B. Nyberg. 1972. Rate of depurination of native DNA. Biochemistry 11:3610-3618.

Li74 Lindahl, T. and B. Nyberg. 1974. Heat-induced deamination of cytosine residues in deoxyribonucleic acid. Biochemistry 13:3405-3410.

Ll81 Lloyd, D.C., and R. J. Purrott. 1981. Chromosome aberration analysis in radiological protection dosimetry. Radiat. Protect. Dosim. 1:19-28.

Lu87a Lundgren, D. L., F. F.Hahn, W. C. Griffith, R. G. Cuddihy, P. J. Haley, and B. B. Boecker 1987. Effects of relatively low-level exposure of rats to inhaled $^{144}CeO_2$. III. Pp. 308-12 in Inhalation Toxicology Research Institute Annual Report 1986-1987, J. D. Sun, and J. A. Mewhinney, eds. U.S. Department of Energy Report LMF-120. Springfield, Va.: National Technical Information Service.

Lu87b Lundgren, D.L., F. F. Hahn, W. C. Griffith, R. G. Cuddihy, F. A. Seiler, and B. B. Boecker. 1987. Effects of relatively low-level thoracic or whole-body exposure of rats to X-rays. I. Pp. 313-317 in Inhalation Toxicology Research Institute Annual Report 1986-1987. J. D. Sun and J. A. Mewhinney, eds. U.S. Department of Energy Report LMF-120. Springfield, Va.: National Technical Information Service.

Ma70 MacMahon, B., and T. F. Pugh. 1970. Epidemiology Principles and Methods. Boston: Little, Brown and Company.

Mc86 McClellan, R. O., B. B. Boecker, F. F. Hahn, and B. A. Muggenburg. 1986. Lovelace ITRI Studies on the toxicity of inhaled radionuclides in beagle dogs. Pp. 74-96 in Life-Span Radiation Effects Studies in Animals: What Can They Tell Us?, R. E. Thompson and J. A. Mahaffey, eds. U.S. Department of Energy Report CONF-830951. Springfield, Va.: National Technical Information Service.

Me88 Metivier, H., R. Masse, G. Rateau, D. Nolibe, and J. Lafuma. 1988. In press. New data on the toxicity of $^{239}PuO_2$ in baboons. Proceedings of the CEC/CEA/DOE-Sponsored Workshop on Biological Assessment of Occupational Exposure to Actinides. Versailles, France, May 30-June 2, 1988.

Mi78 Michaels, H. B., and J. W. Hunt. 1978. A model for radiation damage in cells by direct effect and by indirect effect: A radiation chemistry approach. Radiat. Res. 74:23-24.

Mi85 Mitchell, D. L., C. A. Haipek, and J. M. Clarkson. 1985. (6-4) Photoproducts are removed from the DNA of UV-irradiated mammalian cells more efficiently than cyclobutane pyrimidine dimers. Mutat. Res. 143:109-112.

Mo36 Mottram, J. C. 1936. Factor of importance in radiosensitivity of tumors. Br. J. Radiol. 9:606-614.

Mu28 Muller, H.J. 1928. The effects of X-radiation on genes and chromosomes. Science 67:82.

NCRP80 National Council on Radiation Protection and Measurements. 1980. Influence of dose and its distribution in time on dose-response relationships for low LET radiations. Report No. 64. Washington, D.C.: National Council on Radiation Protection and Measurements.

NCRP84 National Council for Radiation Protection and Measurements (NCRP). 1984. Evaluation of Occupational and Environmental Exposures to Radon and Radon Daughters in the United States. NCRP Report No. 78. Bethesda, Md.: National Council on Radiation Protection and Measurements.

NCRP85 National Council for Radiation Protection and Measurements (NCRP). 1985. General Concepts for the Dosimetry of Internally Deposited Radionuclides. NCRP Report No. 84. Bethesda, Md.: National Council on Radiation Protection and Measurements.

NCRP87a National Council on Radiation Protection and Measurements (NRCP). 1987. Recommendations on limits for exposure to ionizing radiation. Report No. 91. Bethesda, Md.: National Council on Radiation Protection and Measurements.

NCRP87b National Council on Radiation Protection and Measurements (NCRP). 1987. Ionizing Radiation Exposures of the Population of the United States. Report No. 93. Washington, D.C.: National Council on Radiation Protection and Measurements.

NIH85 National Institutes of Health. 1985. Report of the National Institutes of Health Ad Hoc Working Group to Develop Radioepidemiological Tables. NIH Publication 85-2748. Washington, D.C.: Superintendent of Documents, Government Printing Office.

NRC80 National Research Council, Committee on the Biological Effects of Ionizing Radiations (BEIR III). 1980. The Effects on Populations of Exposure to Low Levels of Ionizing Radiation. Washington, D.C: National Academy Press. 524 pp.

NRC88 National Research Council, Committee on the Biological Effects of Ionizing Radiations (BEIR IV). 1988. Health Risks of Radon and Other Internally Deposited Alpha-Emitters. Washington, D.C.: National Academy Press. 602 pp.

Pal84 Palcic, B. and L. D. Skarsgard. 1984. Reduced oxygen enhancement ratio at low doses of ionizing radiation. Radiat. Res. 100:328-339.

Pa86 Park, J.F., G. E. Dagle, H. A. Ragan, R. E. Weller, and D. L. Stevens. 1986. Current status of life-span studies with inhaled plutonium in beagles at Pacific Northwest Laboratory. Pp. 455-470 in Life-Span Radiation Effects Studies in Animals: What Can They Tell Us?, R. E. Thompson and J. A. Mahaffey, eds. U.S. Department of Energy Report CONF-830951. Springfield, Va.: National Technical Information Service.

Pat49 Patt, H. M., E. B. Tyree, R. L. Straube, and D. E. Smith. 1949. Cysteine protection against x-irradiation. Science 110:213.

Ro86 Rothman, K. J. 1986. Modern Epidemiology. Boston: Little, Brown and Co.

Ro84 Rowley, J. D. 1984. Biological implications of consistent chromosome rearrangements in leukemia and lymphoma. Cancer Res. 44:3159-3168.

Ru54 Russell, L. B. 1954. The effects of radiation on mammalian prenatal development. Pp. 861-918 in Radiation Biology, vol. I, part II. A. Hollaender, ed. New York: McGraw-Hill.

Ry82 Rydberg, B., and T. Lindahl. 1982. Nonenzymatic methylation of DNA by the intracellular methyl group donor S-adenosyl-L-methionine is a potentially mutagenic reaction. EMBO J. 1:211-216.

Sa88 Sanders, C. E., K. E. Lauhala, J. A. Mahaffey, and K. E. McDonald. 1988. Low-level $^{239}PuO_2$ lifetime studies. Pp. 31-34 in Pacific Northwest Laboratory Annual Report for 1987 to the DOE Office of Energy Research. Part 1. Biomedical Sciences. U.S. Department of Energy Report PNL-6500 Part 1. Springfield, Va.: National Technical Information Service.

Sa40 Sax, K. 1940. An analysis of X ray induced chromosome aberrations in Tradescantia. Genetics 25: 41-66.

Sc74 Schmidt E., G. Rimpl, and M. Bauchinger. 1974. Dose-Response Relation of Chromosome Aberations in Human Lymphocytes After in Vitro Irradiation with 3-MeV Electrons. Radiat. Res. 57:228.

Sc74c Schroeder, T. M., and J. German. 1974. Bloom's syndrome and Fanconi's anemia: Demonstration of two distinctive patterns of chromosome disruption and rearrangement. Humangenetik 25:299-306.

Sh87 Shimizu,Y., H. Kato, W. J. Schull, D. L. Preston, S. Fujita, and D. A. Pierce. 1987. Lifespan report 11, part 1. Comparison of Risk Coefficients for Site Specific Cancer Mortality Based on the DS86 and T65DR Shielded Kerma and Organ Doses. RERF TR 12-87. Hiroshima: Radiation Effects Research Foundation.

Si66 Sinclair, W. K., and R. A. Morton. 1966. X ray sensitivity during the cell generation cycle of cultured Chinese hamster cells. Radiat. Res. 29:450-474.

Sm84 Smith, H., and G. Gerber, eds. 1984. Lung modelling for inhalation of radioactive materials, Proceedings of a meeting jointly organized by the Commission of the European Communities and the National Radiological Protection Board, Oxford, March 26-28, 1984. Report EUR9384EN.

Sn83 Snipes, M. B., B. B. Boecker, and R. O. McClellan. 1983. Retention of monodisperse or polydisperse aluminosilicate particles inhaled by dogs, rats and mice. Toxicol. Appl. Pharmacol. 69:345-362.

Sn84 Snipes, M. B., B. B. Boecker, and R. O. McClellan. 1984. Respiratory tract clearance of inhaled particles in laboratory animals. Pp 63-71 in Lung Modelling for Inhalation of Radioactive Materials, H. Smith and G. Gerber, eds. Report EUR 9384EN.

St88 Storer, J. B., T. J. Mitchell and R. J. M. Fry. 1988. Extrapolation of the relative risk of radiogenic neoplasms across mouse strains and to man. Radiat. Res. 114:331-353.

Ter63 Terasima, T. and L. J. Tolmach. 1963. Variations in several responses of HeLa cells to X-irradiation during the division cycle. Biophys. J. 3:11-33.

Th86 Thacker, J., R. E. Wilkinson, and D. T. Goodhead. 1986. The induction of chromosome exchange aberrations by carbon ultrasoft X-rays in V79 hamster cells. Int. J. Radiat. Biol. 49:645-656.

Th86d Thompson, R. C., and J. A. Mahaffey, eds. Life-Span Radiation Effects Studies in Animals: What Can They Tell Us? Report CONF-830951. Springfield, Va.: National Technical Information Service.

Th81 Thomson, J. F., F. S. Williamson, D. Grahn, and E. J. Ainsworth. 1981. Radiat. Res. 86(3):559-572, 572-588.

Ul84 Ullrich, R. L. 1984. Tumor induction in BALB/c mice after fractionated or protracted exposures to fission.spectrum neutrons. Radiat. Res. 97:587-597.

Un76 Underbrink, A., A. Kellerer, R. Mills, and A. Sparrow. 1976. Radiat. Environ. Biophys. 13:295.

UN77 United Nations Scientific Committee on the Effects of Ionizing Radiation (UNSCEAR). 1977. Genetic effects of radiation. pp. 425-564. In Sources and Effects of Ionizing Radiation. Report A/32/40. Thirty Second Session, Supplement No. 40. New York: United Nations.

UN86 United Nations Scientific Committee on the Effects of Ionizing Radiation (UNSCEAR). 1986. Genetic Effects of Radiation. Pp. 7-164 in Ionizing Radiation: Sources and Biological Effects. Report A/41/16. Forty First Session, Supplement No. 16. New York: United Nations.

Vi83 Vijayalaxmi, H. J.Evans, J. H. Ray, and J. German. 1983. Bloom's syndrome: Evidence for an increased mutation frequency in vivo. Science 221:851-853.

Vo81 Vogel, H. H., and H. W. Dickson. 1981. Abstracts of the 29th Annual Meeting of the Radiation Research Society, Minneapolis. Radiat. Res. 87(2):453.

Wa87 Ward, J. F. 1988. DNA damage produced by ionizing radiation in mammalian cells: identities, mechanisms of formation and reparability. Prog. Nucleic Acids Res. Mol. Biol. 35:96-128.

Wa88 Ward, J. F., C. L. Limoli, P. Calabro-Jones, and J. W. Evans. 1988. Radiation vs. chemical damage to DNA. Anticarcinogenesi and Radiation Protection, O. F. Nygaard, M. Simic, and P. Cerutti, eds. New York: Plenum.

Wi85 Widom, J., and A. Klug. 1985. Structure of the 300 A chromotin filament: X-ray diffraction frm oriented samples. Cell 43:207-213.

Wi87 Willis, A.E., and T. Lindahl 1987. DNA ligase I deficiency in Bloom's syndrome. Nature 325:355-357.

Wr86 Wrenn, M. E., G. N. Taylor, W. Stevens, C. W. Mays, W. S. S. Jee, R. D. Lloyd, D. R. Atherton, F.W. Bwenger, S. C. Miller, J. M. Smith, L. R. Shabestan, L. A. Woodbury, and B. J. Stover. 1986. DOE life-span radiation effects studies in experimental animals at University of Utah Division of Radiobiology. Pp. 32-52 in Life-Span Radiation Effects Studies in Experimental Animals: What Can They Tell Us?, R. C. Thompson, and J. A. Mahaffey, eds. U.S. Department of Energy Report CONF-830951. Springfield, Va.: National Technical Information Service.

Yu80 Yuhas, J. M., J. M. Spellman, and F. Cullo. 1980. The role of WR2721 in radiotherapy and/or chemotherapy. Pp. 303-308. Radiation Sensitizers, L. Brady, ed. New York: Masson.

2
Genetic Effects of Radiation

INTRODUCTION

Ionizing radiation damages the genetic material in reproductive cells and results in mutations that are transmitted from generation to generation. The mutagenic effects of radiation were first recognized in the 1920s, and since that time radiation has been used in genetic research as an important means of obtaining new mutations in experimental organisms. Although occupational exposure to high levels of radiation has always been of concern, not until during and after World War II was there a concerted effort to evaluate the genetic effects of radiation on entire populations. These efforts were motivated by concern over the effects of extremely large sources of radiation that were being developed in the nuclear industry, of radioactive fallout from the atmospheric testing of atomic weapons and of the rapidly increasing use of radiation in medical diagnosis and therapy. In 1956 the National Academy of Sciences-National Research Council (NAS-NRC) established the Committee on the Biological Effects of Atomic Radiation (denoted the BEAR Committee), which was the forerunner of the subsequent NAS-NRC committees on the Biological Effects of Ionizing Radiation (BEIR committees; of which this BEIR V report is one). A series of reports from the U.N. Scientific Committee on the Effects of Atomic Radiation (UNSCEAR) has also addressed the genetic effects of radiation exposure on populations.

Although there is a continuing need to assess the genetic effects of radiation exposure, for several reasons the perspective has changed somewhat from that in the 1950s. First, it is now clear that the risk of cancer in individuals exposed to radiation is significant and that limiting exposure to radiation to reduce the risk of cancer also limits the genetically significant exposure. Second, the instruments and techniques used in medical radiation have improved significantly, so that the overall doses used in medical diagnoses are reduced and patient exposure in all but the targeted organs is lessened. Third, in regard to the induction of mutations, the greater current risk seems to result from exposure to chemical mutagens in the environment rather than from the exposure of populations to radiation. Despite changed conditions, estimating the genetic effects of radiation remains important for setting exposure standards, both for the general population and for those exposed in their occupations.

There are many difficulties in measuring the genetic effects of exposure of the human population to radiation and other mutagens. This is why, more than 20 years after the BEAR Committee first addressed the issues of radiation exposure, there is still uncertainty and controversy. The following are some of the difficulties and considerations that must be kept in mind.

The genetic effects of radiation are expressed, not in irradiated individuals, but in their immediate or remote offspring. The time lag is great because of the duration of the human life cycle, and massive epidemiologic studies with long-term follow-up are needed to accumulate sufficient data for statistical analysis. Moreover, for risk estimation of exposures that are not uniformly or randomly delivered to the entire population, the age and sex distribution of the exposed population and the different probabilities of having children for members of the population of each age and sex must be taken into account.

The mutations induced by radiation can also occur spontaneously. When humans are exposed to low doses of radiation, it is difficult to estimate what small increment of mutations is induced by radiation above that from spontaneous background radiation. However, radiation has been found to be mutagenic in all organisms studied so far, and there is no reason to suppose that humans are exempt from radiation's mutagenic effects. These mutagenic effects are expected to be harmful to future generations because, in experimental organisms, the majority of new mutations with detectable effects are harmful, and it is assumed that humans are affected similarly. Indeed, the harmful effects of mutations that occur spontaneously in humans are well documented, because many of them result in genetic disease.

The genetic effects of radiation must be detected through the study of certain endpoints, for example, visible chromosome abnormalities, proteins with altered conformations or charges, spontaneous abortions, congenital malformations, or premature death. In addition, radiation induced mutations may affect different endpoints to different degrees. For example, the dose of radiation required to double the incidence of one endpoint need not be the same as that required to double the incidence of a different endpoint.

The BEIR I Committee (NRC72) espoused five general principles of risk estimation. Subsequent committees have generally followed these strictures whenever possible, as has the present committee. They are as follows:

1. Use relevant data from all sources, but emphasize human data when feasible. In general, when data of comparable accuracy exist, place greater emphasis on organisms closest to man.
2. Use data from the lowest doses and dose rates for which reliable data exist, as being more relevant to the usual conditions of human exposure.
3. Use simple linear extrapolation between the lowest reliable dose data and the spontaneous or zero dose rate. In order to get any kind of precision from experiments of manageable size, it is necessary to use dosages much higher than those expected for the human population. Some mathematical assumption is necessary, and the linear model, if not always correct, is likely to err on the safe side.
4. If cell stages differ in sensitivity, weight the data in accordance with the duration of the stage.
5. If the sexes differ in sensitivity, use the unweighted average of data for the two sexes.

Deliberate exposure of humans to radiation without diagnostic or therapeutic justification is unacceptable, and therefore, most genetic studies have had to be carried out in experimental organisms, particularly mice. Such studies raise numerous additional problems of their own, including extrapolation of results obtained under experimental conditions to the conditions relevant to population exposure, such as dose rates, fractionation, and other variables; and extrapolation from an experimental organism such as the mouse, in which radiation effects may be estimated with some confidence, to humans, because organisms differ in radiation sensitivity.

UNSCEAR (UN86) has summarized three principal assumptions that are necessary for extrapolating data from mice and other suitable mammals to humans:

1. The amount of genetic damage induced by a given type of radiation under a given set of conditions is the same in human germ cells and in those of the test species used as a model.
2. The various biological (e.g., sex, germ cell stage, age, etc.) and physical (e.g., quality of radiation, dose rate, etc.) factors affect the magnitude of the damage in similar ways and to similar extents in the experimental species from which extrapolations are made and in humans.

3. At low doses and at low dose rates of low-LET (linear energy transfer) irradiation there is a linear relationship between dose and the frequency of genetic effects studied.

Direct studies of the genetic effects of radiation exposure to human populations have been carried out on the children of the Japanese populations in Hiroshima and Nagasaki who were irradiated in the atomic bombings in August 1945. Results of these careful and very extensive studies, when taken at face value, suggest that humans may be somewhat less sensitive to radiation than mice.

The BEIR I Committee (NRC72) used two methods of estimating genetic effects. One method relied on direct estimates. This method was used whenever possible, for example with reciprocal translocations. The other method was indirect and was used for such endpoints as gene mutation. The indirect method required estimates of the mutation rates, the incidence of genetic disease in the human population, and the extent to which the incidence depends on recurrent mutation, to infer the increased incidence of genetic disease resulting from radiation exposure. Both immediate, first-generation effects and long-term, equilibrium effects were estimated from either the direct or indirect estimates of induced mutation by taking into account the presumed rates of mutant elimination to project the ratio of newly induced genetic damage to that transmitted from previous generations. The BEIR III Committee (NRC80) reviewed and updated the BEIR I report (NRC72). New estimates caused some changes in the previous estimates, and some new methods of estimation were added.

The BEIR V Committee has reviewed and reevaluated the data that are pertinent to the estimation of genetic risks in humans. The present report summarizes the methods and conclusions of previous committees. In deriving new risk figures, it places rather more emphasis on the results of the studies of Japanese atomic-bomb survivors than have previous BEIR reports. However, the committee has also made use of the extensive radiation studies carried out with mice, which are briefly reviewed.

SUMMARY OF CONCLUSIONS

Based on our review of relevant data from humans, other mammals, and mice, the BEIR V Committee believes that the values in Table 2-1 give the current best estimates of risk based on the conclusion that the doubling dose in humans is not likely to be smaller than the approximate 1 Sv (100 rem) obtained from studies in mice. Table 2-1 gives the estimated genetic effects of an average population exposure of 1 rem/30-year generation. Admittedly there are uncertainties, but the calculated risks are based on an impressive body of data and knowledge of radiobiological principles.

As will be reviewed below, attempts to estimate doubling doses from data on Japanese atomic-bomb survivors have consistently led to values larger than those derived from the animal data, and consequently they imply lower risks. Although risks calculated from animal data have large confidence intervals, estimates from those exposed to radiation in Hiroshima and Nagasaki are known with even less precision. In spite of these uncertainties, the data suggest a real difference, with the estimated lower 95% confidence limit of the human data approximating the median of a large number of values obtained in mice. If it is assumed that the apparent difference is real, humans would be less sensitive to radiation induction of mutations in germ cells than mice, and the risks in Table 2-1 should be considered conservative. On the other hand, the human data might be biased too low for reasons that are not presently understood, in spite of all the careful work that has gone into their collection and analysis. The BEIR V Committee is in no better position to decide the issue than were the previous groups and individuals who have grappled with it. Considering the uncertainty, the BEIR V Committee has adopted what it considers a prudent position in basing its risk estimates on the approximate lower 95% confidence limit for humans. This approach, while admittedly conservative, has the advantage of leading to risk estimates that, if anything, are too high rather than estimates that subsequent data may prove to be too low.

The background and methodology for the estimates given in Table 2-1 are provided in the following sections. The material not only provides the background for Table 2-1 but also summarizes the methods and conclusions of previous BEIR, UNSCEAR, and other reports.

It must be emphasized again that virtually all mutations have harmful effects. Some mutations have drastic effects that are expressed immediately, and these are eliminated from the population quite rapidly. Other mutations have milder effects and persist for many generations, spreading their harm among many individuals in the distant future. However, many of the long-term effects are impossible to estimate given present data and understanding, and for this reason the present committee emphasizes the effects of mutations that manifest themselves in the first generation, since these are of immediate concern and can be estimated with some confidence. The effects in the first generation are primarily those caused by simple Mendelian dominant and X chromosome-linked recessive traits because of their high heritabilities. Other kinds of mutations may be more important in the long run and constitute a significant burden for future generations.

Much of the uncertainty in estimating the risks of radiation-induced mutations centers on traits with complex patterns of inheritance that result from the combination of multiple genetic and environmental factors. Risk estimates are determined in part by the degree to which these traits are

TABLE 2-1 Estimated Genetic Effects of 1 rem per Generation[a]

Type of Disorder	Current Incidence per Million Liveborn Offspring	Additional Cases/10^6 Liveborn Offspring/rem/Generation	
		First Generation	Equilibrium
Autosomal dominant			
Clinically severe[b]	2,500[c]	5–20[d]	25[e]
Clinically mild[f]	7,500[g]	1–15[d]	75[e]
X-linked	400	<1	<5
Recessive	2,500	<1	Very slow increase
Chromosomal			
Unbalanced translocations	600[h]	<5	Very little increase
Trisomies	3,800[i]	<1	<1
Congenital abnormalities	20,000–30,000	10[j]	10–100[k]
Other disorders of complex etiology[l]			
Heart disease[m]	600,000		
Cancer	300,000	Not estimated	Not estimated
Selected others	300,000		

[a] Risks pertain to average population exposure of 1 rem per generation to a population with the spontaneous genetic burden of humans and a doubling dose for chronic exposure of 100 rem (1 Sv).

[b] Assumes that survival and reproduction are reduced by 20–80% relative to normal (s = 0.2– 0.8), which is consistent with the range of values in Table 2-2.

[c] Approximates incidence of severe dominant traits in Table 2-2.

[d] Calculated using Equations (2–7), with s = 0.2– 0.8 for clinically severe and s = 0.01– 0.2 for clinically mild.

[e] Calculated using Equation (2-1), with the mutational component = 1.

[f] Assumes that survival and reproduction are reduced by 1–20 percent relative to normal (s = 0.01– 0.2).

[g] Obtained by subtracting an estimated 2,500 clinically severe dominant traits from an estimated total incidence of dominant traits of 10,000.

[h] Estimated frequency from UNSCEAR (UN82,UN86).

[i] Most frequent result of chromosomal nondisjunction among liveborn children. Estimated frequency from UNSCEAR (UN82, UN86).

[j] Based on worst-case assumption that mutational component results from dominant genes with an average s of 0.1; hence, using Equation (2-3), excess cases $<30{,}000 \times 0.35 \times 100^{-1} \times 0.1 = 10$.

[k] Calculated using Equation (2-1), with the mutational component 5–35%.

[l] Lifetime prevalence estimates may vary according to diagnostic criteria and other factors. The values given for heart disease and cancer are round-number approximations for all varieties of the diseases, and the value for other selected traits approximates that for the tabulation in Table 2-4.

[m] No implication is made that any form of heart disease is caused by radiation among exposed individuals. The effect, if any, results from mutations that may be induced by radiation and expressed in later generations, which contribute, along with other genes, to the genetic component of susceptibility. This is analogous to environmental risk factors that contribute to the environmental component of susceptibility. The magnitude of the genetic component in susceptibility to heart disease and other disorders with complex etiologies is unknown.

Table 2-1 *Continued*

Most genes affecting the traits are thought to have small effects, and new mutations would each contribute a virtually insignificant amount to the total susceptibility of the individuals who carry them. However, a slight increase in genetic susceptibility among many individuals in the population may produce, in the aggregate, a significant effect overall. Because of great uncertainties in the mutational component of these traits and other complexities, the committee has not made quantitative risk estimates. The risks may be negligibly small, or they may be as large or larger than the risks for all other traits combined.

determined by mutations, but the mutational component of many of the most common traits is very uncertain. The BEIR V Committee recommends that more research be carried out on such complex disorders to sort out their genetic and environmental causes.

METHODS OF RISK CALCULATION

Table 2-1 is based on the doubling dose method, which is summarized below, along with several other methods that have been used.

The Doubling Dose Method

The doubling dose method is based on the following equation:

$$\text{induced burden} = \text{spontaneous burden} \times (\text{doubling dose})^{-1} \times \text{mutation component} \times \text{dose}. \quad (2\text{-}1)$$

As a hypothetical example, if the spontaneous burden is 20,000 per million liveborn for some class of genetic disease in the human population, the doubling dose is estimated to be 100 rem, and the average mutation component for these diseases is one-half, then, if the parents in each generation are exposed to 1 rem, the induced burden is 100 cases/10^6 liveborn/generation. That is, after the population has reached a new equilibrium between selection and mutation (which is inflated by the added increment of radiation), one expects 100 additional cases of genetic disease in each generation because of the increased radiation.

Although the doubling dose method is based on equilibrium considerations, the method can be used to estimate the effects of an increase in the mutation rate on the first few generations by taking a proportion of the equilibrium damage. For example, for a permanent increase in the mutation rate, the effect of a dominant mutation in the nth generation is $1 - (1 - s)^n$ of the equilibrium damage, where $(1 - s)$ is the fitness of carriers of the dominant gene.

In previous BEIR reports the reciprocal of the doubling dose has been called the *relative mutation risk*, and Equation (2-1) can be written as follows:

$$\text{induced burden} = \text{spontaneous burden} \times (\text{relative mutation risk}) \times \text{mutation component} \times \text{dose}. \quad (2\text{-}2)$$

This was done, in part, to avoid the concept of doubling dose, which is sometimes misunderstood. By definition, the doubling dose is that dose required to induce a number of mutations equal to the spontaneous frequency. However, its use in this report is confined to the range of low doses at which the dose-response curve is essentially linear. We thus have $m = m_0 + aD$, where m_0 is the spontaneous frequency, D is the dose, a is the induction rate, and m is the total mutation frequency (spontaneous plus induced). The doubling dose is then m_0/a and its reciprocal, $a/m_0 = (m - m_0)\, m_0 D$ is the relative mutation risk, that is, the number of mutations induced as a fraction of the spontaneous number per unit dose.

If the sexes differ in doubling dose, then the overall doubling dose is a weighted average of the sex-specific doubling doses. Denoting the male and female sexes as 1 and 2, respectively, and again attending only to the linear part of the dose-response curve, the following equation is obtained:

$$m = m_1 + m_2 + a_1 D_1 + a_2 D_2 \quad (2\text{-}3)$$

where m_1, a_1, D_1 and m_2, a_2, D_2 are the sex-specific spontaneous frequencies (m), induction rates (a), and doses (D) for males and females, respectively. If a population were exposed to $D_1 = DD_1 = m_1/a_1$ and $D_2 = DD_2 = m_2/a_2$, the mutation burden would double. DD_1 and DD_2 are the sex-specific doubling doses for males and females respectively. The common dose to both sexes that will double the mutation rate is:

$$DD = (m_1 + m_2)/(a_1 + a_2) \quad (2\text{-}4)$$

which is the a-weighted average of the sex-specific doubling doses.

Doubling doses from experimental mouse data are usually based on the exposure of a single parent and are sometimes referred to as *gametic*. Doubling doses estimated from the data from Japanese atomic-bomb survivors are sometimes based on joint parental exposure and are referred to as *zygotic*. For example, Neel and Schull (Ne74) have regressed various endpoints such as early infant death and malformations on the sum of the mother's and the father's doses. In this situation the linear part of the response curve can be written as (assuming a mutation component of 1)

$$m = m_1 + m_2 + a(D_1 + D_2). \tag{2-5}$$

An estimate of the doubling dose of $(m_1 + m_2)/a$ is then the summed parental dose that would double the mutation rate. Neel and Schull and collaborators have called this the *zygotic doubling dose.* To convert this to an average, or gametic doubling dose for the sexes, the zygotic doubling dose is divided by 2.

The Direct Method

The direct method of risk calculation was pioneered by Ehling (Eh76a,b) and Selby and Selby (Se77) to estimate first-generation effects for dominant mutations rather than relying on the assumption of the proportionate effects implicit in the doubling dose method.

In the direct method, the induction rate for a specific class of defects in mice (e.g., cataracts and skeletal anomalies) is measured directly by using high-dose-rate radiation, and the results are corrected for dose rate. Then, the proportion of serious dominant genetic disorders in humans that involves similar defects is estimated, and this is used as a proportionality factor to estimate the effect of radiation on all dominant mutations in humans. For example, if the spermatogonial chronic induction rate for skeletal defects in the mouse was 4×10^{-6}/rad/gamete, and in humans about one in five serious dominant disorders involved the skeleton, then the first-generation effect of spermatogonial chronic radiation would be estimated by this method as 20 induced cases/10^6 liveborn/rad.

The committee had little confidence in the reliability of the individual assumptions required by the direct method let alone the product of a long chain of uncertain estimates that follow from these assumptions. Therefore, they did not place heavy reliance on the direct method in making their risk estimates, but used it only as a test of consistency.

The Gene Number Method

In the gene number method, one attempts to estimate the total number of mutations produced by exposure to radiation by using the equation:

$$\text{No. of induced mutations} = \text{No. of genes} \times (\text{induction rate/gene/unit dose}) \times \text{dose}. \tag{2-6}$$

This approach dates back to the BEAR Committee (NRC56) and Muller's elegant concept of "genetic death." BEAR states:

> One way of thinking about this problem of genetic damage is to assume that all kinds of mutations on the average produce equivalent damage, whether as a drastic effect on one individual who leaves no descendants because of this damage, or a wider effect on many. Under this view, the total damage is measured by the number of mutations induced by a given increase in radiation, this number to be multiplied in one's mind by the average damage from a typical mutation.

In other words, each harmful mutation ultimately causes one genetic death, which is either expressed all at once in the death of a single individual or is perhaps spread out as smaller effects over hundreds of individuals and hundreds of generations. One difficulty with this approach is that it is difficult to translate it usefully into societal cost and human suffering. Another problem is that no satisfactory definition or estimate of the total number of mutable genes is available. For these and other reasons, the BEIR V Committee eschewed risk estimates based on gene number.

PREVIOUS ESTIMATES OF HUMAN DOUBLING DOSE

BEAR (1956)

The BEAR Committee (NRC56) concluded that "the actual value of the doubling dose is almost surely more than 5R and less than 100R. It may very well be from 30R to 80R." The exact calculations from which these values, in roentgens, were obtained are not included in the report, except to say that

> the calculations which lead to an estimate of this 'doubling dose' necessarily involve the rates of both spontaneous and radiation-induced mutations in man. Neither of these rates has been directly measured; and the best one can do is to use the excellent information on such lower forms as fruit flies, the emerging information for mice, the few sparse data we have for man—and then use the kind of biological judgement which has, after all, been so generally successful in interrelating the properties of forms of life which superficially appear so unlike but which turn out to be remarkably similar in their basic aspects.

No distinction between acute and chronic dose was made. The doubling dose range given by the BEAR Committee would now be considered to apply to acute radiation. It must be remembered that at the time that the BEAR report was written, neither the dose-rate effect nor the distinction between premeiotic and postmeiotic cell stage response to radiation were known.

BEIR I (1972)

The BEIR I (NRC72) estimate of the doubling dose was given as a range of 20-200 rem, which was determined as follows. A chronic radiation dose to mouse spermatogonia was said to yield about 0.5×10^{-7} recessive mutations/rem/gene. The comparable figure for mouse oocytes was taken to be zero, giving an average of 0.25×10^{-7}. The spontaneous mutation rate was estimated from human dominant and X chromosome-linked mutation data to be in the range 0.5×10^{-6} to 0.5×10^{-5}, giving the doubling dose range of 20-200 rem. The figure of 20 rem was considered as being probably too low after a rough minimum doubling dose was calculated from the data then available from survivors in Hiroshima and Nagasaki.

BEIR III (NRC80)

Although BEIR III (NRC80) subscribed to the general principles of BEIR I (NRC72), it disagreed with the calculation of the doubling dose. Unlike BEIR I, which constructed a hybrid doubling dose based on the induced mutation rate in mice and the spontaneous mutation rate in humans, BEIR III chose to calculate a doubling dose for mice and extrapolate it to humans. The stated objection to the BEIR I method was that it mixed the induced rate of a set of mouse genes preselected for high mutability with an estimate of a human spontaneous rate for more typical genes. BEIR III took as an induced rate 6.6×10^{-8} mutations/locus/rem, from mouse spermatogonia irradiated at 0.009 rem/minute and below. The corresponding spontaneous rate was 7.5×10^{-6}, giving a point estimate of the doubling dose (for chronic radiation) of 114 rem. The committee then doubled and halved this figure to arrive at a final range of 50-250 rem to take into account uncertainties raised by the mouse oocyte data and the data from atomic-bomb survivors in Japan.

Other Estimates Based on Mice

Abrahamson and Wolff's (Ab76) linear-quadratic analysis of the mouse data lead to doubling dose estimates in the range of 43-131 rad. Analyses of data from Russell (Ru77) and Russell and Kelly (Ru82a) on low-dose-rate data in female and male mice, respectively, give a range of 99-160 rad. Finally, Denniston's (De82) analysis of the mouse data using the Lea (1947) model $Y = a + bD + cD^2G$ yielded a point estimate of 109 rad.

The Japanese Data

In contrast to the doubling dose estimates in mice, those derived from the human data have tended to be larger, sometimes by a factor of 3 or more. For example, Schull et al. (Sc81) state:

> In general, human exposure to radiation will not be acute and of the magnitude experienced by the inhabitants of Hiroshima and Nagasaki, but either interrupted

or chronic, and at much lower levels. Under such circumstances, the genetic yield of chronic radiation in mice is approximately one-third that of acute radiation. If mice and people are similar in this respect, the doubling dose for human chronic exposure suggested by these data becomes 468 rems, in contrast to the estimate of 100 rems for low LET, low dose, low-dose-rate exposure recently adopted by a committee of the International Commission on Radiological Protection.

Past committees have been reluctant to make heavy quantitative use of the data from Japan, despite their careful collection and analysis, in part because doubling doses derived from them are highly sensitive to several assumptions. For example, with respect to the two endpoints untoward pregnancy outcome and F_1 mortality, Neel, Schull, and collaborators have usually assumed a spontaneous rate of about 5% and a mutation component of about 5%, giving a spontaneous rate due to mutation of 0.0025. This is the numerator in a doubling dose estimate. However, a problem that these investigators have always been keenly aware of is that the doubling dose estimates are extraordinarily sensitive to these assumptions. For example, if the mutation component of untoward pregnancy outcome were actually 3% rather than 5%, a difference well within the range of plausible values, then the published doubling dose would be 40% too high. On the other hand, if the true mutation component were 7%, the published doubling dose would be 40% too low. Similarly, using 4% rather than 5% as the mutational component decreases the doubling dose by 20%, and using 6% as the mutational component increases the doubling dose by 20%.

Additional uncertainties complicate the estimation of human doubling dose. For example, neither the total spontaneous rate nor the induction rates per rad (which are not significantly different from zero in the Japanese data) are known with much precision. In addition, it is not obvious that the factor of 3 often used to convert the Japanese data from a high to a low dose rate is entirely appropriate. This factor was obtained from irradiation of mouse spermatogonia. Given that mouse data are the only data available on this point, the inference from the Japanese data that the mean radiosensitivity of humans is different from that of mice suggests that the dose rate conversion factor may also differ. Additional uncertainties in interpreting the conversion factor for mice are that it comes from comparison of acute high doses and chronic high doses and not from the more relevant comparison of acute low doses and chronic low doses, and the mouse data are based in part on experiments with radiation of different qualities (x rays, ^{137}Cs gamma rays, and ^{60}Co gamma rays), although radiation quality is unlikely to contribute much to the difference. These issues are admittedly difficult, but the doubling doses quoted for chronic radiation are very sensitive to the conversion factor. Prudence again seems to dictate that risks be based on a lower confidence limit rather than a point estimate.

CALCULATION OF RISK ESTIMATES

The risks in Table 2-1 are based on the assumption of a doubling dose of 100 rem. This is in agreement with the UNSCEAR reports of 1972, 1977, 1982, and 1986. A doubling dose of 100 rem approximates the lower 95% confidence limit for the data from atomic-bomb survivors in Japan, and it is also consistent with the range of doubling doses observed in mice. While it is somewhat arbitrary, the number has the advantage of arithmetic simplicity and is a round number that does not invite an unwarranted assumption of high accuracy. To the extent that the risks in Table 2-1 may be inaccurate, they are to be regarded as probably being too high rather than too low. For purposes of setting radiation standards, it is wiser to estimate risks that might be too large rather than risks that might be too small.

Estimating First Generation and Equilibrium Effects

Dominant Disorders

Several approaches to dominant disorders are possible. BEIR I (NRC72) essentially used the formula:

$$\text{first generation effect} = \text{spontaneous burden} \times (\text{doubling dose})^{-1} \times s, \qquad (2\text{-}7)$$

where $1 - s$ is the assumed average fitness of individuals suffering from dominant disorders. The BEIR I committee (NRC72) assumed a spontaneous burden of 1%, a doubling dose of between 20 and 200 R, and they estimated s as about 1/5, giving a first generation effect of 10 to 100 cases/10^6 liveborn/R. BEIR III (NRC80) assumed the doubling doses to be in the range of 50-250 R and similar estimates for the spontaneous burden and fitness as in BEIR I, from which the formula estimates 8-40 cases/10^6 liveborn/R. (However, BEIR III used the direct method for calculating dominants, see below). Raising the lower bound from 20 to 50R has a significant effect on the estimated risks.

The very different direct method for estimating first-generation effects of dominant disorders was pioneered by Ehling (Eh76a,b) and Selby and Selby (Se77), as described earlier in this chapter. BEIR III (NRC80) invoked the following argument using the data of Selby and Selby (Se77) on the induction of skeletal mutations in mice by gamma irradiation:

$$
\begin{aligned}
\text{risk} = \; & \text{induction rate of skeletal mutations } (37/2646)(600^{-1}) \\
& \times \text{ correction for dose rate and fractionation } (1/3)(1/1.9) \\
& \times \text{ multiplication factor for extrapolating} \\
& \quad \text{skeletal to all dominants } (5 - 15) \\
& \times \text{ correction for seriousness of traits } (0.25 - 0.75) \\
& \times \text{ correction for sex } (1.44) \\
= \; & 5 - 65 \times 10^{-6}
\end{aligned}
$$

This argument gave a risk of 5-65 dominant disorders/10^6 liveborn in the first generation after exposure of the entire population (both sexes) to 1 rem, but the calculation requires the multiplication of several factors of uncertain magnitude. The argument also implies that the average fitness for dominant disorders is 0.675-0.875 (bracketing the value of 0.8 assumed in BEIR I), which is in good agreement with the value of 0.83 calculated from the data of Childs (Ch81) in Table 2-2 (discussed below).

Ehling (Eh78) used data on the induction of cataracts due to a dominant mutation in mice from gamma irradiation to estimate the risk following 1 rem as:

$$
\begin{aligned}
\text{risk} = \; & \text{induction rate per rem } (1.3 \times 10^{-6}) \\
& \times \text{ correction for dose rate and fractionation } (0.3 \times 0.85) \\
& \times \text{ multiplication factor for total dominant damage } (32.4) \\
& \times \text{ extrapolation factor from mouse to human } (1.2) \\
= \; & 14 \times 10^{-6}
\end{aligned}
$$

In these and the previous example the correction factors used for low dose rate, fractionation, and sex were all derived from data using the mouse specific locus system for detecting recessive mutations, which is described in a section on animal studies later in this chapter.

NUREG/CR-4214 (NUR85) gave an estimate of 110 cases of newly induced dominant disorders in 490,000 births after an exposure of approximately 8 R. This corresponds roughly to 30 cases/10^6 liveborn/R.

A somewhat different approach is as follows. Childs (Ch81) has assembled data on some 25 dominant human genetic disorders or groups of disorders, the most severe of which are listed in Table 2-2. The total birth frequencies in Childs' tabulation is given as $5{,}840 \times 10^{-6}$,with an average selection coefficient of about 1/6. Assuming a doubling dose of 100 R, the

TABLE 2-2 Live Birth Frequencies, Reproductive Fitness, and Mutation Rates for Dominant Disorders

Disease (10^6)	Birth Frequency (10^6)	Fitness	Mutation Rate (10^6)
Diseases for which reasonable estimate of mutation rate is possible			
Retinoblastoma	24	0.5	6
Polyposis coli	71	0.8	7
Neurofibromatosis	350	0.5	93
Spherocytosis	220	0.8	22
Huntington disease	300	0.8	5
Myotonic dystrophy	220	0.7	28
Blindness	30	0.3	10
Deaf mutism	69	0.3	24
Cataracts with early onset	40	0.7	6
Aniridia	15	0.9	3
Cleft lip with lip pits	11	0.8	1
Polycystic kidney disease	860	0.8	76
Primary basilar impression	100	0.8	10
Achondroplasia	30	0.2	12
Diaphysial aclasia	50	0.7	8
Osteogenesis imperfecta	40	0.6	9
Osteopetrosis	10	0.8	1
Marfan syndrome	30	0.7	5
Tuberous sclerosis	25	0.2	10
Rare diseases of early onset	130	0.5	30
TOTAL	2,625		366
Diseases for which mutation rate estimate is subject to large uncertainty			
Hypercholesterolemia	2,000	1.0	<20
Porphyria: intermittent acute	15	0.9	1
Porphyria: variegate	15	1.0	<1
Otosclerosis	1,000	1.0	<20
Amelogenesis imperfecta	60	1.0	1
Dentinogenesis imperfecta	125	1.0	<1
TOTAL	3,215		<43
GRAND TOTALS	5,840		409

SOURCE: J. D. Childs (Ch81).

Childs' data give a first generation effect of about 10 dominant cases per million liveborn per R.

Alternatively, one can use Childs' estimates of the spontaneous mutation rates for these disorders, by means of the approximate relation

$$\text{first-generation effect} = 2U/\text{doubling dose},$$

where $U = 409 \times 10^{-6}$ is the total spontaneous mutation rate (Ch81). The estimate is 8 cases/10^6 liveborn/R. The two estimates from Childs' data are not independent, but they demonstrate the consistency of the data.

This approach has the positive feature that it is based on a reasonably well-defined set of diseases that, in fact, constitute a substantial portion of the incidence of dominant disorders in humans.

All these risk estimates for dominant disorders are roughly in agreement and compatible with a doubling dose on the order of 100 rem (1 Sv). The BEIR V Committee has divided the autosomal dominant disorders into categories based on their relative fitness as related to the severity of clinical symptoms. When both categories are combined, the estimate is 6-35 cases of dominant disorders induced in the first generation/10^6 liveborn/rad, with an equilibrium value of 100. The time required to go halfway to equilibrium is about $0.693/s$ generations (Mo82); for s in the range of 0.2-0.8 (clinically severe), this is approximately 4-9 generations, and for s in the range of 0.01-0.2 (clinically mild), it is approximately 4-70 generations.

X Chromosome-Linked Disorders

The dynamics of X chromosome-linked genes are much the same as those of autosomal genes and for this reason they are often included with dominant mutations. Trimble and Doughty (Tr74) give the birth frequency of X-linked disorders as about 400/10^6; Childs (Ch81) cites a value closer to 300/10^6 liveborn. For an X chromosome-linked gene, the proportion of the equilibrium excess of cases that appears in the first generation is approximately $s/(2 + R)$, where $1 - s$ is the fitness of affected males and R is the ratio of male to female mutation rates. In the Childs (1981) compilation, the average value of s is about 0.75. If R is between 3 and 1, the proportion of the equilibrium excess cases occurring in the first generation is between 0.15 and 0.25. For a doubling dose of 100 rem, this implies less than 1 case/10^6 liveborn in the first generation.

Using the same estimates given above, the per-generation excess attained after the population reaches equilibrium between mutation and selection is less than 5 cases/10^6 liveborn/rad. The time required to go halfway to equilibrium is about $0.693(3/s)$, (Mo82), or in this case about 3 generations.

Recessive Disorders

Past BEIR committees have concluded that the increase in disease due to recessive mutations following an increase in the mutation rate from chronic radiation will be too slight or too remote in the future to justify quantitative estimation. Some geneticists disagree (e.g., Neel Ne57). Searle and Edwards (Se86a) have recently addressed whether the induction of recessive mutations significantly increases the mutational burden. The essence of their result is that the first generation effect after a population exposure of 1 R is about $[2\ u/DD]\ \Sigma q$, where u is the average spontaneous mutation rate, DD is the doubling dose, and Σq is the sum of the recessive equilibrium gene frequencies for all recessive disorders. The sum of the q values reflects the meeting of a newly induced mutation with a previous mutation already established in the population. If this sum is taken to be on the order of 1 and the spontaneous mutation rate is taken to be 12×10^{-6}, (Mo81), then for a doubling dose of 100 rem, the first-generation effect is less than 1 recessive case/10^6 liveborn/rem, confirming previous expectations.

The equilibrium between selection and mutation when the mutation rate is increased is attained so slowly that it is relevant only to a hypothetical population existing in the distant future. The time required to go halfway to equilibrium is about $0.693/2\,Qs$ where $Q = (u/s)^{0.5}$ (Mo82). For this reason the present committee has not attempted a quantitative risk estimate for recessive mutations at equilibrium.

Moreover, there are good reasons to believe that the majority of recessive mutations are actually partially dominant in their effects on fitness. For example, in *Drosophila melanogaster*, spontaneous recessive lethal mutations reduce heterozygous viability by 4-5%, but lethal mutations isolated from natural populations cause a 1-2% reduction. Based on allele frequencies, the average recessive lethal allele appears to persist in a *Drosophila melanogaster* population for about 50 generations before it is eliminated by selection, which is far too short a time to be entirely a result of homozygous lethality.

In humans, also, there is some indication that recessive mutations are partially dominant. The evidence comes from consanguineous matings and the often unexpectedly low equilibrium frequencies of recessive genotypes. Whether partial dominance also applies to radiation-induced recessive mutations is less certain, but to the extent that it does, such mutations act like dominant mutations for the purpose of risk calculations.

Translocations

BEIR I (NRC72) estimated a first generation effect of 70 recognized abortions and 12 unbalanced rearrangements born/10^6 liveborn/R. The equlilibium values were only slightly larger. These estimates were based on an estimated mouse spermatogonial induction rate for semisterility of 1.5×10^{-5}/gamete/rad for low dose irradiation, and the conservative assumption is that females would have a similar frequency.

To calculate the risk from induced translocations, BEIR III (NRC80) utilized data from humans and the marmoset (Br75). The frequency of multivalent translocations in the primary spermatocytes of humans and marmoset was taken to be about 7×10^{-4}/rem, based on high dose-rate 250 kV x ray doses of 78 R in humans (371 cells examined) and doses of 25

R, 50 R and 100 R in marmosets (600 cells examined at each dose). The present committee's review of the relevant data suggests that a value of 2×10^{-4}/rem would be more appropriate (see the later section in this chapter on chromosome aberrations in mice and other mammals). In any case, the BEIR III calculation of risk of induced transmitted balanced translocations was $(7 \times 10^{-4})(2/3)(1/2)(0.45/2) = 5.25 \times 10^{-5}$ translocations/rem, where 2/3 is the assumed ratio of the observed incidence of partial sterility to that calculated on the basis of the incidence of multivalent translocations in primary spermatocytes, 1/2 is the correction for dose rate, and 0.45 is the assumed frequency of alternate segregation of which 1/2 yield balanced translocation gametes. To accommodate the uncertainties regarding the dose rate reduction factor, the BEIR III Committee preferred to use the order-of-magnitude range 1.7×10^{-5} to 1.7×10^{-4} translocations/rem.

The corresponding calculation for unbalanced products was $(7 \times 10^{-4})(2/3)(1/2)(0.55)(0.05)(1/4) = 1.6 \times 10^{-6}$ unbalanced zygotes/rem, where 0.55 is the assumed frequency of adjacent segregation, 5% of such translocation gametes are assumed to be capable of producing viable aneuploids, of which 1 in 4 lead to viable zygotes. Again, an order-of-magnitude range was given as 0.5×10^{-6} to 5×10^{-6} unbalanced zygotes/ rem. Multiplying by 2 (assuming females are about as inducible as males) leads to BEIR III's conclusion (Table IV-2 in BEIR III) that fewer than 10 cases/10^6 of induced chromosomal aberrations would appear in the first generation following exposure to 1 rem of radiation.

NUREG/CG 4214

The U.S. Nuclear Regulatory Commission NUREG report (NU85) also used experimental data obtained from marmosets and humans. They took the induction rate of multivalent translocations in spermatogonia irradiated by x rays at and below 100 R (4 data points, one human and three marmosets) as 7.4×10^{-4}. Their calculations were $(7.4 \times 10^{-4})(1/2)(0.4)(1/4) = 3.7 \times 10^{-5}$ balanced translocations/rem, where 1/2 is a dose rate correction, and 0.4 is a relative biological effectiveness (RBE) correction to go from x rays to gamma rays. Again, 1/4 of the segregants were assumed to be balanced translocations. For unbalanced products, the calculation was $(7.4 \times 10^{-4})(1/2)(0.4)(1/2)(1/10) = 7.4 \times 10^{-6}$ unbalanced zygotes/rem, where 1/2 is the frequency of adjacent segregation and 1/10 is the probability of survival. These values are for males. In females, the induced translocations are expected to result from chromatid breaks, so the corresponding calculations were $(7.4 \times 10^{-4})(1/2)(0.4)(1/16) = 9.25 \times 10^{-6}$ balanced translocations/rem, and $(7.4 \times 10^{-4})(1/2)(0.4)(6/16)(1/10) = 5.6 \times 10^{-6}$ unbalanced zygotes/rem.

Comparing the NUREG calculations with the BEIR III results, three differences are seen. BEIR III makes a correction for transmission but NUREG does not (Ge84). NUREG makes a correction for x rays to gamma rays (NCRP80), but BEIR III does not. These differences approximately cancel out each other. Finally, NUREG attempts to calculate explicitly the effect of radiation on oocytes, whereas BEIR III formally assumed that the female rate was equal to the male rate but suspected that the female rate was actually lower.

UNSCEAR 1982

UNSCEAR (UN82), summarizing another UNSCEAR report (UN77), calculated $(7.4 \times 10^{-4})(1/4)(1/10 \text{ to } 1/2)(2)(0.06) = (2.1 \text{ to } 10.5) \times 10^{-6}$ unbalanced zygotes/rem, where, again, the marmoset and human data were used, 1/4 is the conversion factor from multivalents to semisterility and segregation, the range 1/10 to 1/2 is used for dose rate correction, and twice as many unbalanced as balanced gametes are expected, of which about 6% would survive. The result is similar to the previous ones. In addition, UNSCEAR concluded that the female rate could be considerably lower and ". . . should it turn out that the rate of induction in human spermatogonia is more similar to that in the rhesus monkey, the estimates may need revision downward, and consequently the quantitative figures arrived at must be considered provisional at present."

As noted, the BEIR V Committee's review of the relevant data suggests a rate of translocation induction of 2×10^{-4}/rem, with a dose rate effect somewhat larger than previously thought. These revisions imply that previous estimates were somewhat too high. The committee suggests that an appropriate upper limit to the first generation effect caused by unbalanced products arising out of induced reciprocal translocations is less than 5 cases/10^6 liveborn/rad. It does not appear that Robertsonian translocations, which are such a prominent feature of the spontaneous burden in humans, are readily induced by radiation.

Nondisjunction

For a number of years, there has been an unresolved possibility that low doses of radiation, such as those used in diagnostic radiology, might induce chromosome nondisjunctions in exposed women. Most concern has focused on the possible induction of trisomy-21 (Down syndrome). The frequency of Down syndrome is strongly influenced by maternal age, rising to nearly 4% of all live births among women over 40 years of age, and the possibility that radiosensitivity also increases with age must be considered. The issue was addressed in Note 15 of Chapter IV in BEIR III (NRC80), in recent UNSCEAR reports (UN77, UN82, UN86), and in a review by de Boer and Tates (de83). The following provides a brief review of the subject.

Of 13 studies on the Down syndrome in humans discussed by Denniston (De82), 9 were retrospective and 4 prospective. No claim has been made for an effect caused by paternal radiation, but four of the studies found a significant effect caused by maternal radiation (one prospective and three retrospective studies). Of the remaining nine studies in which no statistical significance was attained, five were in the positive direction, two showed no difference, and two were in the negative direction. Overall, looking only at the direction of the data and ignoring whether or not they were statistically significant, there were nine showing positive effects and two showing negative effects. This is significant at the 0.033 level, assuming no effect. However, because of the way some of the data were collected (reliance on subject's memory of past irradiation), there is likely a bias in the positive direction. If, under the hypothesis of no association, the probability of observing data in the positive direction is only as high as 0.53, the sign test for consistency is no longer significant at the 5% level.

No effect on nondisjunction has been seen in the data from survivors of the Hiroshima and Nagasaki bombings (Aw87), and the claim of an effect on the incidence of Down syndrome in a high-background-radiation area of India has been severely criticized on statistical grounds.

Although nondisjunction can be induced with relatively large doses (1 to 6 Gy) of x irradiation in various dictyate oocyte maturation stages in mice (Te85), other studies have concluded that, at low doses, (<1 Gy) nondisjunction is not induced to any significant degree (Sp81, Te82). The positive results obtained by Uchida and Lee (Uc74) at low doses are at variance with results of subsequent studies (Go81, Te82). Therefore, notwithstanding the importance of nondisjunction to the spontaneous burden in humans, it appears that the induction of nondisjunction by low-level irradiation of immature oocytes may not present a serious concern. However, as discussed below in the section on chromosomal nondisjunction in mice, preovulatory oocytes, within three hours of ovulation, are extremely sensitive to the induction of aneuploidy at doses as low as 10 rads (Te82, Te86). Even if this effect occurs in humans, the brevity of the sensitive period would leave the risk estimates essentially unchanged.

Irregularly Inherited Traits

The so-called irregularly inherited disorders are those for which a genetic component has been established or seems likely, but which do not give simple Mendelian ratios. Irregular inheritance poses a serious problem to risk estimation. Although these traits constitute a significant portion of the total genetic burden in human populations, their response to an increase in the mutation rate from radiation is not predictable with any great confidence because of the uncertainty in their mode of inheritance.

An important concept relevant to irregularly inherited traits is the mutation component. If the incidence (I) of a condition can be written as $I = a + bu$, where u is the mutation rate and a and b are constants, then the mutation component of the condition is $M = bu/(a + bu)$. M is the proportion of the incidence attributable to recurrent mutation, and $a/(a + bu)$ is the part attributable to other causes. If the mutation rate is increased from u to $u(1 + k)$, the incidence eventually increases from I to $I(1 + Mk)$.

The heritability of a trait is a measure of that part of the total phenotypic variability that can be ascribed to genetic variability in the population. The ratio of the total genetic variance to the total phenotypic variance is called the "broad-sense heritability"; the ratio of the "additive" genetic variance (only part of the total genetic variance) to the total phenotypic variance is called the "narrow-sense heritability." For a trait maintained by balance between directional selection and mutation, if both broad-sense and narrow-sense heritability are high, then M is high. If both are low, then M is low. If the broad-sense heritability is high and the narrow-sense heritability is low, M cannot be predicted unless the specific mode of inheritance is known; however, any increase in the incidence following an increase in the mutation rate should be very slow (Cr81).

Trimble and Doughty (Tr74) estimated that about 9% of all liveborn humans are seriously handicapped at some time during their lifetimes by genetic disorders of complex etiology, either congenital abnormalities, anomalies that are expressed later, or consitutional and degenerative diseases. Their estimate is somewhat indirect. They adjusted data based on incidences prior to age 20 to account for disorders appearing later in life. BEIR III accepted this estimate and combined it with their own doubling dose range of 50-250 R and mutation component range of 5-50% to estimate an equilibrium excess of 20-900 induced cases of irregularly inherited disorders/R/10^6 liveborn. No first generation effect was estimated.

Estimating the equilibrium effect on irregularly inherited disorders due to an increase in the mutation rate raises several problems:

1. The mutation components are not known for these disorders, even approximately. Many of the traits are genetically and environmentally heterogeneous—a mixture of simple Mendelian etiologies, multifactorial threshold factors, and purely environmental causes. To the extent that the traits are accurately described by a multifactorial threshold model, the mutation component is undoubtedly low and the approach to equilibrium is very slow. To the extent that the traits include a simple Mendelian component, the mutation component is high and the approach to equilibrium

depends on the exact nature of the model (e.g., dominant versus recessive, overdominance versus mutation-selection balance). The BEIR III Committee dealt with these uncertainties as well as they could and considered a range of mutation component between 5 and 50%.

2. Irregularly inherited disorders are diverse in terms of the nature of the defects represented (e.g., anencephaly versus varicose veins), severity (e.g., cleft lip versus club foot), time of action (birth to old age), and so on. This diversity makes it difficult to present a single overall measure of impact on the population. For example, the spontaneous frequency is determined by the rather arbitrary definition of what constitutes a serious disorder rather than one that is clinically significant.

3. Irregularly inherited disorders—even those with a substantial mutation component—have a slow rate of approach to equilibrium following a change in the mutation rate. Measures, such as excess number of cases per generation at equilibrium, are virtually meaningless because the very slowness of the approach may mitigate the seriousness of the threat to the population. The potential impacts cannot be quantified because the increased genetic load is spread out over so many generations into the future in an environment that is totally unpredictable at the present time.

Since the BEIR III report (NRC80), new information on the spontaneous incidence and the genetic nature of irregularly inherited disorders has become available (Cz84a, Cz84, UN86, Cz88). For purposes of the present discussion, it will be convenient to divide the irregularly inherited disorders into isolated congenital abnormalities and all others.

Congenital Abnormalities: Table 2-3 lists nine congenital abnormalities with an estimated combined birth incidence in Hungary of about 5%, and estimates of the heritabilities both of their liabilities and of the traits themselves. All such tabulations are somewhat vague in the diagnostic criteria used to identify the traits, and the high incidence of congenital dislocation of the hip in Hungary is so exceptional as to suggest overreporting. The BEIR V Committee estimates the birth incidence of congenital abnormalities at 20,000-30,000/10^6 liveborn (Table 2-1), which is consistent with the data in Table 2-3 when the high value for congenital dislocation of the hip is discounted.

The distinction between the heritability of a trait's liability, assuming a threshold model, and the heritability of the trait itself is crucial, because the mutation component is more related to the heritability of the trait than to the heritability of liability. In the threshold model it is assumed that underlying each trait is a quantitative variable called liability, which is normally distributed and the result of many genetic and environmental terms of small effect. Individuals with a value of liability above a threshold are affected; those below the threshold are normal. By observing the

TABLE 2-3 Selected Isolated Congenital Abnormalities[a]

Trait	Birth Incidence (10^3)	Heritability of Liability[b]	Heritability of Trait[c]	MZ Concordance of Trait[d]
Anencephaly/spina bifida	2.9	35–70	2–7 (4)	5
Cleft lip (cleft palate)	1.0	70–90	4–8 (10)	30
Pyloric stenosis	1.5	60–90	3–9 (12)	65
Ventricular septal defect	1.5	35–80	1–7 (3)	20
Congenital dislocation of hip	28.0	60–80	17–27 (50)	50
Talipes equinovarus	1.3	65–95	4–10 (11)	30
Congenital inguinal hernia	11.4	40–60	6–11 (18)	50
Simple hypospadias	4.4	45–85	4–13 (9)	50
Undescended testicles	13.5	35–65	5–13 (11)	15
TOTAL	54.1			

[a] All values except birth incidences are rounded to the nearest 5%.
[b] Ranges are ±1 standard deviation, i.e., an approximate 68% confidence interval.
[c] Range obtained from the liability heritabilities from the formula $h_T^2 = 2(q - p)/(1 - p)$, where $q = p^z$ and $z = \tan\{[(D/4)(1 - 0.5h_L^2)]\,[1 + (0.5h_L^2)^5]\}$; values in parentheses were obtained directly from sib recurrence risks, q (Czeizel and Tusnady, 1984).
[d] The MZ twin concordances, C, yield maximum estimates of the broad sense heritability of each trait through the formula $H_T^2 = (C - p)/(1 - p)$.

SOURCE: Modified from A. Czeizel and K. Sankaranarayanan (Cz84).

population incidence of a trait p and the recurrence risk for relatives of affected individuals q, an estimate of the narrow heritability of liability can be obtained (Fa65, Sm70, Cu72). These estimates depend not only on the accuracy of the estimates of p and q but also on the assumptions of the threshold model.

Alternatively, the disorder itself can be thought of as a quantitative trait taking either of two values: 0 for normal and 1 for affected. An estimate of the narrow heritability of the trait is obtained from relatives by the formula $Rh_T^2 = (q - p)/(1 - p)$, where R is the coefficient of relationship. An approximate relation between the heritability of liability h_L^2 and the heritability of the trait h_T^2 is given in footnote c in Table 2-3. The concordance between monozygotic (MZ) twins may be considered as an approximate maximum estimate of the broad-sense heritability of the disorder.

In Table 2-3 all numbers except the birth incidences and heritabilities of traits have been rounded to the nearest 5%. The incidences, liability heritabilities, and MZ twin concordances are from Cz84a. All estimates, especially those from the twin data, are inflated to an unknown extent by environmental correlations. The twin data also yield very unstable estimates because of small sample size.

In general, the estimates of trait heritabilities from sibling data do not differ much from those in the entire data of Czeizel and Tsunady (Cz84a) or those derived indirectly by using estimates of liability heritabilities from the threshold model. (One exception is congenital dislocation of the hip.) In rough terms, the heritabilities of the traits themselves are about 1/10 those of the liabilities.

At face value, the MZ concordances suggest that broad-sense heritabilities are much larger than narrow-sense heritabilities. This discrepancy is more likely caused by environmental correlations peculiar to twins rather than to a large amount of dominance and epistatic variance. In any event, whether the mutation components of these disorders are closer to 5 or 50% (the BEIR III range), the uniformly low narrow heritabilities would indicate that the approach to equilibrium following a rise in the mutation rate would be very slow indeed. On the other hand, to the extent that any of these disorders includes a significant proportion of cases with a simple monogenic origin (which have a mutation component of 1), the overall mutation component would be increased.

The risk estimates for this category of traits are listed in Table 2-1. The equilibrium value is based on Equation (2-1) with the assumption that the mutation component of the traits is between 5 and 35%. The upper limit of 10 for the first-generation effect is based on the worst-case assumption that the mutational component is due entirely to dominant genes.

Other Disorders of Complex Etiology: The data in Table 2-4 are taken from a recent set of data from Hungary presented in preliminary form by UNSCEAR (UN86). The table shows (1) large total lifetime prevalence (over 30%) and (2) large estimated heritabilities of liability based on a multifactorial threshold model. However, the heritabilities of the traits themselves are much smaller (see preceding section).

If anything, the disorders in Table 2-4 are even more heterogeneous than the congenital abnormalities in Table 2-3. In Table 2-4, lifetime prevalences rather than birth frequencies are given. Many of the disorders have a rather late age of onset. The total lifetime prevalence for the selected disorders tabulated is about 30%. Assuming independence, approximately 27% of individuals suffer from at least one of these diseases sometime during their lifetimes.

The heritabilities in Table 2-4 again pertain to liability calculated from the Hungarian data and with the assumption of a multifactorial threshold etiology. The narrow heritabilities of the traits themselves are approximately 1/10 of these values (see preceding section). To the extent that these disorders are heterogeneous and confounded with monogenic or simple Mendelian disorders whose equilibrium frequencies result from a balance between mutation and selection, the mutation components would be elevated. On the other hand, several of the disorders are known to be correlated with variation in the HLA histocompatibility complex (e.g.,

TABLE 2-4 Selected Diseases of Complex Etiology

Disease	Lifetime Prevalence per 10^4	Liability Heritability
Grave's disease	65	0.47
Diabetes mellitus	407	0.65
Diabetes mellitus (IDDM)	20	0.30
Gout	18	0.50
Schizophrenic psychoses	85	0.80
Affective psychoses: unipolar	500	0.60
Affective psychoses: bipolar	100	0.90
Multiple sclerosis	4	0.58
Epilepsy	60	0.50
Glaucoma	160	0.32
Allergic rhinitis	360	0.43
Asthma	249	0.70
Peptic ulcer	460	0.65
Idiopathic proctocolitis	3	0.60
Cholelithiasis	94	0.63
Coeliac disease	13	0.80
Calculus of the kidney	90	0.70
Atopic dermatitis	60	0.50
Psoriasis	39	0.75
Systemic lupus erythematosus	4	0.90
Rheumatoid arthritis	131	0.58
Ankylosing spondylitis	19	0.79
Scheuermann disease	50[a]	0.56
Adolescent idiopathic scoliosis	41	0.88
TOTAL	3,032	

[a] Includes only the 5% of cases identified by radiographic screening that are deemed to be of clinical significance.

ankylosing spondylitis, rheumatoid arthritis, psoriasis, coeliac disease, and diabetes). To the extent that population variation in the HLA complex is caused by balancing of selection, the mutation components of these disorders would be reduced correspondingly.

As in the case of the congenital abnormalities, data on twins generally show substantially higher concordances in monozygotic (MZ) than dizygotic (DZ) twins, testifying to a likely significant genetic component in these disorders. The general pattern is that the broad-sense heritabilities of the traits are considerably larger than the narrow-sense heritabilities. Consequently, the mutation components are indeterminant without further information, but it seems likely that any change in the frequencies of these diseases caused by a change in the mutation rate would be attained very slowly.

The data in Table 2-4 are for selected diseases and do not include data for cancer and heart disease, which are the most common diseases with complex etiologies. Cancer and heart disease are listed separately in Table 2-1, and the lifetime prevalence figures are approximations in round numbers for the prevalence of all varieties of the diseases. By enumerating heart disease in Table 2-1, the committee makes no implication that radiation can induce heart disease in exposed individuals. The effect of radiation on this and other diseases with complex etiologies (with the exception of cancer) is through new mutations that may increase the susceptibilities of their carriers to the onset of the diseases. From a genetic point of view, the mutational component of diseases with complex etiologies results from a number of genes, usually with small individual effects, that in combination determine susceptibility to environmental factors causing the disease. In the case of heart disease, for example, these environmental factors include diet and tobacco smoking. Any individual mutation is extremely unlikely to tip the balance between a person's health and disease. Rather, each new mutation is an additional genetic risk factor that combines with other genetic and relevant environmental risk factors. For the individual, a new mutation may contribute a marginally insignificant amount to the overall risk, but for the population, the small individual effects are cumulative and may become very significant.

For diseases with complex etiologies, the lifetime prevalences sum to greater than 100%, which means that few individuals escape them completely, and many suffer from more than one. Since the prevalence is one component of the risk estimate (Equation 2-1), this factor is very large. However, the prevalence factor is offset in part by an unknown, but presumably low, mutational component. Unfortunately, the mutational component is not known even to its order of magnitude, and for this reason, as well as other complexities enumerated in the preceding section on congenital abnormalities, the committee has not estimated risks for this category of traits. While the risks could be negligible, they could also be as large or larger than all the other entries in Table 2-1 combined.

BACKGROUND DATA FROM HUMANS

Three key sets of background data for humans concern the genetic burden resulting from spontaneous mutation, the rate of spontaneous mutation, and the data from survivors of the bombings of Hiroshima and Nagasaki. These are briefly reviewed below.

The Spontaneous Genetic Burden

Table 2-5 shows estimates of spontaneous frequencies of genetic disorders. The estimates used by the BEIR V Committee are also summarized. The categories of disorders are autosomal dominant, X chromosome-linked

TABLE 2-5 Estimated Spontaneous Burden (per 1,000 live births)

Source	Dom.	X-linked	Rec.	Chrom.	Congen. Abn.	Other Multifact.
Stevenson (1959)	30.7	0.2	1.0	—	10.1	10.3
UNSCEAR (1966)	9.5	0.4	2.1	4.2	25.0	15.0
BEIR (1972)	10.0	0.4	1.5	5.0	15.0	25.0
Trimble and Doughty (1974)	0.8	0.4	1.1	2.0	42.8	47.3
UNSCEAR (1977)	10.0[a]		1.0	4.0	90.0[b]	
Carter (1977)[c]	7.0	0.4	2.5	6.0	24.4	—
BEIR (1980)	10.0[a]		1.1	6.0	90.0[b]	
Childs (1981)	5.8	0.3	—	—	—	—
UNSCEAR (1982)	10.0[a]		2.5	6.3	43.0	—
Czeizel and Sankaranarayanan (1984)		—	—	—	59.7	—
UNSCEAR (1986)	10.0[a]		2.5	6.3	60.0	600
This committee	10.0[d]	0.4	2.5	4.4[e]	20–30	1,200[f]

NOTE: Abbreviations: Dom., dominant; Rec., recessive, Chrom., chromosomal abnormality; Congen. Abn., congenital abnormality; other multifact., other multifactorial trait.

[a] Dominant and X-linked combined.
[b] Congenital abnormalities and "other multifact." categories combined.
[c] Chromosomal abnormalities from Evans (1977).
[d] Divided into 2.5 clinically severe and 7.5 clinically mild.
[e] Divided into 0.6 unbalanced translocations and 3.8 trisomies (includes sex chromosome trisomies).
[f] Includes heart disease, cancer, and other selected disorders (Table 2-4). Note that the total exceeds 100%. The genetic component in many of these traits is unknown. To the extent that genetic influences are important, the effects are through genes that have small individual effects but that act cumulatively among themselves and in combination with environmental factors to increase susceptibility.

recessive, autosomal recessive, chromosomal abnormalities, congenital abnormalities, and other multifactorial traits. The last category is made up of a group of disorders for which the exact mode of inheritance is unknown. Some may prove to be monogenic in origin; others are undoubtedly threshold traits, for example, the congenital abnormalities. Five entries in Table 2-5 are based on original data: those of Stevenson (St59), Trimble and Doughty (Tr74), Carter (Ca77), Czeizel and Sankaranarayanan (Cz84), and Childs (Ch81). The remaining entries are consensus estimates of committees based largely on data from the first four studies listed in the table. A discussion of the main points presented in Table 2-5 follows.

The most dramatic discrepancy is between the data of Stevenson and those of Trimble and Doughty with respect to autosomal dominant disorders. The Stevenson estimate of 30.7/1,000 live births is inflated by the incorporation of a number of traits that are now known not to be autosomal dominant, of traits of inconsequential clinical importance, or both (Tr77). The 10 most frequent traits in the Stevenson list make up about 70% of the total frequency, and most of these fall in the above categories of inappropriateness. On the other hand, the value of 0.8/1,000 from Trimble and Doughty is undoubtedly an underestimate because it is based on studies of individuals from birth to 21 years of age. Consequently, the estimate does not include serious genetic diseases due to single dominant genes that are manifested later in life. It can be seen from Table 2-5 that committees have chosen a middle course, with an estimate of about 10/1,000, often lumping dominant and X chromosome-linked traits together because of their similar responses to an increase in the mutation rate.

Over the years the estimated frequencies of recessive disease and chromosomal abnormalities have increased somewhat. Estimates of congenital abnormalities have increased substantially. Like the autosomal dominant traits, the estimate for congenital abnormalities is highly dependent on the definition of "serious." The value of 60/1,000 from Cziezel and Sankaranarayanan, which was also used by UNSCEAR (UN86), is so high, in part, because of the unusually high frequency of congenital dislocation of the hip in Hungary. The surprisingly high value of 600/1,000 for lifetime prevalence of other multifactorial disorders given by UNSCEAR (UN86) includes such entities as diabetes mellitus, gout, schizophrenia, affective psychoses, epilepsy, glaucoma, hypertension, varicose veins, asthma, psoriasis, ankylosing spondylitis, and juvenile osteochondrosis of the spine. Disorders with such high frequencies are, of course, not strictly independent, but the message, nevertheless, is that virtually all humans suffer from ill health at some time in their lives, and ill health can usually be attributed in part to genetic factors.

Estimating Spontaneous Mutation Rates

Table 2-6 gives some representative mutation rates estimated in humans. These values are consistent with the values given more than 25 years ago (Pe61, Cr61).

It is well recognized that published mutation rates are probably a biased estimate of all mutation rates, because it is more likely that those loci with higher natural rates will be studied. A simple correction for this bias is to use the harmonic mean of the studied loci.

From the data collected by Vogel and Rathenberg (Vo75) and Childs (Ch81) (Table 2-6), the harmonic means for dominant and X chromosome-linked traits are both about 8×10^{-6} if the Von Hippel-Lindau syndrome is omitted from the dominant traits, or 3×10^{-6} if the Von Hippel-Lindau syndrome is included. On the other hand, for X chromosome-linked traits

TABLE 2-6 Selected Mutation Rates

	Mutation Rate (10^6)		
Trait	Vogel and Rathenberg (Vo75)	Childs (Ch82)	Morton (Mo81)
Autosomal dominant			
Achondroplasia	6–13	12	10
Aniridia	3–5	3	3
Dystrophia myotonica	8–11	28	10
Retinoblastoma	5–12	6	8
Acrocephalosyndactyly	3–4	—	4
Osteogenesis imperfecta	7–13	9	10
Tuberous sclerosis	6–11	10	8
Neurofibromatosis	44–100	93	73
Intestinal polyposis	13	7	13
Marfan syndrome	4–6	5	5
Polycystic kidneys	65–120	76	92
Multiple exostoses	6–9	8	8
Von Hipple-Lindau syndrome	0.2	—	1
Pelger anomaly	—	—	6
Spherocytosis	—	—	22
Microphthalmos	—	—	6
Waardenburg's syndrome	—	—	4
Nail-patella syndrome	—	—	2
Huntington disease	—	—	2
Multiple teangiectasia	—	—	2
TOTAL: fairly reliable	—	366	—
including uncertain	—	409	—
X-linked recessives			
Hemophilia A	32–57	36	13
Hemophilia B	2–3	3	1
Duchenne MD	43–105	60	88
Incontinentia pigmenti	6–20	—	13
Orofaciodigital syndrome	5	—	5
Lesch-Nyhan syndrome	—	—	2
TOTAL	—	140	—

SOURCE: Crow and Denniston (Cr85).

from the data of Stevenson and Kerr (Table 2-7), a supposedly far less biased sample, the median is about 0.1×10^{-6} and the mean is about 3×10^{-6}. The Morton estimates give harmonic means of 4×10^{-6} for dominant traits and 3×10^{-6} for X chromosome-linked traits. Cavalli-Sforza and Bodmer (Ca71) plotted the cumulative frequency of published rates against the log mutation rate and found the plot to be approximately linear, suggesting that the log-normal distribution is a good distribution for describing mutation rates. From the fitted line they estimated the median

TABLE 2-7 Mutation Rates for X-Linked Recessives

Mutation Rate (10^6)	Frequency of Traits with Mutation Rate
50	1
20–49	1
10–19	1
5–9	2
1–4	9
0.1–0.9	11
<0.1	24
TOTAL	49

SOURCE: Stevenson and Kerr (St67).

to be 0.16×10^{-6} and the mean to be about 7×10^{-6}. All of these estimates are derived from overlapping sets of data.

In sum, the spontaneous per-locus mutation rate for dominant and X chromosome-linked traits has a mean of approximately 5×10^{-6} and a median perhaps an order of magnitude lower.

The mutation rate of autosomal recessives is much less certain. Morton (Mo81) has examined this problem in detail. Using the harmonic mean argument, he derives an estimate of 12×10^{-6} clinically detectable mutations/locus/generation. In this regard, Neel (Ne57) commented that "it is entirely conceivable that the loci thus far selected for study in man are those at which a high proportion of all possible alleles results in readily detectable effects, but at which the per locus mutation rate is fairly representative of the human species." In that case the arithmetic mean of 22×10^{-6} is more appropriate.

The Hiroshima-Nagasaki Data

A pregnancy termination study (Ne56) analyzed some 75,000 births, of which 38,000 had at least one parent who was exposed to radiation. No significant effects on still births, birth weight, congenital abnormalities, infant mortality, childhood mortality, leukemia, or sex ratio were found. A significant distortion of the sex ratio had been reported (Ne53), but the effect subsequently disappeared. In 1960 the pregnancy termination study was augmented with additional children of survivors and controls. A cohort, the F_1 mortality sample, was created, consisting of (1) all infants who were liveborn in the two cities between May 1946 and December 1958, one or both of whose parents were within 2,000 meters of the hypocenter, (2) an age-matched and sex-matched group of children with one parent who was more than 2,500 meters from the hypocenter and the other parent who was the same distance from the hypocenter or who was not exposed at all, and (3) an age-matched and sex-matched group of children neither of whose parents were exposed. No statistically significant effects of radiation have been demonstrated to date (Ne74, Sc81, Sc81a, Sa82).

A cytogenetic study of the children of exposed parents was begun in 1968 (Aw75). Ten metaphase preparations are routinely examined from each child. No significant effect has been demonstrated (Aw87).

The investigation of rare electrophoretic variants in children born to proximally and distally exposed parents was begun in 1972 as a pilot study and was begun in earnest in 1976 (Ne80). Each child is examined for rare electrophoretic variants of 28 proteins of the blood plasma and erythrocytes, and since 1979, a subset of the children is further examined for deficiency variants of 10 erythrocytic enzymes. If the variant is not found in either parent and a discrepancy in biological parentage can be excluded, a mutation has been identified. Among the children of proximally exposed parents, the equivalent of 667,404 locus tests have been done, yielding three probable mutations. The corresponding value for the comparison groups is three mutations in 466,881 tests. The point estimate of the mutation rate is higher in the control population, but the difference is not significant (Ne88).

Table 2-8 provides the lower 95% confidence limits of doubling dose estimated for various endpoints in the data from the Japanese atomic-bomb survivors summarized by Schull et al. (Sc81a). Other data from the studies of the atomic-bomb survivors give comparable results (Ne74, Sa82). Prior to calculating the doubling doses from the regression coefficients, negative regression coefficients were set equal to zero. In all cases, following Schull et al. (Sc81), a spontaneous rate of 0.0025 was used in the calculation. Schull et al. stated ". . . during the interval covered by this study, characterized by an infant and childhood mortality of about 7 percent, we could assume that approximately one in each 200 liveborn infants die before reaching maturity because of mutation (point or chromosomal) in the previous generation. . . . We still believe that this estimate is valid, but to err on the conservative side we will reduce the figure to one in 400 and apply it not only to the survival data but also to the data on untoward pregnancy outcomes." All lower 95% confidence limits shown are gametic doubling doses, assuming an equal contribution by the mother and father when necessary. The lower 95% confidence limits in Table 2-8 are for chronic radiation (low dose); that is, the acute doubling doses derived directly from the published regression coefficients have all been arbitrarily multiplied by a factor of 3 obtained from mouse data. As emphasized earlier, the factor of 3 is based on acute single doses in mice that are much greater than those experienced in Hiroshima or Nagasaki, and the factor of 3 cannot be applied to the Japanese data with great confidence. Although

TABLE 2-8 Estimated Lower 95% Confidence Limits of Doubling Dose from Chronic Radiation for Malformations, Stillbirths, Neonatal Death, and All Untoward Pregnancy Outcomes—Hiroshima and Nagasaki Data

Group	Malformations	Stillbirths	Neonatal Death	All Untoward Outcomes
All groups	96	124	90	60
	(62)	(129)	(115)	(44)
Only mother exposed	277	32	23	29
	(63)	(40)	(79)	(28)
Only father exposed	65	344	56	41
	(49)	(136)	(45)	(29)
Combined	119	64	35	36
	(63)	(76)	(69)	(35)
Both mother and	41	73	75	37
father exposed	(51)	(82)	(101)	(36)

NOTE: Data are the lower 95% confidence limits of the doubling dose adjusted for concomitant sources of variation (and, in parentheses, the lower 95% confidence limit for unadjusted data). The spontaneous rate of the endpoint was assumed to be 0.0025 throughout. For acute doubling doses, divide by 3. Calculations are for RBE = 1. For all estimates adjusted for concomitant sources of variation, the range is 23–344, the median is 62, and the mean is 86. For all estimates unadjusted for concomitant sources of variation, the range is 28–136, the median is 62, and the mean is 67.

the Committee believes that the factor of 3 may overestimate the risks, this point is arguable. Conceivably, the true correction factor for the dose rate in humans at the relevant doses could be as small as 1 or as large as 5. Use of the smaller number would bring estimates of human doubling doses more in line with the range of values observed in mice.

Data based on the revised dosimetry system, DS86, were not available to this committee in the detail necessary for doubling dose estimates at the time the report was being prepared. However, while the committee's calculations are based on the old T65DR dosimetry system, reanalysis based on the revised DS86 dosimetry seems to present essentially the same results (Ot87). The various entries in Table 2-8 are not independent, because they are derived from different subsets of the data for which different methods of analysis, removing different sets of concomitant variables by regression (e.g. inbreeding, parental ages, year of birth), were used. Most of the confidence limits have not been published as such by the investigators who are most familiar with the data (although the estimated limits are based on published regression coefficients), and the lower 95% confidence limits given in Table 2-8 are included here simply to give a general qualitative impression. All estimates were calculated by using regression coefficients, none of which are significantly different from zero, and all estimates depended heavily on estimated gonadal doses, the estimated spontaneous rate (about 5%) and the estimated mutation component (about 5%).

Table 2-8 provides the *minimum* doubling dose estimates, based on the one-sided 95% confidence intervals, assuming that the spontaneous rate and correction for low dose are known without error. These estimates tend to be more stable than point estimates, because the minimum estimates are more closely bounded below by zero. The values are somewhat scattered, in part because of the small sample size. The medians of the 95% confidence limits for both the adjusted and unadjusted data are about 60 rem, and the mean for the adjusted data is 86 rem. Rather than take the estimates literally and impute to them more accuracy than is warranted, the committee has rounded the estimate to the nearest 100 rem and used this as an approximate lower 95% confidence limit for the human doubling dose. The calculations in Table 2-1 are based on this 100 rem minimum doubling dose. It is noteworthy that the range 50-100 rem includes the majority of the minimum estimates in Table 2-8.

BACKGROUND DATA FROM MICE AND OTHER MAMMALS

Over the years the mouse has been the main source of experimental information regarding the genetic effects of radiation in mammals, and previous committees have relied heavily on mouse data to substantiate their estimates. The mouse radiation studies are briefly reviewed here to demonstrate their general consistency and to show that the mouse doubling dose is on the order of 100 rads.

Summarizing the mouse results as a whole, the following qualitative and semiqualitative conclusions are drawn primarily from Russell (Ru60) and subsequent papers:

1. Radiation-induced mutation rates are higher in mice than in *Drosophila melanogaster* (this original finding, in a sense, stimulated much of the subsequent work on mice because of its obvious greater relevance to estimating radiation risks in humans).
2. For specific locus mutations induced in the spermatogonial stage, there is no significant change in mutation rate with time after irradiation (i.e., the risk does not decrease with time after exposure).
3. Radiation-induced mutation rates differ markedly from locus to locus.
4. Mutations induced in spermatogonia and postspermatogonial stages differ with respect to absolute frequency and relative frequencies among loci and by radiation quality.
5. A significant proportion of mutations detected in the specific locus test (see below) have proved to be recessive lethals.

6. Some of the recessive lethal mutations have had a heterozygote effect dramatic enough to be identified in specific individuals.
7. Dominant effects on viability are demonstrable in the first-generation progeny of irradiated males.
8. Chronic irradiation is considerably less effective in inducing mutations in both spermatogonia and oocytes. This dose rate effect appears to be greater in females than in males.
9. A significant proportion of radiation-induced mutations in the specific locus test are small deletions.
10. The immature mouse oocyte is highly sensitive to cell killing.

A detailed summary of quantitative results in the mouse and other mammals is provided in Tables 2-9 and 2-10. Standard errors are not given because they tend to reflect experimental factors more than they do the true level of biological uncertainty. Rates have also been rounded so as not to imply greater precision than that which may actually exist. Although there is a significant amount of recognized genetic and nongenetic variance in the mutation rates, the uncharacterized variance is likely to be greater than that identified and measured under laboratory conditions. The uncertainties in the data base may be troublesome, but the existence of significant genetic and nongenetic variance is an intrinsic property of mammalian populations.

Table 2-9 summarizes estimates of spontaneous mutation rates for various endpoints, and Table 2-10 summarizes the estimated induced mutation rates per rad for the same endpoints for high and low dose rates of low-LET radiation exposure and for fission neutrons. Comparing the values for low and high dose rates in Table 2-10 for the endpoint recessive visible mutations (specific locus tests), the conversion factor for acute to chronic radiation is 22/7, or very nearly 3. This is the factor often used previously to convert acute doses to chronic doses in humans. It was argued earlier that application of any such conversion factor from mice to humans might warrant some skepticism, notwithstanding the fact that mice are the only mammal in which relevant data exist. Table 2-10 shows, however, that a conversion factor of 5-10 in mice could be defended just as easily. The evidence cited below suggests that the conversion factor may differ according to the particular endpoint. In any event, if the highest conversion factors are applied to the data from the Japanese atomic-bomb survivors, they imply a human doubling dose of greater than 1,000 rem. This value might be taken as a possible upper limit of the human doubling dose, and risk values based on it can be obtained from Table 2-1 by dividing the tabulated values by 10.

Table 2-11 provides estimated doubling doses for chronic radiation exposure primarily in mice. Values in parentheses are based on high dose rates and have been converted to chronic dose rates by using the factor

TABLE 2-9 Estimated Spontaneous Mutation Rates (Primarily Mouse)

Genetic Endpoint and Sex	Spontaneous Rate
Dominant lethal mutations	
Both sexes	$2 \times 10^{-2} - 10 \times 10^{-2}$/gamete
Recessive lethal mutations	
Both sexes	3×10^{-3}/gamete
Dominant visible mutations	
Male	
Skeletal	3×10^{-4}/gamete
Cataract	2×10^{-5}/gamete
Other	8×10^{-6}/gamete
Female	8×10^{-6}/gamete
Recessive visible mutations (7-locus tester stock)	
Male	8×10^{-6}/locus
Female	$2 \times 10^{-6} - 6 \times 10^{-6}$/locus
Reciprocal translocations (observed in meiotic cells)	
Male	
Mouse	$2 \times 10^{-4} - 5 \times 10^{-4}$/cell
Rhesus	8×10^{-4}/cell
Heritable translocations	
Male	$1 \times 10^{-4} - 10 \times 10^{-4}$/gamete
Female	2×10^{-4}/gamete
Congenital malformations (observed in utero in late gestation)	
Sexes combined	$1 \times 10^{-3} - 5 \times 10^{-3}$/gamete
Aneuploidy (hyperhaploids)	
Female	
Preovulatory oocyte	$2 \times 10^{-3} - 15 \times 10^{-3}$/cell
Less mature oocyte	$3 \times 10^{-3} - 8 \times 10^{-3}$/cell

range 5-10. The medians for all endpoints are summarized at the bottom of Table 2-11. The direct estimates strongly suggest a doubling dose of about 100 rads. The indirect and combined estimates also support this value, but are slightly higher, possibly because the conversion factor 5-10 is somewhat too high. Overall, considering the uncertainties in the value of the conversion factor, the data are in excellent agreement with the proposed chronic doubling dose of 100 rad in mice.

Taking the values in Table 2-11 at face value for the endpoint of congenital malformations, and making no assumptions about the mutational component of this category of traits, the doubling dose for exposed males is at the high end of the range. This endpoint is, arguably, the most closely analogous to the kinds of endpoints in the study of Japanese atomic-bomb survivors, and it is again consistent with the view that the doubling dose obtained from the study of humans in Japan may well be greater than the median of all studies of mice.

TABLE 2-10 Estimated Induced Mutation Rates per Rad (Primarily Mouse)

Genetic Endpoint, Cell Stage, and Sex	Low-LET Radiation Exposure: High Dose Rate	Low-LET Radiation Exposure: Low Dose Rate	Fission Neutrons (Any Dose Rate)
Dominant lethal mutations			
Postgonial, male	10×10^{-4}/gamete	5×10^{-4}/gamete	75×10^{-4}/gamete
Gonial, male	10×10^{-5}/gamete	2×10^{-5}/gamete	40×10^{-5}/gamete
Recessive lethal mutations			
Gonial, male	1×10^{-4}/gamete		
Postgonial, female	1×10^{-4}/gamete		
Dominant visible mutations			
Gonial, male	2×10^{-5}/gamete		
Skeletal	5×10^{-7}/gamete		
Cataract	$5-10 \times 10^{-7}$/gamete		
Other	$5-10 \times 10^{-7}$/gamete	1×10^{-7}/gamete	25×10^{-7}/gamete
Postgonial, female	$5-10 \times 10^{-7}$/gamete		
Recessive visible mutations (specific locus tests)			
Postgonial, male	65×10^{-8}/locus		
Postgonial, female	40×10^{-8}/locus	$1-3 \times 10^{-8}$/locus	145×10^{-8}/locus
Gonial, male	22×10^{-8}/locus	7×10^{-8}/locus	125×10^{-8}/locus
Reciprocal translocations			
Gonial, male			
Mouse	$1-2 \times 10^{-4}$/cell	$1-2 \times 10^{-5}$/cell	$5-10 \times 10^{-4}$/cell
Rhesus	2×10^{-4}/cell		
Marmoset	7×10^{-4}/cell		
Human	3×10^{-4}/cell		
Postgonial, female Mouse	$2-6 \times 10^{-4}$/cell		
Heritable translocations			
Gonial, male	4×10^{-5}/gamete		
Postgonial, female	2×10^{-5}/gamete		
Congenital malformations			
Postgonial, female	2×10^{-4}/gamete		
Postgonial, male	4×10^{-5}/gamete		
Gonial, male	$2-6 \times 10^{-5}$/gamete		
Aneuploidy (trisomy)			
Postgonial, female			
Preovulatory oocyte	6×10^{-4}/cell		
Less mature oocyte	6×10^{-5}/cell		

The studies on which the data in Tables 2-9 to 2-11 are based are summarized briefly in the following discussion.

The Mouse and Other Laboratory Mammals: A Summary of Present Knowledge

The BEIR V Committee decided to include a brief summary of the present knowledge of the genetic effects of ionizing radiation in laboratory mammals. Such a summary was not included in previous BEIR reports (NRC72, NRC80), although many critical issues were discussed in a series of notes or appendices to the chapters on genetic effects. Prior committees deferred to the excellent detailed reviews of radiation genetics published by the United Nations (UN72, UN77), and the present committee continues that tradition to include the most recent documents (UN82, UN86). The thorough reviews of mutation induction in mice by Searle (Se74) and by Selby (Se81) are also recommended as excellent sources of information. We believe, however, that present and future users of the BEIR committee reports could benefit from a concise summary that identifies the scope and limitations of our understanding.

The information is presented under several general headings of genetic endpoints and under each endpoint includes the information that can contribute either to the projection of radiation-induced genetic risks to humans or, if not directly appropriate for such use, to a better appreciation of the range of information available from studies with experimental animals.

Dominant Mutations

By definition, mutations in this category are detected in the immediate F_1 progeny of the irradiated generation. Tests for heritability are straightforward, unless the method of detection requires sacrificing the animals, as in the case of mutations affecting the skeletal system, in which the animals under scrutiny must be bred prior to final evaluation to prevent the loss of any potential new mutations. Information in this general category falls into three subclasses: mutations causing (1) skeletal abnormalities, (2) abnormalities of the lens, and (3) all other dominant mutations. All data have been obtained from the study of mice.

Skeletal Abnormalities

In the original studies by Ehling (Eh65, Eh66), the mutation rate for single doses of x rays was estimated to be about 1×10^{-5}/gamete/R for spermatogonia and about 3×10^{-5}/gamete/R for the postspermatogonial cell stages. Both values were corrected for control occurrences.

TABLE 2-11 Estimated Doubling Doses for Chronic Radiation Exposure (Primarily Mouse)

Genetic Endpoint and Sex	Doubling Dose (rad)[a]
Dominant lethal mutations	
Both sexes	40–100
Recessive lethal mutations	
Both sexes	(150–300)
Dominant visible mutations	
Male	
Skeletal	(75–150)
Cataract	(200–400)
Other	80
Female	(40–160)
Recessive visible mutations	
Postgonial, male	
Postgonial, female	70–600
Gonial, male	114
Reciprocal translocations	
Male	
Mouse	10–50
Rhesus	(20–40)
Heritable translocations	
Male	(12–250)
Female	(50–100)
Congenital malformations	
Female, postgonial	(25–250)
Male, postgonial	(125–1,250)
Male, gonial	(80–2,500)
Aneuploidy (hyperhaploids)	
Female	
Preovulatory oocyte	(15–250)
Less mature oocyte	(250–1,300)
Median (mouse, all endpoints, both sexes)	
Direct estimates	70–80
Indirect estimates	(150)
Overall	100–114

[a] Values not in parentheses are based on the spontaneous rate divided by the induced rate/rad for the low dose rate; values in parentheses are based on the spontaneous rate divided by the induced rate/rad at the high dose rate, multiplied by 5–10 to correct for the dose rate effect.

A major study by Selby and Selby (Se77) gave a spermatogonial rate of 2.3 × 10^{-5}/gamete/R of ^{137}Cs gamma rays. The exposure involved a 100 R and a 500 R exposure separated by 24 hours. This type of fractionation procedure is often used in mouse genetics to augment the yield of mutations per unit dose while avoiding excessive cell killing (Ru62). These data were used in the BEIR III report as an integral part of the risk analysis for dominant disabilities. However, the mutation rate was adjusted for both dose rate and dose fractionation factors for that application.

Abnormalities of the Lens (Cataracts)

All available data are from studies by Ehling and colleagues (Eh85) and Graw et al. (Gr86b). For x- and gamma-irradiated spermatogonia, the mutation rate ranges between about 3 × 10^{-7} and 13 × 10^{-7}/gamete/R. Both single and split doses (24-hour interval) were used, but no consistent variation related to exposure factors was seen. Limited information on postspermatogonial stages indicates a rate per gamete that is two- to fivefold greater than that for spermatogonia.

All Other Dominant Mutations

This is a heterogeneous class of mutations that includes, but is not limited to, changes in growth rate, coat color, limb and tail structure, hair texture, eye and ear size, congenital malformation incidence, and histocompatibility. For most traits, detection can be done nondestructively by consistent evaluation of the F_1 progeny. The study of malformations requires prenatal observation, and the data in this subclass, although limited, will be presented later in this chapter in the section on complex traits.

Efforts to determine a mutation rate for histocompatibility loci have been essentially negative. No significant increase in mutation frequency was noted for either x-irradiated sperm or spermatogonia (Du81, Ko76). The failure to detect significant increases suggests that these loci are either much less mutable or more liable to lethal mutation than expected on the basis of known mutation rates for specific recessive visible mutations in mice.

The balance of the quantitative data on dominant visible mutations is from the Medical Research Council Radiobiology Unit, Harwell, United Kingdom (Lu71, Se74). The spontaneous rate is about 8 × 10^{-6}/gamete/generation, and the induced rate for single doses of x rays to spermatogonia is about 5 × 10^{-7}/gamete/R. A study using protracted ^{60}Co gamma rays compared with fission neutrons (mean energy of about 0.7 MeV) gave spermatogonial mutation rates of 1.3 × 10^{-7}/gamete/rad for gamma rays and 25.5 × 10^{-7}/gamete/rad for neutrons resulting in an RBE value of 20 (Ba66). Dominant visible mutations were also scored in a study on x-irradiated females exposed to single doses of 200, 400, and 600 rad (Ly79). The induced rates were between 5×10^{-7} and 10×10^{-7}/gamete/rad.

These data on dominant mutations have usually been considered to be minimum estimates because of incomplete ascertainment of all classes of mutation events. Other studies reported by Searle and Beechey (Se85, Se86) with a different marker stock suggests that the rate may be as high as about 3×10^{-6}/gamete/rad of x rays (1,000 rad given in two 500-rad doses with a 24-hour interval), which implies that the value of 5×10^{-7} may be low by a factor of 3-6.

In summary, data on dominant visible mutations have yielded rates that vary by a factor of 20 for comparable types of exposure, but this range is no more than that observed for other genetic endpoints. Although the data are limited in the range of doses and exposure factors used, they demonstrate dose rate and LET factors or ratios that agree closely with those observed in more extensive studies with other endpoints.

Dominant Lethal Mutations

Data for this category of genetic events have been largely ignored in the analysis of genetic risks, because dominant lethal mutation rates have been used principally to measure damage induced in the meiotic and postmeiotic cell stages. Damage in these stages has been considered to be only transient and of limited concern for human populations. In addition, most of the mutations would be eliminated early in gestation, and many would be eliminated prior to implantation (see Note 14 in NRC80). This class of injury now requires some consideration because (1) the endpoint has been used for broad comparisons of dose rate and LET factors, (2) the category has been broadened to include the results of extensive retrospective analyses of data on litter size changes and preweaning mortality from earlier genetic studies (UN86), and (3) the concern about continuous low levels of environmental or occupational exposure requires that consideration be given to damage that is being induced continuously in the meiotic and postmeiotic cell stages.

Dominant lethal mutations, generally called simply dominant lethals, are scored among the first-generation progeny of an irradiated generation, essentially by their absence. Compared with appropriate controls, a deficiency in the number of offspring is measured at any time from conception to weaning age, which is at about 21 days of age in mice, the species for which most data have been obtained. Lethal mutations that express themselves between conception and implantation in the uterine wall (preimplant losses) are not as reliable a measure as those that occur between implantation and birth (postimplant losses) or as those that are manifested as postnatal reductions in litter size at any time from birth to weaning.

Dominant lethals are attributed to the induction of one or more major chromosome or chromatid aberrations that interfere with the complex sequence of cell and tissue differentiations that occur during organogenesis and fetal growth. The chromosome imbalances that typify these lethal mutations are usually selectively eliminated during mitotic cell division, so they do not persist in the stem cell population. Rates of induction are sensitive to cell stage in gametogenesis, with the highest rates occurring in the postgonial stages.

Postgonial Stages

There is a remarkable uniformity among the results of many individual studies that used high-dose-rate, low-LET irradiation of male mice that were then bred for the first 4 to 5 weeks after exposure. A rate of about 10×10^{-4}/gamete/rad has generally been observed (Eh71, Sc71, Gr79, Gr84, Ki84). Although control values vary among different genetic strains of mice, these values range only between about 0.025 and 0.1/gamete.

Dose rate has only a small influence on the mutation rate in the postgonial stages, and the small amount of repair implied by this dose rate effect is probably due to induced unscheduled DNA synthesis. The mutation rate drops to about 5×10^{-4}/gamete/rad at low dose rates. For the high-LET radiations, such as fission neutrons and 5-MeV alpha particles, the RBE value is about 5 (Gr79, NCRP87). Protracted exposure to neutrons appears to act in the opposite manner seen for low-LET radiation exposure and the mutation rate for lethal mutations increases at low total doses (less than 10 rads of neutrons) by about 50%, so the neutron/gamma RBE value increases to about 15.

Data for irradiated females are sparse, but a study by Kirk and Lyon (Ki82) for the period from 1 to 28 days postirradiation indicates that the rate varies with time but averages about the same as that seen for the male, about 10×10^{-4}/gamete/rad. Data from the same institution involving guinea pigs, rabbits, and golden hamsters suggest that mice may have a higher rate than other species for lethal mutations induced in males, but a similar rate exists for all species when compared with dominant lethals induced in irradiated females (Ly70, Co75).

Age does not appear to influence the induced mutation rate for dominant lethals, although the control rate may increase. It should be noted that when male mice are periodically scored for induction rate after continuous or repeated exposure to gamma rays or neutrons, a steady state value for the postgonial cell stages develops that is essentially equal to the sum of values for all injuries accumulated during the 5-week postgonial period (Gr86a).

Stem Cell (Gonial) Stage

In a strict genetic sense, most dominant lethals cannot persist in the stem cell population. Although they are induced in these cells, most are quickly eliminated at cell division because of lethal chromosome imbalance. Balanced chromosome aberrations do persist, however, and are transmitted through the series of mitotic divisions occurring in the proliferative phase of gametogenesis. Balanced translocations induced in the stem cell segregate chromosomally unbalanced gametes during the meiotic divisions. These unbalanced gametes behave like the dominant lethals induced directly in postgonial stages and their induction rates reflect the induction rates for the translocations themselves. For example, a translocation-bearing spermatocyte will produce the expected four spermatids, but on average, two spermatids will carry unbalanced chromosome sets and act as lethal mutations. One spermatid will be balanced and viable (the transmission of the original aberration), and the other will be chromosomally normal and viable.

Lüning and Searle (Lu71) summarized the available data to about 1970, in which the average rate for dominant lethals induced in spermatogonia by high-dose-rate, low-LET radiation was about 9×10^{-5}/gamete/R. More recent data give values between 7×10^{-5} and 10×10^{-5}/rad for both x rays and gamma rays. A significant dose-rate effect has been seen for gamma radiation; the rate drops to about 3×10^{-5}/rad for weekly exposures to 1.4×10^{-5}/rad for continuous, low-intensity gamma radiation exposure (Gr79, Gr83). No dose-rate effect was seen for single versus weekly neutron exposures in these studies. The fission neutron-induced rate is about 40×10^{-5}/gamete/rad, which gives RBE values of 4 to 5 for single doses, 10 to 15 for weekly exposures, and 25 or greater for continuous irradiation.

The 1986 UNSCEAR report (UN86) summarized data originally taken in the form of litter size reductions at birth, at weaning, or both, which is essentially a neonatal to postnatal measure of dominant lethals induced in spermatogonia. The data are from Selby and Russell (Se85), Lüning (Lu72), and Searle and Papworth (UN86). The data from Searle were from a study published in 1966 by Batchelor et al. (Ba66), and the analysis by Selby used data collected by the Russells at the Oak Ridge National Laboratory (ORNL), Oak Ridge, Tenn., in the 1950s. The UNSCEAR analysis made several adjustments to the findings to make them consistent with regard to the response to low-dose rate and low-LET radiations. The rates from the three sets of data were 11×10^{-6}, 19×10^{-6}, and 24×10^{-6}/R or equivalent, which is not significantly different from the value of

14×10^{-6} given previously for losses measured in utero. These results were surprisingly similar, and the variation among the values is certainly within the limits of experimental error for the type of measurements involved. These mutation rates predominantly reflect the chromosomally unbalanced gametes segregating from balanced translocations. Higher rates would normally be expected for observations made at weaning compared with those made in utero, but no study has examined this type of lethality longitudinally over the full 6-week period from conception to weaning.

In summary, dominant lethal mutations show consistent rates among different studies. For postgonial stages, it is about 10×10^{-4}/gamete/rad for high-dose-rate, low-LET exposure. Low-dose-rate exposure reduces the value by a factor of 2. For spermatogonial stages, the high-dose-rate value is about 1×10^{-4}. The dose-rate factor is about 7, and the low-dose-rate value lies between 10×10^{-6} and 25×10^{-6}/gamete/rad, depending upon method of ascertainment. RBE values for fission neutrons are between 5 and 15 at a high dose rate and 20 to 40 at a low dose rate. Continuous exposure induces a steady equilibrium rate reflecting the high sensitivity of the postgonial cell stages.

Recessive Autosomal and Sex-Linked Lethal Mutations

Mutation rates in this classical category of genetic injury have been somewhat elusive in mammalian genetics because, until recently (Ro83), no chromosome inversion stocks were available to facilitate the detection and isolation of new mutations. The methods that have been used, for example, the Haldane swept-radius procedure or the outcross-backcross test (Ha56), are not efficient, as they require a series of test generations and close attention to the sampling variance of litter size. The majority of the available data have been reviewed by Lüning and Searle (Lu71) and Searle (Se74).

Recessive Autosomal Lethal Mutations

The reviews noted previously gave an estimated mutation rate for spermatogonia exposed to high-dose-rate x irradiation of about 1×10^{-4}/gamete/R. This value has been confirmed by Lüning and Eiche (Lu75). A test with 14.5-MeV neutrons by Lüning et al. (Lu75a) yielded a mutation rate in the same range. More recently, a study by Lüning and Eiche (Lu82) with x-irradiated adult and fetal female mice has produced mutation rate estimates in the range of 0.8×10^{-4} to 1.3×10^{-4}/gamete/rad (maturing oocytes) and no indication of a significant difference in mutagenic sensitivity for oogonia. A multigeneration study with x-irradiated rats (Ta69) has given mutation rates for recessive lethals varying from 1×10^{-4} to 1.6×10^{-4}/gamete/R, depending on the age at which the litter size was measured. The lowest value was at birth and the highest was at 69 days of age.

There are limited data from studies using an inversion of a major portion of chromosome 1 of mice (Ro83). Two lethal mutations were detected in 364 gametes tested after exposure to 892 R of x rays. These data were from exposed postgonial cells. The rate, 6.2×10^{-6}/gamete/R, relates to about 3.5% of the genome. Assuming this portion is representative of the whole genome, the rate multiplies up to 1.8×10^{-4}/gamete/R, which is a reasonable expectation for postgonial cells compared to data available from spermatogonia.

Sex-Linked Lethal and/or Detrimental Mutations

Efforts by Auerbach et al. (Au62), Schröder (Sc71), and Grahn et al. (Gr72) to determine the mutation rate for sex-linked lethal and/or detrimental mutations were uniformly unsuccessful, although Grahn et al. generated an unproven estimate of 8.5×10^{-5}/X chromosome/R. Recently, the discovery and use of a large inversion of the X chromosome has succeeded in providing a proven estimate (Ly82). The inversion scores 85% of the X chromosome. An x ray dose of 500 rad + 500 rad (24 hour interval) to the spermatogonia gave a mutation rate of 3.7×10^{-6}/X chromosome/rad.

In summary, the recessive autosomal lethal mutation rate is about 1×10^{-4} to 2×10^{-4}/gamete/rad, for both sexes. There are no data on the influence of dose rate or the effects of fission neutrons. The sex-linked lethal mutation rate is probably no more than 4×10^{-6}/X-chromosome/rad and may be one-half this value if one allows for the possible augmenting effect of the split-dose exposure regime used to obtain the only available estimate.

Recessive Visible Mutations

The data in this category are all from studies in which the specific locus test system in mice was used. Experiments in which this test procedure was used have been performed for about 40 years in several major laboratories. The data base is extensive. In a few instances, the data are complex and even controversial, but for the most part, data from this test are both uncomplicated and quantitative. They have, as a result, provided the principal basis for understanding the effects of most physical and biological variables that influence the mutation rate. The previous BEIR Committee reports (NRC72, NRC80) and all UNSCEAR reports (UN58, UN62, UN66, UN72, UN77, UN82, UN86) have relied heavily on the data obtained from the results of this test. Due to the scope of the data and the availability of many detailed reviews and summaries, this overview only presents the principal mutation rates that define the importance of the major influencing variables. The variables are as follows:

Physical variables:

1. Total dose
2. Dose rate
3. Fractionation pattern
   Size of dose increment
   Interval between doses
4. LET

Biological variables:

1. Cell stage
2. Sex
3. Age at exposure
   Age at breeding test (time since last exposure)
4. Test stock or locus at risk

The test procedure uses a genetic marker stock that carries, in the homozygous state, a number of easily identifiable recessive mutations with known viabilities and locations in the genome. An irradiated wild-type male or female is crossed to the multiple recessive test stock and a new mutation at any of the marker loci can be detected in the F_1 progeny. Subsequently, a series of test matings can be performed to ensure allelism, to test for viability, and to establish the new mutant stock for any additional detailed genetic analysis. Principally, however, the detection of a mutation in the F_1 progeny can be considered unequivocal evidence for the occurrence of a new mutation.

Several tester stocks have been developed, but nearly all data are from one stock developed at ORNL by Russell (Ru51). This stock consists of seven recessive visible mutants: six coat color mutants and one structural (ears) mutant. A second tester stock was developed at Harwell, United Kingdom (Ly66), and was used only briefly. It carries six recessive mutants, one common to the ORNL line (a color mutant), four other coat color mutants, and one structural (skeletal) mutant. A third stock has been developed in the Soviet Union (Ma76) from Ehling (Eh78), but it apparently has not been used in radiation studies. Recently, a fourth stock carrying three pairs of closely linked mutants has been developed at Harwell by Searle and colleagues (Se85a, Se86). Where data are available, the differences among the stocks would seem to devolve to differences among the loci themselves, not to the different genetic backgrounds (Fa87).

The intrinsic value of the specific locus test system is in the clarity of the endpoint and its utility for testing concomitant variables quantitatively. Nevertheless, the reader should be cautioned to appreciate that data principally based on only seven loci should not be presumed to represent the full genome of the mouse, let alone the genomes of other mammals, including that of humans.

As noted, the data from the specific locus test are too extensive to be presented in detail. This overview is, therefore, limited to the principal estimates that define the influence of the major physical and biological variables. The interested reader can find detailed information from the UNSCEAR reports, collectively, from the reviews by Green and Roderick (Gr66), Searle (Se74), and Selby (Se81), and from more topical summaries by Russell et al. (Ru58), Russell (Ru65, Ru77), Russell and Kelly (Ru82a,b), Ehling and Favor (Eh84), Batchelor et al. (Ba66), and Lyon et al. (Ly72a).

Studies with Males

The spontaneous mutation rate for the seven-locus tester stock is between 8×10^{-6} and 8.5×10^{-6}/locus on the basis of pooled data from the three principal laboratories (ORNL, Harwell, and Neuherberg) that involve observations on over 800,000 control progeny. The best estimate presently available is $8.1 \pm 1.2 \times 10^{-6}$/locus (Ru82b). This value is not cell-stage specific and can be used for comparisons with data from any study. It seems likely, on the basis of the characteristics of the spontaneous events, that they have occurred predominantly in the stem cells.

The induced rate for spermatogonia exposed to single doses of low-LET radiation delivered at high dose rates is generally considered one of the baseline values. The present best estimate is $21.9 \pm 1.9 \times 10^{-8}$/locus/rad (Ru82b) at single doses of x rays between 300 and 700 rads. Above this dose level, the mutation rate drops sharply to less than 10×10^{-8}/locus/rad, a phenomenon attributed to the overriding effect of cell killing.

The data for postgonial cell stages are not as complete, but the rate per locus per rad is two- to threefold greater than for spermatogonia and reaches a level of about 65×10^{-8} to 70×10^{-8} among progeny conceived during the first 4 weeks after exposure to 300 rad of x rays (data from Russell in Se78).

The other important baseline value for spermatogonia is for the response to low-dose-rate, low-LET radiations (in this instance ^{137}Cs and ^{60}Co gamma rays). The rate is $7.3 \pm 0.8 \times 10^{-8}$/locus/rad for total doses between 35 and 900 rad (Ru82a). The dose-rate factor is 3.0 ± 0.4. This value of 3 is low in comparison with the effect for specific locus mutations in oocytes and for translocations induced in both sexes. (See discussions earlier in this chapter of the application of this factor to the human data obtained from survivors of the atomic bombings of Hiroshima and Nagasaki.) Russell et al. (Ru58) noted that there is little or no dose-rate effect for cells exposed at postgonial stages.

The rate for fission neutron doses below 100 rads is between 100 and 150 × 10^{-8}/locus/rad (Ru65, Ba66, Se67). Dose rate has no influence, and the derived rate depends upon the dose-response model used. Above 100 rad the response to a single neutron dose drops significantly below that which is expected, a finding comparable to that seen with high doses (about 1,000 rad) of x rays. Neutron dose protraction causes the mutation rate at these higher doses to rise above the single-dose value to a level consistent with a linear projection from the lower doses. This is the so-called reversed dose-rate effect reported by Batchelor et al. (Ba67), a phenomenon sometimes seen in other neutron radiobiology studies. Unfortunately, there are no data available for doses below about 50 rad, so the mutation rate at low neutron doses (less than 10 rad) is unknown. It could be as high as about 200 × 10^{-8}, as judged from the responses seen for other genetic and somatic endpoints. RBE values are 5 to 7 at high dose rates and up to 20 or more at low dose rates.

The response to an internally deposited alpha-emitter, ^{239}Pu, is intermediate to those of gamma rays and neutrons, with a rate of 18 × 10^{-8} at low dose rates and an RBE value of 2 to 3 [data from Russell in Report 89 from the National Council on Radiation Protection and Measurements (NCRP87)].

Dose fractionation studies have presented an interesting phenomenon in terms of the mutation rates induced in spermatogonia. Russell (Ru62) reported a highly significant augmentation of the mutation rate when 1,000 R was delivered in two 500 R increments separated by a 24-hour interval. The observed rate was about double the rate expected on the basis of linear extrapolation from the responses at 300 R and 600 R. A shorter interval or a greater number of fractions did not duplicate this finding, while a 15-week interval produced an additive response to the two increments. Russell also demonstrated that the augmentation phenomenon occurred with a total dose of 600 R given in 100-R and 500-R fractions 24 hours apart (Ru64). Cattanach and Moseley (Ca74) have extended the information to include intervals of 4 and 7 days and found the two 500-R doses to be roughly additive. In further studies, Cattanach and Jones (Ca85) tested fractions of 100 R + 900 R and found the results to be subadditive, so that dose size and dose interval are both factors in this type of response. The augmentation effect was also reported by Lyon and Morris (Ly69) for both specific locus and dominant visible mutations induced in the six-locus Harwell tester stock. It has been assumed that this augmentation effect is a general one and would be seen with all other genetic endpoints. It is not seen for the induction of translocations however (Ca74), so the assumption of universality may not be appropriate.

Studies with Female Mice

Data from female mice are not as extensive as those for male mice and are limited by the fact that most data are from mature and maturing oocytes. The adult female may be fertile only for about 6 weeks following a single exposure of 100 rad or more because of the killing of oocytes at their resting stage in the process of oogenesis, the dictyate stage. For those circumstances in which fertility does continue, no significant increase in the mutation rate has been seen for conceptions occurring 7 weeks or later after irradiation (Ru77, Ly79). This observation incorporates data from many experiments that have provided a total of 325,000 offspring and only 4 observed mutants. This approximates the spontaneous mutation rate.

The procedure for estimating the induced mutation rate for maturing oocytes has involved some controversy, which was discussed in Note 9, Chapter IV of BEIR III (NRC80) and by UNSCEAR (UN77, UN82). In simple terms, the controversy arose from differences in the interpretation applied to the mutation rate data that would account for (1) the observed nonlinear response to single doses and (2) a vanishingly small mutation response to low-dose-rate exposures. The alternative interpretations concerned the emphasis placed upon a more classical cytogenetic model for the mutational event (Ab76) compared with that on the existence of complex repair mechanisms (Ru58, Ly79). At present, the issue is moot, because, as Denniston (De82) noted in a review of genetic risk estimates, curve-fitting cannot resolve the controversy, given the lack of adequate data.

The spontaneous mutation rate estimated in the female has been an integral part of the noted controversy, because, of the eight spontaneous mutations reported by Russell over a series of studies (Ru77), two occurred as single events and six occurred in one cluster. Lyon et al. (Ly79) concur with Russell (Ru77) in the position that the cluster should be treated as one event, for a total of three events, giving a spontaneous rate of 2.1 $\times 10^{-6}$/locus. Upper and lower estimates would be 1.4 $\times 10^{-6}$ and 5.6 $\times 10^{-6}$ respectively, depending upon the assumptions that either three or eight events would be used. The assumption of two events was included in the analysis presented by Lyon (Ly79), but this assumption is not favored by either Russell or Lyon.

The response of mature oocytes to single doses of x rays delivered at 50 R/minute or greater is distinctly non-linear, concave upward, over the dose range of 50 R to 600 R. For progeny conceived during the first week after exposure, a linear-quadratic equation gives a linear term of 39 $\times 10^{-8}$/locus/rad (Ly79). Data from the first full 6 weeks, while less complete than those for the first week, indicate that the nonlinear response persists and the mutation rate (linear term) remains high, with the possibility it can approach a value of 50 $\times 10^{-8}$/locus/rad (Se74,

NRC80). Protracted exposures delivered either as continuous low-dose-rate exposures or multiple-increment fractionated exposures give a linear response over the dose range of 200 R to 600 R. The mutation rate is between 1.1×10^{-8} and 3.0×10^{-8}/locus/rad, depending on certain assumptions concerning the spontaneous rate and the use of data from older females (Ru77). This is clearly below the value for males by a factor of 2 or more, while the high-dose-rate value is greater than the value for males by nearly a factor of 2. The dose-rate factor is therefore at least 10 for females, compared with only 3 for males.

The limited data for fission neutrons give a mutation rate of about 145×10^{-8}/locus/rad, as derived from the data of Russell (Ru72) for single doses of 30, 60 and 120 rad. Assuming no dose-rate effect for neutrons, the RBE value would be 5 at high dose rates of low-LET radiation and 50 or greater at low dose rates.

Other Variables

Age

Age may influence the response in two ways: from variation in age at exposure and age at testing, which may also be confounded with elapsed age since the last exposure. For young adult male mice exposed for 12 weeks and then mated for the following 18 months, there was no age-related variance in the mutation rate; the rate remained essentially constant (Ba66). In a study reported by Russell and Kelly (Ru82a) four groups of males were each exposed to radiation for 8 weeks. The first group began receiving radiation at 9 weeks of age, and the three subsequent groups began exposure at 90-day intervals. No significant dependence on age was observed. Thus, for adult male mice, age does not appear to influence the mutation rate.

For female mice, the elapsed time since last exposure is critical because of the sensitivity of the dictyotene oocyte to the lethal effects of exposure. The mutation rate in the first week is usually somewhat lower than that in the second through sixth weeks, whereupon the rate drops to zero (Ru77). One set of data reported by Russell (Ru63) suggested that older females (6 to 9 months of age compared with those 2 to 4 months of age) had a significantly higher mutation rate in their second litters but not in their first litters. This seems to have been an isolated observation that has not been confirmed.

Some data are also available on the response of male and female mice exposed during prenatal, neonatal, and juvenile age periods. The data are from a mixture of experiments and conditions. Searle and Phillips (Se71) exposed mice to 108 rad of fission neutrons over a 1 week period prior to day 12 of gestation and then test-mated the animals as young adults. Mutation rates were 42 × 10^{-8}/locus/rad for the males and 58 × 10^{-8}/locus/rad for the females. Both values are only about one-third those found with irradiated adults.

At 17.5 days of fetal life, a single dose of 200 R produced mutation rates of 21 × 10^{-8}/locus/R for males and 7 × 10^{-8}/locus/R for females (Ca60). Exposure of newborn mice to single doses of x rays induced mutation rates of 13.7 × 10^{-8}/locus/rad for males (Se73) and about 10 × 10^{-8}/locus/rad for females (Se80). Selby (Se73a) exposed male mice at the ages of 2, 4, 6, 8, 10, 14, 21, 28, and 35 days and the mutation rates tended to dichotomize into the two periods of 2-6 days compared to 8-35 days. The rate was 17.5 × 10^{-8}/locus/rad at 2-6 days and 30.6 × 10^{-8}/locus/rad at 8-35 days. As the average rate for single exposures is about 22 × 10^{-8}/locus/rad, the Committee considers that none of the values from birth through 5 weeks of age differed significantly from those for adults. However, the newborn males do seem to have a lower rate. In general, prenatal, newborn, and juvenile animals of both sexes appear to be less sensitive than their adult counterparts.

Tester Stock or Locus at Risk

The two Harwell stocks have produced mutation rates about one-third the value seen with the ORNL stock (Ly66, Ly69, Se85a, Se86). This variation probably reflects differences in the loci at risk in the three stocks rather than an effect of the background genotype. Favor et al. (Fa87) have tested six of the seven loci in the ORNL stock in two unrelated inbred backgrounds, the BALB/c and DBA/2 mouse strains. Mutation rates were identical with those found in the hybrid tester stock.

The observed frequency of mutations among the seven loci varies by at least 30-fold, and 50% or more of the induced mutations have occurred at only two loci, the brown (*b*) and piebald (*s*) loci. Less than 20% were at the agouti (*a*), dilute (*d*), and short-ear (*se*) loci, while the remaining 25% occurred at the albino (*c*) and pink-eye (*p*) loci. Only the ORNL stock has tested the *b* and *s* loci, and only the *a*, *d*, and *se* loci have been common loci for the several stocks. It would be expected therefore that the overall mutation rates for the different stocks should differ by at least a factor of 2.

Chromosome Aberrations

In 1964 a new procedure became available for making cytological preparations of mammalian spermatocytes in meiosis that permitted reliable screening for the occurrence of chromosome and chromatid aberrations (Ev64). The technique soon became widely used, and much quantitative data have since been collected on the cytogenetic effects of radiation exposure of the male germ line and, to a lesser extent, the female germ line. Most of the quantitatively useful data involve the induction of balanced or symmetrical chromosome translocations. These translocations are of concern because they produce an increase in prenatal losses through the segregation of chromosomally unbalanced germ cells during gametogenesis. They also perpetuate themselves by segregating chromosomally balanced but translocation-bearing gametes (the heritable translocation).

The kinetics of translocation induction and the genetic consequences of their occurrence were discussed in Notes 3 and 14, Chapter IV, BEIR III (NRC80). A principal concern was the risk that a small number of carriers of an unbalanced chromosome set segregating from a balanced translocation heterozygote would survive to birth and thus add to the frequency of severe physical or mental abnormalities among the offspring of irradiated parents. There was no discussion in the BEIR III report of the parameters and variables influencing the induction of the original translocations. Because the induction rates for translocations depend on many important variables, such as LET and other exposure parameters, and data are now also available from a number of mammalian species other than mice, the major aspects of translocation induction rates will be summarized here. It is not possible to provide a detailed summary because the data are too diverse and because many investigations have used the translocation endpoint for the study of mechanisms of damage and repair, which goes beyond our immediate interests. The following overview identifies only the major variables and the magnitude of their influence on the rate of translocation induction. Detailed reviews of the original studies can be found in UNSCEAR reports (UN72, UN77, UN82, UN86), and in Leonard (Le71), Adler (Ad82), and van Buul (Bu83). Much of the information comes from a series of studies from Harwell (Cattanach, Lyon, Searle), ORNL (Brewen, Preston), Mol, Belgium (Leonard and colleagues), and the Soviet Union (Pomerantseva and colleagues).

Male Mice

In many respects, the variables that influence translocation induction and their effects are similar to those that influence the specific locus mutation rate. Similar to the specific locus test data, a baseline value is seen for the rate of translocations induced in spermatogonia by exposure to single-dose, high-dose-rate, low-LET radiations. Although there is some variation among different mouse strains and hybrids, the average induced linear rate is 1×10^{-4} to 3×10^{-4} cells with translocations/rad over a dose range up to about 300 rad. The spontaneous rate also varies among different mouse strains and hybrids, but generally ranges between 2×10^{-4} and 2×10^{-3} cells with translocations. Under ideal conditions for collection and scoring, the response to x rays or gamma rays is nonlinear and shows a classical linear-quadratic dose-response relationship up to about 600 rads (Pr73). At higher doses, the response levels off and later drops. This is attributed to cell killing.

After a single exposure, the rate tends to remain unchanged for about 3-6 months, followed by a modest (20-30%) decrease in value (Le70, Al85). The decrease may not always be detected because there may also be a general increase in the spontaneous frequency of aberrations in older animals (Mu74, Pa83).

Several studies have examined the influence of dose rate, and the results have been consistent in demonstrating that there is a steady decline in the basic induction rate per rad as the dose rate decreases from about 100 rad/minute down to about 0.1 rad/minute. The induction rate declines by a factor of 10, down to about 1.5×10^{-5}/rad. An absolute minimum rate would probably be about 1×10^{-5}/rad (Se76, Br79a). At low dose rates, the response is linear over wide dose ranges (greater than 1,000 rad), and it is this linear regression on dose that steadily declines as the dose rate declines. In other words, there does not appear to be a single linear term in a series of linear-quadratic equations.

Several studies with fission neutrons have also provided generally consistent results (Gr84). The response peaks at about 100 rad and then drops sharply when single doses are used. Up to 100 rad, the response may be either linear or nonlinear with a negative dose-squared term. The RBE value for linear terms at low doses is about 5. When neutron doses are protracted or are given in repeated small fractions, the response is either equal to or greater than the response to low single doses (Gr83). The augmentation of response is probably no greater than about 25% at low doses. At doses above 100 rad, there is no decline in response, so the augmentation factor ranges from about 2 at 100 rad to 5 or more at 150 rad. The RBE value for protracted exposures, neutrons versus gamma rays, varies with dose rate in low-LET radiation exposures, but approaches 50 at the lowest dose rates (Gr86a).

Studies with alpha-emitters have not given consistent results. Nevertheless, the response is no greater than that seen with fission neutrons and it may be less (Se76, Gr83). High-energy neutrons are also less effective than fission neutrons.

Dose fractionation has been used extensively to study cell stage sensitivity, cell synchronization and repair, and the interaction of mutagenic and lethal actions (see, for example, Cattanach and colleagues Ca74, Ca76). For the purposes of this report, the findings can be reduced to a few general observations. For split doses with intervals of less than 1 day, variable responses are seen that are usually subadditive. Intervals of 18 to 36 hours yield responses that are generally additive for the two doses. It is important to note that superadditivity or augmentation of injury is not observed with the 24-hour interval as is observed in the specific locus test (Ca74). With intervals of days to weeks, subadditive responses are seen to at least a 3-week interval in some studies and up to 6 weeks in others. Eight-week intervals produce clear additivity of the individual doses, even when exposures are repeated beyond only a single pair of doses (Pr76). With long intervals between doses, the decline in response seen for high single doses does not occur.

When small dose increments (less than 50 rad) are given at daily or weekly intervals, additivity exists, but the rate of response is less than that seen for comparable single doses, and the magnitude of this drop in response depends on the size of the dose increment, the dose interval, and the instantaneous dose rate (Ly70a, Ly70b, Ly72, Ly73; Gr86b, Gr88). As there are no generalized formulations to describe or predict responses to repeated exposures, most analyses are empirical. Lyon has made the suggestion that some resistance to subsequent exposures may even be induced, although such an effect would have to be short-lived (less than 1 week). In any event, the responses to repeated low doses are not greater than the effect of single doses and are not less than the response to low-dose-rate (less than 0.1 rad/minute) continuous or near-continuous exposures.

The cell stage in spermatogenesis is an important factor, although the data are not as clear or complete as they are for spermatogonia (Ad82). Spermatocyte stages, spermatids, and spermatozoa are more sensitive than spermatogonia, with spermatids being the most sensitive, according to data from F_1 male progeny derived from irradiated sires. The damage induced in spermatocytes and scored at first metaphase is complex, because rearrangements involve both chromosomes and chromatids. Fragments and deletions are also seen from the exposure of spermatocytes. Results from different studies are not consistent, but generally, the rates of induction for translocations are about two- to fourfold greater than they are for stem cells. Dose-rate factors are limited because meiotic and postmeiotic stages have a limited repair capacity. Fission neutrons may have high RBE values, comparable to those for stem cell exposures, because of their efficiency in producing chromosome or chromatid breaks and fragments. Alpha particles, on the other hand, are not as efficient as neutrons because of their extremely dense ionization track (Gr83).

Female Mice

The data from adult female mice are quite limited in comparison with those from male mice because the information is largely restricted to mature and maturing oocytes that can be screened for only the first

6 weeks after exposure to radiation. However, a recent report by Griffin and Teage (Gr88a) has shown that significant increases in structural and numerical chromosome abnormalities can be induced in immature oocytes by low-dose-rate gamma irradiation to total doses of 1, 2, or 3 Gy.

Data have been obtained by both cytological and breeding tests. Because the oocyte stage is exposed, the responses involve chromatid as well as chromosome aberrations and include interchanges, fragments, and deletions. Direct comparison with males is difficult because the stage at which the oocyte rests during postnatal life, the dictyotene stage, has no exact parallel in spermatogenesis.

Irradiated oocytes express cytogenetic damage in complex ways, and chromosome fragments make up 30-50% of the total damage (Ca77). Fragments would usually be lost in the next cell division, so that deletions or deficiencies would occur in the zygotes formed from the resulting gametes. Induction rates either for total cytogenetic damage or for rearrangements alone are generally similar for cells of both sexes that are in comparable stages. Response kinetics for single doses are nonlinear, with a strong positive quadratic (dose-squared) term evident at doses above 200 rad. For rearrangements, the rate below 200 rad for oocytes is 1×10^{-4} to 2×10^{-4}/rad during the first week after exposure (Br79). The rate rises to about 6×10^{-4}/rad during the second and third weeks, a response pattern comparable to that seen for specific locus mutations in oocytes.

Also comparable to the specific locus test data is the observation that a significant dose-rate effect exists: reducing the dose rate from about 100 to about 0.04 rad/minute reduces the effectiveness by a factor of 7-10 (Br77). Evidence of repair capability is also seen in the results of split-dose studies with short intervals of 90 minutes to 1 day.

Age is another factor for females. The spontaneous frequency and induced rates of common chromosome aberrations are higher in female mice of about 1 year of age or greater (Se85a).

Although several attempts have been made to detect aberrations induced by neutron irradiation, no clear evidence has been obtained (Se74a). Aberrations are certainly induced in oocytes by neutrons, however, because there is clear evidence of an increase in the frequency of dominant lethal mutations, which are attributable to complex cytogenetic damage.

Mammals Other Than Mice

At least eight mammalian species have been screened for the induction of reciprocal translocations in spermatogonia by single doses of low-LET radiation. At least six different inbred or F_1 hybrid strains of mice have been studied, along with three other small laboratory mammals (guinea pigs, rabbits, and hamsters) and several primate species, including rhesus monkeys and humans. The basic dose-response curve is similar for all species. There is an initial linear increase with dose, a plateauing of the response, and then a decrease in the induction rate as cell killing intervenes. The response for mice peaks (for single doses of low-LET radiation) at about 600 rad, but for all other species the maximum response is at 300 rad or less. The initial linear coefficients fall within the limits of about 0.8×10^{-4} and 3.5×10^{-4} translocations/cell/rad for all the species except for the marmoset, *Saguinus fuscicollis* (Ma85). In this species, the rate of response is estimated to be 7.4×10^{-4}/rad (Br75). The limited data available for humans give a rate of about 3.4×10^{-4}/rad, which is near the high end of the range (Br75). The highest value for mice, however, is about 2.6×10^{-4}/rad, which is not significantly below the human value. The response of the rhesus monkey is 0.86×10^{-4}/rad (Bu83, Bu86). However, recent studies with two species of *Macaca* indicate that the best value for this genus is about 2×10^{-4}/rad (Ad88).

As noted earlier, both the BEIR III (NRC80) and UNSCEAR (UN77) committees used a value of 7×10^{-4}/rad as a reasonable estimate of the human response to low single doses of x rays or gamma rays. That value was derived from a combination of data from marmosets and humans at doses of 100 rad or less. Only one datum point, at 78 rad, was taken from the human data, while three data points, at 25, 50, and 100 rads, were taken from the marmoset data. A control value of zero events, which was the case for both species, was used to complete the analysis. In this manner the value of 7×10^{-4} was derived from a merged data set from two species, a practice not commonly used in extrapolation modeling. In more recent UNSCEAR reports, more emphasis was placed on direct estimates from studies with rhesus and crab-eating monkeys. These two primate species produced the maximum difference in the rates of response noted previously (0.86×10^{-4} to 7.4×10^{-4}). In addition, UNSCEAR (UN86) also noted some preliminary (unpublished data) dose-rate data with the crab-eating monkey that suggest a factor of 10 reduction in effectiveness for a dose rate of 0.002 rad/minute compared with a dose rate of 25 rad/minute. The factor of 10 is similar to that seen in mice.

Other Aberrations

Irradiation of the meiotic stage in gametogenesis in either sex has demonstrated that chromosome and chromatid breaks, leading to the formation of fragments, deletions, and dicentrics, are readily induced. Rates of induction are not consistent among different studies and are dependent on the exact cell stage in gametogenesis and on the quality of the cytological preparations. On average, following administration of single doses of x rays or gamma rays, the rate of other aberrations would probably be about equal to the rate of rearrangements alone, which was noted above to be at least two- to fourfold greater than the rate of induction of stem cells. While the chromosomally unbalanced gametes that would result from these other aberrations would be eliminated early in fetal life and would not contribute to the transmissible genetic burden, they would increase the frequency of reproductive failures early in gestation.

Finally, chromosome inversions have been induced experimentally in mice and have been characterized in order to be used in other studies. The rate of induction by radiation is not clear, but it probably does not exceed 4×10^{-5}/gamete/rad for cells exposed at postmeiotic stages (Ro71).

Complex Traits

Complex traits are difficult to study in the laboratory, and therefore mutation rates or comparable coefficients of induced risk have not been available for use in genetic risk assessments. Nevertheless, the data from animal studies on complex traits carried out over the past decade have achieved some modest success, and the summary of information in this category will be presented in terms of two classes of traits. The first class includes traits that have provided some opportunity for rate analysis, and the second includes traits for which evidence exists of a response to increased mutation pressure, but not of sufficient quality or repeatability to yield a risk coefficient.

Traits with Quantifiable Rates of Induction

Congenital Abnormalities

The frequency of congenital malformations, including small stature or reduced growth rate, in the first-generation progeny of x-irradiated male and female mice has been evaluated in late gestation (No82, Ki82, Ki84, Ru86). Irradiated oocytes yield consistent dose-response data between 100 and 500 rad. The rate is 1×10^{-4} to 2×10^{-4}/gamete/rad, but it is slightly lower among progeny conceived in the first postirradiation week. For male mice, the average response to doses between 100 and 500 rad is 4×10^{-5}/gamete/rad for the postmeiotic cell stages of sperm and spermatids, while irradiated spermatogonia yielded a value of 2×10^{-5} to 3×10^{-5}/gamete/rad (Ki84). Initially, Nomura (No82) did not see a significant response for the exposed male parent, but recent data (No88) suggest a rate of about 6×10^{-5}/gamete/rad for spermatogonia. Rutledge et al. (Ru86) observed a yield of 0.5×10^{-5} to 2×10^{-5}/gamete/rad for spermatogonia exposed to 2,000 rads given in four increments of 500 rad each separated by 4-week intervals.

The genetic basis of the observed malformations has not been fully ascertained. Recent studies suggest a major proportion could be due to dominant mutations with a high penetrance that are expressed and lost in the first generation. A small number with a low penetrance may persist into later generations (Ly88, No88). The spectrum of induced abnormalities appears to be typical for mice. About one-half of the traits are classified as dwarfism, which is defined as a body size smaller than 75% of the average of all littermates. Reduced stature has also been seen as a common expression for some specific-locus mutants (piebald, *s*, for example) and has been successfully evaluated for heritability in recent studies on dominant visible traits (Se86). Nevertheless, the observation that dwarfism constitutes about 50% of all abnormalities urges some caution in the use of these data as a surrogate for human malformations.

Heritable Translocations

Balanced reciprocal translocations are generally transmissible to subsequent generations. Their frequency should theoretically be about one-fourth the induction rate in spermatogonia. In laboratory studies with mice, the value of one-fourth has been achieved only at a dose of 150 rad, the lowest dose used by Generoso et al. (Ge84). Ford et al. (Fo69), in their detailed cytogenetic evaluation of the transmissibility of balanced translocations, concluded that only about one-half of the expected number would be found in the F_1 progeny (that is, only one-eighth of the induced frequency rather than one-fourth). It is reasonable to expect the value of one-fourth to pertain to balanced translocations induced at all low doses (less than 50 rad) and low dose rates of low-LET radiations.

The experimentally derived rate induced by single or split doses of x rays delivered at high dose rates was estimated to be 34×10^{-6}/gamete/rad by Lüning and Searle (Lu71) and 39×10^{-6}/gamete/rad by Generoso et al. (Ge84). The spontaneous rates were given as 1×10^{-3}/gamete by Lüning and Searle and about 1×10^{-4}/gamete by Generoso et al. Pomerantseva et al. (Po76) observed a rate of 31×10^{-6}/gamete/rad following three exposures to 300 rad of gamma rays with a 4-week interval between exposures, but no control estimate was given. A rate of 15×10^{-6} to 30×10^{-6}/gamete/rad has been observed for 5-MeV alpha particles from gonadal burdens of ^{239}Pu (Ge85) suggesting an RBE of 1 or less for this high-LET radiation. There are no substantive data from neutron irradiations, but RBE values should mimic those seen for reciprocal translocation induction and, therefore, should range from about 5 to 45.

Data from irradiated female mice are extremely limited, but the summaries given in UNSCEAR reports of 1977 and 1982 suggest a value no greater than about one-half that seen for males (15×10^{-6}/gamete/rad following a single x-ray dose of 300 rad).

Chromosomal Nondisjunction

Relevant human data on the possible induction of nondisjunction by radiation was discussed earlier in this chapter. Studies of mice by Tease (Te82, Te85) are pertinent and quantitative. The preovulatory oocyte of the female mouse is sensitive to the induction of nondisjunction (specifically, hyperhaploidy) at low doses of x rays (10, 25, and 50 rad). The rate is 6×10^{-4} to 7×10^{-4}/cell/rad, and there is no influence of age on this rate of induction, although the intercept increases 10-fold in 1-year-old females compared with that in 90-day-old females. Less mature oocytes (those scored between 9.5 and 23.5 days after exposure) were significantly less sensitive and gave linear rates of 5×10^{-5} to 7×10^{-5}/rad over the dose range of 100 to 600 rad. The mechanisms of induction are not clear, but the frequencies of all structural aberrations observed in preovulatory oocytes were considered sufficient to account for the majority of nondisjunction events (Te86). Thus, in mice at least, age is not a factor in radiosensitivity. The preovulatory oocyte is sensitive to low doses, but less mature oocytes are quite resistant.

Multilocus Deletions

The specific-locus test has provided useful data on the characteristics or phenotypic manifestations of mutations induced by different radiation qualities and in different germ cell stages. Many of the new mutations apparently involve a deletion of a small portion of the chromosome where the marker gene is located, although some would also appear to be at least the equivalent of an intragenic mutation. Deletions that clearly involve more than the specific locus (multilocus deletions) have recently attracted more attention in genetic risk analysis because they will generally have deleterious effects on the heterozygous carriers and are nearly always lethal when they are homozygous (Ru87, de87). The deleterious manifestations in the heterozygote include reductions in viability, growth rate, and fertility and are seen in a variety of organisms, including, in addition to the mouse, both *Drosophila melanogaster* and *Neurospora* species (Sc87).

Russell and Rinchik (Ru87) have presented information on the characteristics of about 300 radiation-induced mutations involving the *d*, *se*, and *c* loci in mice. The frequency of intragenic mutations is small; only 15% of the spontaneous mutations and 11% or less of the induced mutations are in this class. Depending on the cell stage and radiation quality, about 25% to 75% of induced mutations are multilocus deletions, while less than 5% are seen in controls. The balance, from about 25 to 80%, are classed as viable null mutations that could be intragenic mutations, single-gene deletions, or multilocus deletions. About 25% of the low-LET mutations induced in spermatogonia (with no apparent dose-rate effect), and about 55% of those induced in postgonial stages and oocytes result from multilocus deletions. With neutrons, the figures are 35% in spermatogonia and over 70% in postgonial cells and oocytes respectively. Thus, a minimum induction rate for this deleterious class of mutations would be one-fourth (1.8×10^{-8}/rad) of the rate for spermatogonia exposed to low dose rate, low-LET radiations (7.3×10^{-8}/locus/rad). The rate would be about the same for females, allowing for their lower mutation rate, but higher probability of giving rise to the multilocus deletion.

It is likely that this class of detrimental mutations overlaps with mutations that are characterized as producing congenital malformations, dominant visible mutations, and possibly, heritable translocations. These latter categories have induced rates per gamete, the multilocus class is per locus, so there is no simple means of distinguishing them.

Traits Acknowledged To Be Influenced by New Mutations but Lacking Sufficient Data for Risk Analysis

Several studies have endeavored to determine the impact of an increased mutation rate on the general fitness of a population, where fitness incorporates a variety of generally quantitative or continuously distributed traits. The biological components of fitness include all aspects of viability and reproduction, from conception to death. Some specific attributes are evaluated categorically, such as dominant lethal mutations, congenital malformations, and litter size. Many attributes, however, do not lend themselves to the type of rate or risk analysis necessary for the modeling of projected risks to human populations, even though fitness traits are important for the survival and reproduction of a species. Radiation-induced mutations and the concomitant increases in the genetic variance have been used successfully to improve productivity for several economic crops, but in the field of mutation genetics, the quantitative analysis of fitness attributes, in general, has been unsuccessful.

Excellent summaries have been given in a symposium edited by Roderick (Ro64), in a review article by Green (Gr68), and in a tabular review by UNSCEAR (UN72). The issue was also discussed in BEIR III, Note 12 of Chapter IV (NRC80).

The summarized studies dealt with early mortality, growth, reproduction, and long term survival. More recent studies have dealt with growth rate and stature (Se86) (discussed earlier in this chapter in the section on dominant visible mutations) and with the induction of changes in the susceptibility to spontaneous or induced tumors in the mouse (No82, No83).

According to Nomura's data, an increased prevalence of tumors is observed on the basis of a one-time sampling of the F_1 population at 8 months of age. The increase is from about 5% in the control to 25% at a 504-rad dose to cells in postmeiotic stages in males, spermatogonia, or oocytes. There was no shift in the spectrum of tumor types, and 90% were pulmonary adenomas, which is a common neoplasm in some strains of mice. The one-time sample leaves unanswered the question of whether the increased frequency is due to a shift in the time of appearance or is due to a real increase in the total number of tumors over the mouse's lifetime. Previous studies of this type gave negative results (Ko65), although there was evidence of reduced life expectancy in the F_1 progeny of irradiated parents in an early study by Russell (Ru57). As life expectancy in the mouse can be closely related to age, rate, and type of tumor occurrence, Russell's results could have indicated an induced change in death rates from tumors; however, the results of Russell's 1957 study have not been confirmed.

Summary of Data on Mice and Other Laboratory Mammals

Tables 2-9 and 2-10 summarize the data on eight genetic endpoints that have reasonably representative mutation rates. All these data have been derived from studies that were specifically directed toward the particular endpoint; thus, the rates for multilocus mutations are not included because of their indirect derivation. Standard errors are not given because they tend to reflect experimental factors more than they do the true level of biological uncertainty. Most rates have been rounded so as not to imply greater precision than that which may actually exist.

The available data are predominantly from studies in which high-dose-rate exposures with low-LET radiations were used. This reflects the availability or unavailability of appropriate facilities to carry out low-dose-rate irradiations or irradiations with high-LET sources. It also probably reflects the shifting level of interest from radiation mutagenesis to chemical mutagenesis over the past 15-20 years. The effect of this shift has been to leave large gaps in our matrix of information.

For the high-dose-rate, low-LET radiations, mutation rates per gamete or per cell generally fall in the range of 10^{-5} to 10^{-4}/rad, although there are several exceptions. Higher rates are seen for dominant lethal mutations induced in postgonial cells of male mice, for translocations induced in the spermatogonia of one marmoset species, and for aneuploidy induced in the preovulatory oocyte of female mice. Lower rates pertain to dominant visible mutations; however, except for skeletal and cataract mutations, these are recognized to be systematically underestimated. Rates per locus are in the range of 10^{-8} to 10^{-7}.

Low-dose-rate exposures cause the mutation rate to drop by a factor of 5 or greater, and a factor of 10 accommodates the range of values, with one notable exception. The dose-rate factor for the male specific-locus mutation rate is only 3. This is a firmly established value. The reason for this rather low dose-rate factor is not clear, although it is not dissimilar from some values derived from other radiobiological studies on tumorigenesis and life shortening (NCRP Report 64, 1980). RBE values for fission neutron exposures are about 5 for high-dose-rate comparisons and range from 15 to 50 for low dose rates.

Spontaneous mutation rates (Table 2-9) are understandably less well known than the induced rates; this appears to be largely a matter of inadequate sampling statistics. The values for the specific locus test are well defined, although even here they are not free of controversy because of the occurrence of clusters of events. For other endpoints, such as translocations in mice, the range of values often reflects genetic diversity and not uncertainty per se. On this point, the committee notes that there is considerable diversity in the spontaneous rates among all the known specific recessive and dominant genes in mice and humans.

The estimated doubling doses derived from Tables 2-9 and 2-10 are summarized in Table 2-11. Considering all endpoints together, the direct estimates of doubling dose for low dose rate radiation have a median value of 70-80 rad, indirect estimates based on high-dose rate experiments have a median of 150 rad, and the overall median lies in the range of 100 to 114 rad. These estimates support the view that the doubling dose for low-dose-rate, low-LET radiation in mice is approximately 100 rad for various genetic endpoints. This contrasts with the results of the human data obtained from the study of Japanese atomic-bomb survivors, as discussed earlier in this chapter, which suggest that the value of 100 rad represents an approximate lower 95% confidence limit for the human doubling dose.

REFERENCES

Ab76 Abrahamson, S., and S. Wolff. 1976. Re-analysis of radiation induced specific locus mutations in the mouse. Nature 264:715-719.

Ab85 Abrahamson, S. 1985. Risk estimates for genetic effects. In: Assessment of Risk from Low Level Exposure to Radiation and Chemicals: A Critical Overview. A. D. Woodhead et al., eds. New York, Plenum Press.

Ad82 Adler, I. D. 1982. Male germ cell cytogenetics. Pp. 249-276 in Cytogenetic Assays of Environmental Mutagens, T. C. Hsu, ed. Totowa, N.J.: Allanheld, Osmun.

Ad88 Adler, I. D., and C. Erbelding. 1988. Radiation-induced translocations in spermatogonial stem cells of *Macaca fascicularis* and *Macaca mulatta*. Mutat. Res. 198:337-342.

Al85 Alavantic, D., and A. G. Searle. 1985. Effects of postirradiation time interval on translocation frequency in male mice. Mutat. Res. 142:65-68.

Au62 Auerbach, C., D. S. Falconer, and J. H. Isaacson. 1962. Test for sex-linked lethals in irradiated mice. Genet. Res. 3:444-447.

Aw75 Awa, A. A. 1975. Review of thirty years study of Hiroshima and Nagasaki atomic bomb survivors. II. Biological effect. B. Genetic effects. 2. Cytogenetic study. J. Radiat. Res. 16(Suppl):75-81.

Aw87 Awa, A. A. et al. 1987. Cytogenetic study of the offspring of atomic bomb survivors, Hiroshima and Nagasaki. In Cytogenetics, G. Obe and A. Basler, eds. Berlin, Heidelberg: Springer-Verlag.

Ba66 Batchelor, A. L., R. J. S. Phillips, and A. G. Searle. 1966. A comparison of the mutagenic effectiveness of chronic neutron- and gamma-irradiation of mouse spermatogonia. Mutation Res. 3: 218-229.

Ba67 Batchelor, A. L., R. J. S. Phillips, and A. G. Searle. 1967. The reversed dose-rate effect with fast neutron irradiation of mouse spermatogonia. Mutation Res. 4:229-231.

Br75 Brewen, J. G., R. J. Preston, and N. Gengozian. 1975. Analysis of X-ray induced chromosomal translocations in human and marmoset spermatogonial stem cells. Nature 253:468-470.

Br77 Brewen, J. G., H. S. Payne, and I. D. Adler. 1977. X-ray induced chromosome aberrations in mouse dictyate oocytes II. Fractionation and dose rate effects. Genetics 87:699-708.

Br79 Brewen, J .G., and H. S. Payne. 1979. X-ray stage sensitivity of mouse oocytes and its bearing on dose-response curves. Genetics 91:149-161.

Br79a Brewen, J. G, R. J. Preston, and H. E. Luippold. 1979. Radiation-induced translocations in spermatogonia. III. Effect of long-term chronic exposures to gamma-rays. Mutat. Res. 61:405-409.

Bu83 van Buul, P. P. W. 1983. Induction of chromosome aberrations by ionizing radiation in stem cell spermatogonia of mammals. Pp. 369-400 in Radiation-Induced Chromosome Damage in Man, T. Ishihar and M. Sasaki, eds. New York: Alan R. Liss.

Bu86 van Buul, P. P. W., J. F. Richardson, Jr., and J. H. Goudzwaard. 1986. The induction of reciprocal translocations in Rhesus monkey stem-cell spermatogonia:.Effects of low doses and low dose rates. Radiat. Res. 105:1-7.

Ca60 Carter, T. C., M. F. Lyon, and R. J. S. Phillips. 1960. Genetic sensitivity to x-rays of mouse foetal gonads. Genet. Res. 1:351-355.

Ca71 Cavalli-Sforza, L. L., and W. F. Bodmer. 1971. The Genetics of Human Populations. San Francisco: W.H. Freeman.

Ca74 Cattanach, B. M., and H. Moseley. 1974. Sterile period, translocation and specific locus mutation in the mouse following fractionated X-ray treatments with different fractionation intervals. Mutat. Res. 25:63-72.

Ca76 Cattanach, B. M., C. M. Heath, and J. M. Tracey. 1976. Translocation yield from the mouse spermatogonial stem cell following fractionated X-ray treatments: Influence of unequal fraction size and increasing fraction interval. Mutat. Res. 35:257-268.

Ca77 Caine, A., and M. F. Lyon. 1977. The induction of chromosome aberrations in mouse dictyate oocytes by X-rays and chemical mutagens. Mutat. Res. 45:325-331.

Ca77a Carter, C. O. 1977. Monogenic disorders. J. Med. Genet. 14:316-320.

Ca85 Cattanach, B. M., and C. Jones. 1985. Specific-locus mutation response to unequal 1+9 Gy X-ray fractionations at 24-h and 4-day fraction intervals. Mutat. Res. 149:105-118.

Ch81 Childs, J. D. 1981. The effect of a change in mutation rate on the incidence of dominant and X-linked recessive disorders in man. Mutat. Res. 83:145-158.

Co75 Cox, B. D., and M. F. Lyon. 1975. X-ray induced dominant lethal mutations in mature and immature oocytes of guinea-pigs and golden hamsters. Mutat. Res. 28:421-436.

Cr61 Crow, J. F. 1961. Mutation in man. Prog. Med. Genet. 1:1-26.

Cr81 Crow, J. F., and C. Denniston. 1981. The mutation component of genetic damage. Science 212:888-893.

Cr85 Crow, J. F., and C. Denniston 1985. Mutation in human populations. In Advances in Human Genetics, H. Harris and K. Hirschhorn, eds. New York: Plenum.

Cu72 Curnow, R. N. 1972. The multifactorial model for the inheritance of liability to disease and its implications for relatives at risk. Biometrics 28:931-946.

Cz84 Czeizel, A., and K. Sankaranarayanan. 1984. The load of genetic and partially genetic disorders in man. I. Congenital anomalies: estimates of detriment in terms of years of life lost and years of impaired life. Mutat. Res. 128:73-103.

Cz84a Czeizel, A., and G. Tusnady. 1984. Aetiological Studies of Isolated Common Congenital Abnormalities in Hungary. Budapest: Akademiai Kiado.

Cz88 Czeizel, A. et al. In press. The load of genetic and partially genetic diseases in man. II. Some selected common multifactorial diseases: Estimates of population prevalence and of detriment in terms of years of life lost and impaired life. Mutat. Res.

De82 Denniston, C. 1982. Low level radiation and genetic risk estimation in man. Annu. Rev. Genet. 16:329-355.

de83 de Boer, P., and A. D. Tates. 1983. Radiation-induced nondisjunction. Pp. 229-325 in Radiation-Induced Chromosome Damage in Man, T. Ishihar and M. Sasaki, eds. New York: Alan R. Liss.

de87 deSerres, F. J. 1987. Specific-locus studies with two-component heterokaryons of *Neurosoora crassa* predict differential recovery of mutants in mammalian cells. Pp. 55-69 in Mammalian Cell Mutagenesis: Banbury Report 28. New York: Cold Spring Harbor Laboratory.

Du81 Dunn, G. R., and H. I. Kohn. 1981. Some comparisons between induced and spontaneous mutation rates in mouse sperm and spermatogonia. Mutation Res. 80:159-164.

Eh65 Ehling, U. H. 1965. The frequency of X-ray induced dominant mutations affecting the skeleton of mice. Genetics 51:723-732.

Eh66 Ehling, U. H. 1966. Dominant mutations affecting the skeleton in offspring of X-irradiated male mice. Genetics 54:1381-1389.

Eh71 Ehling, U. H. 1971. Comparison of radiation- and chemically-induced dominant lethal mutations in male mice. Mutation Res. 11:35-44.

Eh76 Ehling, U. H. 1976a. Die Gefährdung der menschlichen Erbanlagen im technischen Zeitalter. Fortsch. Röntgenstr. 124:166-171.

Eh76a Ehling, U. H. 1976b. Estimation of the frequency of radiation-induced dominant mutations. ICRP, CI-TG14, Task Group on Genetically Determined Ill-Health.

Eh78 Ehling, U. H. 1978. Specific-locus mutations in mice. Pp. 233-256 in Chemical Mutagenesis: Principles and Methods for Their Detection, Vol. 5, A. Hollaender and F. J. deSerres, eds. New York: Plenum.

Eh84 Ehling, U. H., and J. Favor. 1984. Recessive and dominant mutations in mice. Pp. 389-428 in Mutation, Cancer and Malformation, E. H. Y. Chu and W. M. Generoso, eds. New York: Plenum.

Eh85 Ehling, U. H. 1985. The induction and manifestation of hereditary cataracts.

Pp. 345-367 in Assessment of Risk from Low-Level Exposure to Radiation and Chemicals, A. V. Woodhead, C. J. Shellabarger, V. Bond, and A. Hollaender, eds. New York: Plenum.

Ev64 Evans, E. P., G. Breckon, and C. E. Ford. 1964. An air-drying method for meiotic preparations from mammalian testes. Cytogenetics 3:289-294.

Ev77 Evans, H. J. 1977. Chromosome mutations in human populations. Pp. 57-65 in Conference on Mutations: Their Origin, Nature and Potential Relevance to Genetic Risk in Man. Zentrallarboratorium fur Mutagenitatsprufung.

Fa65 Falconer, D. S. 1965. The inheritance of liability to certain diseases, estimated from the incidence among relatives Ann. Hum. Genet. 29:51-76.

Fa87 Favor, J., A. Neuhauser-Klaus, and U. H. Ehling. 1987. Radiation-induced forward and reverse specific locus mutations and dominant cataract mutations in treated strain BALB/c and DBA/2 male mice. Mutat. Res. 177:161-169.

Fo69 Ford, C. E., A. G. Searle, E. P. Evans, and B. J. West. 1969. Differential transmission of translocations induced in spermatogonia of mice by irradiation. Cytogenetics 8:447-470.

Ge84 Generoso, W. M., K. T. Cain, N. L. A. Cachiero, and C. V. Cornett. 1984. Response of mouse spermatogonial stem cells to X-ray induction of heritable reciprocal translocations. Mutat. Res. 126:177-187.

Ge85 Generoso, W. M., K. T. Cain, N. L. A. Cachiero, and C. V. Cornett. 1985. ^{239}Pu induced heritable translocations in male mice. Mutat. Res. 152:49-52.

Go81 Golbus, M. S. 1981. The influence of strain, maternal age, and method of maturation on mouse oocyte aneuploidy. Cytogenet. Cell Genet. 31:84-90.

Gr66 Green, E. L., and T. H. Roderick. 1966. Radiation Genetics. Pp. 165-185 in Biology of the Laboratory Mouse, 2nd ed., E. L. Green, ed. New York: McGraw-Hill.

Gr68 Green, E. L. 1968. Genetic effects of radiation on mammalian populations. Annu. Rev. Genet. 2:87-120.

Gr72 Grahn, D., W. P. Leslie, F. A. Verley, and R. A. Lea. 1972. Determination of the radiation-induced mutation rate for sex-linked lethals and detrimentals in the mouse. Mutat. Res. 15:331-347.

Gr79 Grahn, D., B. H. Frystak, C. H. Lee, J. J. Russell, and A. Lindenbaum. 1979. Dominant lethal mutations and chromosome aberrations induced in male mice by incorporated ^{239}Pu and by external fission neutron and gamma irradiation. Pp. 163-184 in Biological Implications of Radionuclides Released from Nuclear Industries, Vol. I. Vienna: International Atomic Energy Agency.

Gr83 Grahn, D. 1983. Genetic risks associated with radiation exposures during space flight. Adv. Spac Res. 3:161-170.

Gr83a Grahn, D., C. H. Lee, and B. F. Farrington. 1983. Interpretation of cytogenetic induced in the germ line of male mice exposed for over one year to ^{239}Pu alpha particles, fission neutrons, or ^{60}Co gamma-rays. Radiat. Res. 95:566-583.

Gr84 Grahn, D., B. A. Carnes, B. H. Farrington, and C. H. Lee. 1984. Genetic injury in hybrid male mice exposed to low doses of ^{60}Co gamma-rays or fission neutrons. I. Response to single doses. Mutat. Res. 129:215-229.

Gr86 Grahn, D., B. A. Carnes, and B. H. Farrington. 1986a. Genetic injury in hybrid male mice exposed to low doses of ^{60}Co gamma-rays or fission neutrons. II. Dominant lethal mutation response to long-term weekly exposures. Mutat. Res. 162:81-89.

Gr86a Grahn, D., J. F. Thomson, B. A. Carnes, F. S. Williamson, and L. S. Lombard. 1986b. Comparative biological effects of low dose, low dose-rate exposures to fission neutrons from the Janus reactor or to Cobalt-60 gamma rays. Pp. 285-396 in Proc. Workshop on Californium-252 Brachy-Therapy and Fast Neutron Beam Therapy, Y. Maruyama, J. L. Beach, and J. M. Feola, eds. Nuclear Science Applications, Vol 2, Sect. B. New York: Harwood Academic.

Gr86b Graw, J., J. Favor, A. Neuhauser-Klaus, and U. H. Ehling. 1986. Dominant cataract and recessive specific locus mutation in offspring of X-irradiated male mice. Mutat. Res. 159:47-54.

Gr88 Grahn, D., and B. A. Carnes. 1988. Genetic injury in hybrid male mice exposed to low doses or ^{60}Co gamma-rays or fission neutrons III. Frequencies of abnormal sperm and reciprocal translocations measured during and following long-term weekly exposures. Mutat. Res. 198:285-294.

Gr88a Griffin, C. S., and C. Tease. 1988. Gamma ray-induced numerical and structural chromosome anomalies in mouse immature oocytes. Mutat. Res. 202:209-213.

Ha56 Haldane, J. B. S. 1956. The detection of autosomal lethals in mice induced by mutagenic agents. J. Genet. 54:327-342.

Ki82 Kirk, M., and M. F. Lyon. 1982. Induction of congenital anomalies in offspring of female mice exposed to varying doses of X-rays. Mutat. Res. 106:73-83.

Ki84 Kirk, K. M., and M. F. Lyon. 1984. Induction of congenital malformations in the offspring of male mice treated with X-rays at pre-meiotic and post-meiotic stages. Mutat. Res. 125:75-85.

Ko65 Kohn, H. I., M. L. Epling, P. H. Guttman, and D. W. Bailey. 1965. Effect of paternal (spermatogonial) X-ray exposure in the mouse: Lifespan, X-ray tolerance, and tumor incidence of the progeny. Radiat. Res. 25:423-434.

Ko76 Kohn, H. I., and R. W. Melvold. 1976. Divergent X-ray-induced mutation rates in the mouse for H and "7-1ocus" groups of loci. Nature 259:209-210.

Le47 Lea, D. E. 1947. Actions of Radiations on Living Cells. Cambridge: Cambridge University Press.

Le70 Leonard, A., and Gh. Deknudt. 1970. Persistence of chromosome rearrangements induced in male mice by X-irradiation of premeiotic germ cells. Mutat. Res. 9:127-133.

Le71 Leonard, A. 1971. Radiation-induced translocations in spermatogonia of mice. Mutat. Res. 11:71-88.

Lu71 Lüning, K. G., and A. G. Searle. 1971. Estimates of the genetic risks from ionizing irradiation. Mutat. Res. 12:291-304.

Lu72 Lüning, K. G. 1972. Studies of irradiated mouse populations IV. Effects on productivity in the 7th-18th generations. Mutat. Res. 14:331-344.

Lu75 Lüning, K. G., and A. Eiche. 1975. X-ray-induced recessive lethal mutations in the mouse. Mutat. Res. 34:163-174.

Lu75a Lüning, K. G., C. Ronnback, and W. Sheridan. 1975. Genetic effects of acute and chronic irradiation with 14MeV neutrons. Acta Radiol. 14:401-416.

Lu82 Lüning, K. G., and A. Eiche. 1982. X-ray induced recessive lethal mutations in adult and foetal female mice. Mutat. Res. 92:169-180.

Ly66 Lyon, M. F., and T. Morris. 1966. Mutation rates at a new set of specific loci in the mouse. Genet. Res. 7:12-17.

Ly69 Lyon, M. F., and T. Morris. 1969. Gene and chromosome mutation after large fractionated or unfractionated radiation doses to mouse spermatogonia. Mutat. Res. 8:191-198.

Ly70 Lyon, M. F. 1970. X-ray induced dominant lethal mutation in male guinea-pigs, hamsters and rabbits. Mutat. Res. 10:133-140.

Ly70a Lyon, M. F., T. Morris, P. Glenister, and S. E. O'Grady. 1970a. Induction of translocations in mouse spermatogonia by X-ray doses divided into many small fractions. Mutat. Res. 9:219-223.

Ly70b Lyon, M. F., R. J. S. Phillips, and P. Glenister. 1970b. Dose-response curve for the yield of translocations in mouse spermatogonia after repeated small radiation doses. Mutat. Res. 10:497-501.

Ly72 Lyon, M. F., D. G. Papworth, and R. J. S. Phillips. 1972. Dose-rate and mutation frequency after irradiation of mouse spermatogonia. Nature 238:101-104.

Ly72a Lyon, M. F., R. J. S. Phillips, and P. Glenister. 1972. Mutagenic effects of repeated small radiation doses to mouse spermatogonia. II. Effect of interval between doses. Mutat. Res. 15:191-195.

Ly73 Lyon, M. F., R. J. S. Phillips, and P. Glenister. 1973. The mutagenic effect of repeated small radiation doses to mouse spermatogonia. III. Does repeated irradiation reduce translocation yield from a large radiation dose? Mutat. Res. 17:81-85.

Ly79 Lyon, M. F., R. J. S. Phillips, and G. Fisher. 1979. Dose-response curves for radiation-induced gene mutations in mouse oocytes and their interpretation. Mutat. Res. 63:161-173.

Ly82 Lyon, M. F., R. J. S. Phillips, and G. Fisher. 1982. Use of an inversion to test for induced X-linked lethals in mice. Mutat. Res. 92:217-228.

Ly88 Lyon, M. F., and R. Renshaw. 1988. Induction of congenital malformation in mice by parental irradiation: transmission to later generations. Mutat. Res. 198:277-283.

Ma76 Malashenko, A. M. 1976. Investigation of the effect of diethylsulphate applied at low doses in laboratory mice using the method of specific loci. Genetika 12:163-165.

Ma85 Matsuda, Y., I. Tobari, J. Yamagiwa, T. Utsugi, M. Okamoto, and S. Nakai. 1985. Dose-response relationship of gamma-ray-induced reciprocal translocations at low doses in spermatogonia of the crab-eating monkey (Macaca fascicularis). Mutat. Res. 151:121-127.

Mo81 Morton, N. E. 1981. Mutation rates for human autosomal recessives. Pp. 65-89 in Population and Biological Aspects of Human Mutation, E. B. Hook and I. H. Porter, eds. New York: Academic.

Mo82 Morton, N. E. 1982. Outline of Genetic Epidemiology. Basel: S. Karger.

Mu74 Muramatsu, S. 1974. Frequency of spontaneous translocation in mouse spermatogonia. Mutat. Res. 24:81-82.

NRC56 National Academy of Sciences. 1956. Biological Effects of Atomic Radiation. Report of the Committee on the Effects of Atomic Radiation on Agriculture and Food Supplies. Washington, D.C.

NRC72 National Research Council, Advisory Committee on the Biological Effects of Ionizing Radiation. The Effects on Populations of Exposure to Low Levels of Ionizing Radiations. Washington, D.C.: National Academy of Sciences.

NRC80 National Research Council. 1980. Advisory Committee on the Biological Effects of Ionizing Radiation. The Effects on Populations of Exposure to Low Levels of Ionizing Radiation: 1980. Washington, D.C.: National Academy of Sciences.

NCRP80 National Council on Radiation Protection and Measurements (NCRP). 1980. Influence of dose and its distribution in time on dose-response relationships for low-LET radiations. NCRP Report No. 64. Washington, DC: NCRP.

NCRP87 National Council on Radiation Protection and Measurements (NCRP). 1987. Genetic effects from internally deposited radionuclides. NCRP Report No. 89. Bethesda, Md.: NCRP.

Ne53 Neel, J. V., Morton, N. E., Schull, W. J., McDonald, D. J., Kodani, M., et al. 1953. The effect of exposure of parents to the atomic bombs on the first generation offspring in Hiroshima and Nagasaki (preliminary report). Japan J. Genet. 28:211-218.

Ne56 Neel, J. V., and W. J. Schull. 1956. Studies on the potential effects of the atomic bombs. Acta Genet. 6:183-196.

Ne57 Neel, J.V. 1957. In Effect of Radiation on Human Heredity. Geneva: World Health Organization.

Ne74 Neel, J. V., Kato, H., and W. J. Schull. 1974. Mortality in the children of atomic bomb survivors and controls. Genetics 311-326.

Ne80 Neel, J. V., C. Satoh, H. B. Hamilton, M. Otake, K. Goriki, T. Kageoka, M. Fujita, S. Neriishi, J. Asakawa. 1980. Search for mutations affecting protein structure in children of atomic bomb survivors. Proc. Natl. Acad. Sci. USA 77:4221-4225.

Ne84 Neel, J. V. et. al. 1981. A search for mutations affecting protein structure in children of proximally and distally exposed atomic bomb survivors: preliminary report. RERF TR 5-80.

Ne88 Neel, J. V. et. al. 1988. Search for mutations altering protein charge and/or function in children of atomic bomb survivors: final report. Am. J. Hum. Genet. 42:663-676.

No82 Nomura, T. 1982. Parental exposure to X rays and chemicals induces heritable tumours and anomalies in mice. Nature 296:575-577.

No83 Nomura, T. 1983. X-ray induced germ-line mutation leading to tumours: its manifestation by post-natally given urethane in mice. Mutat. Res. 121:59-65.

No88 Nomura, T. 1988. X-ray and chemically induced germ-line mutation causing phenotypical anomalies in mice. Mutat. Res. 198:309-320.

NUR85 NUREG/CR-4214 1985. Health Effects Model for Nuclear Power Plant Accident Consequence Analysis, J. S. Evans, D. W. Moeller, and D. W. Cooper, eds. U.S. Nuclear Regulatory Commission.

Ot87 Otake, M., H. Yoshimaru, and W. J. Schull. 1987. Severe mental retardation among the prenatally exposed survivors of the atomic bombing of Hiroshima and Nagasaki. A comparison of the T65DR and DS86 dosimetry systems. Radiation Effects Research Foundation. RERF TR 16-87.

Pa83 Pacchierotti, F., U. Andreozzi, A. Russo, and P. Metalli. 1983. Reciprocal translocations in ageing mice and in mice with long-term low-level ^{239}Pu contamination. Int. J. Radiat. Biol. 43:445-450.

Pe61 Penrose, L. S. 1961. Mutation. Pp. 1-18 in Recent Advances in Human Genetics, L. S. Penrose, ed. London: Churchill.

Po76 Pomerantseva, M. D., L. K. Ramaiya, and G.A. Vilkina. 1976. Mutagenic effect of various types of radiation in the germ cells of male mice. X. Frequency of recessive lethal mutation and reciprocal translocations in the spermatogonia of mice subjected to fractionated gamma irradiation. Genetika 12:56-63.

Pr73 Preston, R. J., and J. G. Brewen. 1973. X-ray induced translocations in spermatogonia. I. Dose and fractionation responses in mice. Mutat. Res. 19:215-223.

Pr76 Preston, R. J., and J. G. Brewen. 1976. X-ray induced translocations in spermatogonia. II. Fractionation responses in mice. Mutat. Res. 36:333-344.

Ro64 Roderick, T. H., ed. 1964. Proc. Symp. Effects of Radiation on the Hereditary Fitness of Mammalian Populations. Genetics 50:1019-1217.

Ro71 Roderick, T. H. 1971. Producing and detecting paracentric chromosomal inversions in mice. Genetics 76:109-117.

Ro83 Roderick, T. H. 1983. Using inversions to detect and study recessive lethals and detrimentals in mice. Pp. 135-167 in Utilization of Mammalian Specific Locus Studies in Hazard Evaluation and Estimation of Genetic Risk, F. J. deSerres and W. Sheridan, eds. New York: Plenum.

Ro87 Russell, L. B., and E. M. Rinchik. 1987. Genetic and molecular characterization of genomic regions surrounding specific loci of the mouse. Pp. 109-121 in Mammalian Cell Mutagenesis: Banbury Report 28. New York: Cold Spring Harbor Lab.

Ru51 Russell, W. L. 1951. X-ray induced mutations in mice. Cold Spring Harbor Symp. Quant. Biol. 16:327-336.

Ru57 Russell, W. L. 1957. Shortening of life in the offspring of male mice exposed to neutron radiation from an atomic bomb. Proc. Natl. Acad. Sci. USA 43:324-329.

Ru58 Russell, W. L., L. B. Russell, and E.M. Kelly. 1958. Radiation dose rate and mutation frequency. Science 128:1546-1550.

Ru60 Russell, W. L. 1960. Recent findings in mammalian radiation genetics. In 9th (1959) International Congress of Radiology, Munich. Stuttgart: Georg Thieme.

Ru62 Russell, W. L. 1962. An augmenting effect of dose fractionation on radiation-induced mutation rate in mice. Proc. Natl. Acad. Sci. USA 48:1724-1727.

Ru63 Russell, W. L. 1963. The effect of radiation dose rate and fractionation on mutation in mice. Pp. 205-217 in Repair from Genetic Radiation Damage, F. H. Sobels, ed. New York: Pergamon.

Ru64 Russell, W. L. 1964. The nature of the dose-rate effect of radiation on mutation in mice. Japan J. Genet. 40(Suppl.):128-140.

Ru65 Russell, W. L. 1965. Studies in mammalian radiation genetics. Nucleonics 23:53-62.

Ru72 Russell, W. L. 1972. The genetic effects of radiation. Pp. 487-500 in Peaceful Uses of Atomic Energy, 4th International Conference, Vol. 13, 1971. Vienna: International Atomic Energy Agency.

Ru77 Russell, W. L. 1977. Mutation frequencies in female mice and the estimation of genetic hazards of radiation in women. Proc. Natl. Acad. Sci. USA 74:3523-3527.

Ru82 Russell, W. L., and E. M. Kelly. 1982. Specific locus mutation frequences in mouse stem cell spermatogonia at very low radiation dose rates. Proc. Natl. Acad. Sci. USA 79:539-541.

Ru82a Russell, W. L., and E. M. Kelly. 1982. Mutation frequencies in male mice and the estimation of genetic hazards of radiation in men. Proc. Natl. Acad. Sci. USA 79:542-544.

Ru86 Rutledge, J. C., K. T. Cain, L. A. Hughes, P. W. Braden, and W. M. Generoso. 1986. Differences between two hybrid stocks of mice in the incidence of congenital abnormalities following X-ray exposure of stem-cell spermatogonia. Mutat. Res. 163:299-302.

Sa82 Satoh, C., A. A. Awa, J. V. Neel, W. J. Schull, H. Kato, H. B. Hamilton, M. Otake, and K. Goriki. 1982. Genetic effects of atomic bombs. Pp. 267-276 in Human Genetics, Part A: The Unfolding Genome. New York: Alan R. Liss.

Sc71 Schröder, J. H. 1971. Attempt to determine the rate of radiation-induced recessive sex-linked lethal and detrimental mutations in immature germ cells of the house mouse (*Mus musculus*). Genetics 68:35-57.

Sc71a Schröder, J. H., and O. Hug. 1971. Dominante Letalmutationen in der Nachkommenschaft Bestrahlter Mannlicher Mause. I. Untersuchung der Dosiswirkungsbeziehung und des Unterschiedes Zwischen Ganzund Teilkorperbestrahlung bei Meiotischen und PostMeiotischen Keimzellenstadien. Mutat. Res. 11:215-245.

Sc81 Schull, W. J., M. Otake, and J. V. Neel. 1981. Genetic effects of the atomic bombs: a reappraisal. Science 213:1220-1227.

Sc81a Schull, W. J., M. Otake, and J. V. Neel. 1981. Hiroshima and Nagasaki: a reassessment of the mutagenic effect of exposure to ionizing radiation. Pp. 227-303 in Population and Biological Aspects of Human Mutation, E. B. Hook and I.H. Porter, eds. New York: Academic.

Se67 Searle, A. G. 1967. Progress in mammalian radiation genetics. Pp. 469-481 in Radiation Research 1966. Amsterdam: North Holland.

Se71 Searle, A. G. and R. J. S. Phillips. 1971. The mutagenic effectiveness of fast neutrons in male and female mice. Mutat. Res. 11:97-105.

Se74 Searle, A. G. 1974. Mutation induction in mice. Pp. 131-207 in Advances in Radiation Biology, vol. 4, J. T. Lett, H. Adler, and M. R. Zelle, eds. New York: Academic Press.

Se74a Searle, A. G., and C. V. Beechey. 1974. Cytogenetic effects of x-rays and fission neutrons in female mice. Mutat. Res. 24:171-186.

Se76 Searle, A. G., C. V. Beechey, D. Green, and E.R. Humphreys. 1976. Cytogenetic effects of protracted exposures to alpha-particles from plutonium-239 and to gamma-rays from cobalt-60 compared in male mice. Mutat. Res. 41:297-310.

Se85 Searle, A. G., and C. V. Beechey. 1985. A specific locus experiment with mainly dominant visible results. Genet. Res. 45:224 (Abst).

Se85a Searle, A. G., and C. V. Beechey. 1985. The influence of mating status and age on the induction of chromosome aberrations and dominant lethals in irradiated female mice. Mutat. Res. 147:357-362.

Se86 Searle, A. G., and C. V. Beechey. 1986. The role of dominant visibles in mutagenicity testing. Pp. 511-518 in Genetic Toxicology of Environmental Chemicals, Part B: Genetic Effects and Applied Mutagenesis. New York: Alan R. Liss.

Se86a Searle, A. G. and J. H. Edwards. 1986. The estimation of risks from the induction of recessive mutations after exposure to ionizing radiation. J. Med. Genet. 23:220-226.

Se78 Sega, G. A., R. E. Sotomayor, and J. G. Owens. 1978. A study of unscheduled DNA synthesis induced by x-rays in the germ cells of male mice. Mutat. Res. 49:239-257.

Se73 Selby, P. B. 1973. X-ray-induced specific-locus mutation rate in newborn male mice. Mutat. Res. 18:63-75.

Se73a Selby, P. B. 1973. X-ray-induced specific-locus mutation rate in young male mice. Mutat. Res. 18:77-88.

Se77 Selby, P. B., and P. R. Selby. 1977. Gamma-ray-induced dominant mutations that cause skeletal abnormalities in mice. I. Plan, summary of results and discussion. Mutat. Res. 43:357-375.

Se80 Selby, P. B., S. S. Lee, and E. M. Kelly. 1980. Probable dose-rate effect for induction of specific locus mutations in oocytes of mice irradiated shortly before birth. Genetics 94(Suppl):94-95 (Abst).

Se81 Selby, P. B. 1981. Radiation genetics. In H. L. Foster, J. D. Small, and J. G. Fox, eds. The Mouse in Biomedical Research. vol. 1. New York: Academic Press.

Se85b Selby, P. B., and W. L. Russell. 1985. First-generation litter-size reduction following irradiation of spermatogonial stem cells in mice and its use in risk estimation. Environ. Mutagen. 7:451-469.

Sm70 Smith, C. 1970. Heritability of liability and concordance in monozygotic twins. Ann. Hum. Genet. 34:85-91.

Sp81 Speed, R. M., and A. C. Chandley. 1981. The response of germ cells of the mouse to the induction of non-disjunction by X-rays. Mutat. Res. 84:409-418.

St59 Stevenson, A. 1959. The load of hereditary defects in human populations. Radiat. Res. 1:306-325.

St61 Stevenson, A. C. 1961. The load of hereditary defects in human populations. Radiat. Res. 1(Suppl.):306-325.

St67 Stevenson, A. C., and C. B. Kerr. 1967. On the distributions of frequencies of mutation to genes determining harmful traits in man. Mutat. Res. 4:339-352.

Ta69 Taylor, B. A., and A. B. Chapman. 1969. The frequency of x-ray-induced dominant and recessive lethal mutations in the rat. Genetics 63:455-466.

Te82 Tease, C. 1982. Similar dose-related chromosome non-disjunction in young and old female mice after X-irradiation. Mutat. Res. 95:287-296.

Te85 Tease, C. 1985. Dose-related chromosome non-disjunction in female mice after X-irradiation of dictyate oocytes. Mutat. Res. 151:109-119.

Te86 Tease, C., and G. Fisher. 1986. X-ray-induced chromosome aberrations in immediately preovulatory oocytes. Mutat. Res. 173:211-215.

Tr74 Trimble, B. K., and J. H. Doughty. 1974. The amount of hereditary disease in human populations. Ann. Hum. Genet. 38:199-223.

Tr77 Trimble, B. K., and M. E. Smith. 1977. The incidence of genetic disease and the impact on man of an altered mutation rate. Can. J. Genet. Cytol. 19:375-385.

Uc74 Uchida, I. A., and C. P. V. Lee. 1974. Radiation-induced nondisjunction in mouse oocytes. Nature 250:601-602.

UN58 United Nations Scientific Committee on the Effects of Atomic Radiation (UNSCEAR). 1958. Report A.3838. General Assembly Official Records, 13th Sess., Suppl. No. 17. New York: United Nations.

UN62 United Nations Scientific Committee on the Effects of Atomic Radiation (UNSCEAR). 1962. Report A.5216. General Assembly Official Records, 17th Sess., Suppl. No. 16. New York: United Nations.

UN66 United Nations Scientific Committee on the Effects of Atomic Radiation (UNSCEAR). 1966. Report A.8314. General Assembly Official Records. 21st Sess. Suppl. No. 14. New York: United Nations.

UN72 United Nations Scientific Committee on the Effects of Atomic Radiation (UNSCEAR). 1972. Genetic effects of radiation. Pp. 199-564 in Ionizing Radiation: Levels and Effects. Report A/8725. Twenty-Seventh Session, Supplement No. 25. New York: United Nations.

UN77 United Nations Scientific Committee on the Effects of Atomic Radiation (UNSCEAR). 1977. Genetic effects of radiation. Pp. 425-564 in Sources and Effects of Ionizing Radiation. Report A/32/40. Thirty Second Session, Supplement No. 40. New York: United Nations.

UN82 United Nations Scientific Committee on the Effects of Atomic Radiation (UNSCEAR). 1982. Genetic effects of radiation. Pp. 425-569 in Ionizing Radiation: Sources and Biological Effects. Report A/37/45. Thirty Seventh Session, Supplement No. 45. New York: United Nations.

UN86 United Nations Scientific Committee on the Effects of Atomic Radiation (UNSCEAR). 1986. Genetic effects of radiation. Pp. 7-164 in Ionizing Radiation: Sources and Biological Effects. Report A/41/16. Forty First Session, Supplement No. 16. New York: United Nations.

Vo75 Vogel, F., and R. Rathenberg. 1975. Spontaneous mutation in man. Pp. 223-318 in Advances in Human Genetics, Vol. 5, H. Harris and K. Hirschhorn, eds. New York: Plenum.

Vo84 Vogel, F. 1984. Clinical consequences of heterozygosity for autosomal-recessive diseases. Clin. Genet. 25:381-415.

3

Mechanisms of Radiation-Induced Cancer

BACKGROUND

Carcinogenesis is viewed as a multistep process in which two or more intracellular events are required to transform a normal cell into a cancer cell. The concept that carcinogenesis involves more than one step is derived from three main lines of evidence: (1) the rate of mortality from cancer increases as a power function of age, (2) a long latent period typically intervenes between exposure to a known carcinogen and the appearance of cancer, and (3) three distinct and separate stages have been identified in experimental carcinogenesis: initiation, promotion, and progression.

The fact that the cumulative incidence of cancer increases approximately as the seventh power of age during adult life prompted early investigators to postulate the existence of seven successive events, or steps, in the conversion of a normal cell into a cancer cell; these events were thought to involve mutational changes in the broadest sense (Ar54). This concept failed to recognize, however, the high rates of somatic mutation that such a seven-stage model would require, the dynamic state of the target cells, and the peculiar age distributions typical for the cancers occurring during childhood. If the kinetics of target cells and the possible growth advantage of preneoplastic cells are taken into account, the age distributions of pediatric and adult cancers can be explained in terms of just two rate-limiting mutational steps (e.g., see Mo81), although other events that might be associated with tumor progression or tumor metastasis are not excluded. In a tumor that has grown to a population of 10^6 cells, even events that occur only rarely in each cell division can be expected to occur with a high probability in the total cell population. Models that account for all of the complex factors involved in the mechanisms of carcinogenesis have not yet been developed to the point where they can be used realistically for risk estimation, especially in view of the fact that the sparsity of data available makes it difficult to choose among the various possibilities. In Chapter 4 of this report, therefore, descriptive empirical models are used to arrive at cancer risk estimates.

MECHANISMS

The mechanisms by which radiation may produce carcinogenic changes are postulated to include the induction of: (1) mutations, including alterations in the structure of single genes or chromosomes; (2) changes in gene expression, without mutations; and (3) oncogenic viruses, which, in turn, may cause neoplasia. Although controversy persists as to the relative importance of these hypothetical mechanisms in the induction of carcinogenesis, they are not mutually exclusive, since different mechanisms may be involved at successive stages in carcinogenesis.

The somatic mutation theory of carcinogenesis, proposed by Boveri in 1914 (Bo14), has received further support from the high correlation between the carcinogenicity and the mutagenicity of different agents. In a few types of cancer (e.g., retinoblastoma), moreover, the same specific gene mutation or deletion is found both in familial and nonfamilial cases, as noted in Chapter 1, suggesting that the mutation or the deletion of the gene plays a causative role, as discussed below.

It is possible, on the other hand, that premalignant or malignant alterations do not necessarily result from changes in gene or chromosome structure per se, but from changes in gene expression. Support for this concept comes from evidence that nuclei transplanted from cancer cells into enucleated ova or blastocysts can produce apparently normal organisms or tissues in various species, including mice (Br77). Nevertheless, altered gene expression does not exclude the possibility that premalignant cells might undergo mutation during their conversion to cancer cells.

Initiation, Promotion, and Progression in Carcinogenesis

The following generalizations about the process of carcinogenesis are noteworthy: (1) The effects of radiation and chemical carcinogens which lead to cancer are dose dependent and generally irreversible; (2) the carcinogenic process is dependent on cell proliferation; (3) the changes that initiate carcinogenesis in a cell are passed on to daughter cells; (4) the subsequent events in carcinogenesis can be profoundly influenced by various noncarcinogenic factors; and (5) tumors tend to become increasingly malignant with time through the stepwise outgrowth of progressively more malignant subpopulations of tumor cells.

It is now widely accepted that initiation, the first step in malignant cell transformation, begins the carcinogenic process, while in most cases promotion is required to complete the process (Co83). This concept of carcinogenesis as a two-stage process was suggested originally by studies of tumor induction in mouse skin in which a dose of chemical carcinogen that was too small to cause a detectable increase in the incidence of tumors was found to induce a high incidence of tumors if it was followed by repeated administration of a suitable promoting agent, an agent that did not cause tumors when administered alone (Bo74a, Be75). A synergistic interaction between the initiating effects of radiation (or various chemicals) and specific promoting agents is now known to occur in many different organs and cell systems (Mo64, Pe85, Ja86, Ke84a). In these studies, it was observed that promotion caused a higher incidence of cancer with a shortened latent period (Ry71). It has been widely assumed that a similar two-stage mechanism involving initiation and promotion exists for radiation carcinogenesis.

Whereas most initiating agents, including radiation, are carcinogenic by themselves in a single exposure if they are administered in a sufficiently large dose, promoting agents must be given repeatedly over long periods of time, during which successive phases of promotion may be distinguishable (Pe85). Different promoting agents, moreover, may act at different stages of promotion. By the same token, different agents that inhibit promotion may act at different stages in the process (Pe85).

The term *tumor progression* has been used traditionally to denote the acquisition of increasingly malignant properties within an established cancer, presumably via genetic instability. However, the term has also come to be used to denote the conversion of a benign growth into a malignant growth. In either case, the process reflects the proliferation of a subpopulation of cells within a tumor. This subpopulation of cells expands and overgrows the less aggressive cells. Radiation has been shown to be capable of enhancing the process of progression (Ja87). Other clastogenic agents such as hydroxyurea (Hah86) may also be progression agents for carcinogenesis (Personal Communication, Dr. Henry Pitot). Similarly, initiation-promotion-initiation experiments, in which promotion is followed by a second initiation step brought about by the administration of an initiator, have been found to increase the final incidence of malignant, as opposed to benign tumors (Mo81, He83). While initiation is thought by some investigators to result from mutational events, promotion appears to involve non-mutational effects on the kinetics of intermediate-stage cells.

The first step in the initiation of carcinogenesis, whether by radiation or a chemical carcinogen, has been observed to be an event that occurs in a large percentage of treated cells (Ke85a, Cl86a, Cl86b, Wa88). The frequency with which this event can be produced experimentally far exceeds the frequency of mutations at any one gene locus, contradicting the notion that the initiating event is a specific single-locus mutation. Instead, initiation more likely appears to be an event that increases the genomic instability of the cells in subsequent rounds of cell division (Cl86b, Wa88, Ke84b). Although much experimental data has suggested that the first event in radiation and chemical carcinogenesis is a widespread, nonmutagenic type event, the same data has suggested that later events in the carcinogenic process appear to behave like mutations. Thus the notion that mutagenic events may occur in carcinogenesis still has widespread support, as indicated elsewhere in this report.

The hypothesized high-frequency initiating event could conceivably be a change in gene expression (for example, see Fa80) of a type that might occur in a large proportion of irradiated cells (Sc85); in *Escherichia coli*, for example, radiation induces an error-prone DNA repair system (the SOS system) which leads to mutations that would otherwise occur only rarely (Wi76). Although the SOS system is activated for only a short period of time, other radiation-induced systems may be activated for longer periods; for example, recombinational events in yeast continue to occur for many generations after irradiation (Fa77). In this connection, it is noteworthy that SOS functions are also activated by a protease (Li80a) but are suppressed by protease inhibitors (Me77), which also suppress radiation-induced recombination in yeast (Wi84) and radiation-induced malignant cell transformation in vitro (Ke85b). Many other agents that enhance or suppress carcinogenesis in vivo exert similar effects on malignant cell transformation in vitro (Ke84a); these include retinoids (vitamin A derivatives), antiinflammatory steroidal agents, antioxidants, vitamins, protease inhibitors, and other substances (Sl80, Pe85, Wa85, Ke84a).

After exposure to a carcinogen, proliferation of the exposed cells is essential to their subsequent neoplastic transformation. Tissue irritation, which stimulates cell division, was recognized long ago to increase the probability of tumor development; for example, following carcinogen treatment of the skin or liver, wounding of the skin or partial hepatectomy enhances tumor formation in the skin or liver, respectively (Su73). Similarly, the carcinogenic effects of ^{210}Po alpha radiation on the lung of the hamster are enhanced by repeated instillation of saline into the airway, which stimulates proliferation of pulmonary epithelial cells (Li78, Sh82). Likewise, cigarette smoke, which contains small amounts of many known carcinogenic agents (such as ^{210}Po) and which is a potent irritant, appears to potentiate the effects of inhaled radon and its daughter products in uranium miners (Lo44, Lu71, Sa84). Proliferation is thought to play a role in the fixation of radiation damage which leads to malignant transformation in the expression of that damage and in the promotional phase of cancer development.

The mechanism of tumor promotion is still obscure. Promoters such as phorbol esters are known to interrupt intercellular communication in some cell populations (Tr82), and they have traditionally been thought to be nonmutagenic (Ma83) and thus to act through effects on gene expression (Bo74). Recently, however, some such agents have been found to produce chromosome aberrations (Em81), aneuploidy (Pa81), sister chromatid exchanges (Ki78, Na79), and single-strand breaks in DNA (Bi82). Many promoting agents, moreover, induce free radicals in cells (Go81, Fi85). These free radicals can, in turn, damage DNA. It is noteworthy, therefore, that free radical-generating agents can act as tumor promoters (Ke86) and that inhibitors of free radical reactions can suppress tumor promotion in some systems (Sl83).

Radiation itself also can enhance tumor promotion, tumor progression, and the conversion of benign growths to malignant growths (Ja87). To the extent that the effects of radiation are mediated by free radicals (Li77), which can also mediate the effects of promoting agents (Co83), sequential exposures to radiation may serve to promote tumorigenesis through mechanisms similar to those of chemical promoting agents.

Natural hormones also may promote carcinogenesis in irradiated individuals. However, it is not yet clear how comparable the effects of hormones are compared to the effects of the classical promoting agents. Hormonal promotion conceivably may be mediated through physiological effects on the proliferation and differentiation of cells (Cl86a,b, Wa88). It may also be mediated through autocrine growth factors or their receptors, such as those that may be under the influence of certain oncogenes (Sp85). In some cases, hormones may actually suppress tumor promotion by inducing differentiation in cells that are at risk.

Other factors capable of having a highly significant effect on the various stages of carcinogenesis include age, sex, genetic constitution, capacity to repair DNA, carcinogen metabolism, immunologic status, and dietary factors such as caloric intake (Su73).

Radiobiological Factors Affecting Oncogenic Transformation

During the past two decades, much information has been gathered about radiation carcinogenesis from experimental systems in which cultured mammalian cells are transformed to a malignant state by exposure to radiation. In vitro transformation assays have been used extensively to study the carcinogenic effects of radiation in a highly quantitative fashion and in a defined environment. One major advantage of such in vitro systems is that the effects of radiation on specific target cells can be studied directly without the presence of extraneous factors, which complicate carcinogenesis in vivo. In addition, transformation assays are extremely sensitive, allowing detection of the carcinogenic effects of radiation at doses below those at which statistically significant carcinogenic effects have been observed in animal and human studies. It has been observed by many investigators that radiation-induced transformation in vitro can be modified in the same way as radiation-induced cancer in animals, with the yields of malignant cells varying similarly in response to different characteristics of the radiation (such as total dose, dose rate, fractionation pattern, linear energy transfer (LET), etc.) and many other modifying factors, as described below. It is widely inferred that the processes involved in radiation-induced transformation in vitro are similar to those involved in carcinogenesis in vivo, and that results from in vitro studies are applicable to radiation-induced cancer in vivo. In vitro transformation systems also offer an approach to studying radiation carcinogenesis that is less expensive and less time-consuming than animal experiments.

Dose Response

Commonly used in vitro transformation assays can be divided into two broad classes. First, there is the use of short-term cultures of embryo cells, with clonal assays in which transformed clones can be identified after an incubation period of about 14 days. The transformation frequency and the surviving fraction can then be assessed from the same culture dishes.

Second, there are assays with established cell lines (such as 3T3, 10T1/2, Rat 2) that have become immortal. These are focal assays, and for transformed foci to become identifiable, the culture must be continued for some weeks after the normal cells have reached confluence. Cell survival and transformation frequency cannot be assessed from the same culture dishes. Results can be expressed as transformation frequency per surviving cell, but because the transformation frequency observed is a function of the number of viable cells seeded per culture dish, the data can also be expressed in terms of the number of foci per dish or the fraction of culture dishes bearing foci.

These in vitro assays, based on rodent fibroblasts, have been used widely because they are highly quantitative. Ideally, assays based on human epithelial cells would be more relevant, but, although transformation in human cells has been demonstrated as a result of exposure to radiation or chemicals, quantitative assays are not available.

In recent years, in vivo transformation assays also have been developed for thyroid and mammary cells in rats. Cells are irradiated in situ in the thyroid or mammary gland and are subsequently excised and transplanted

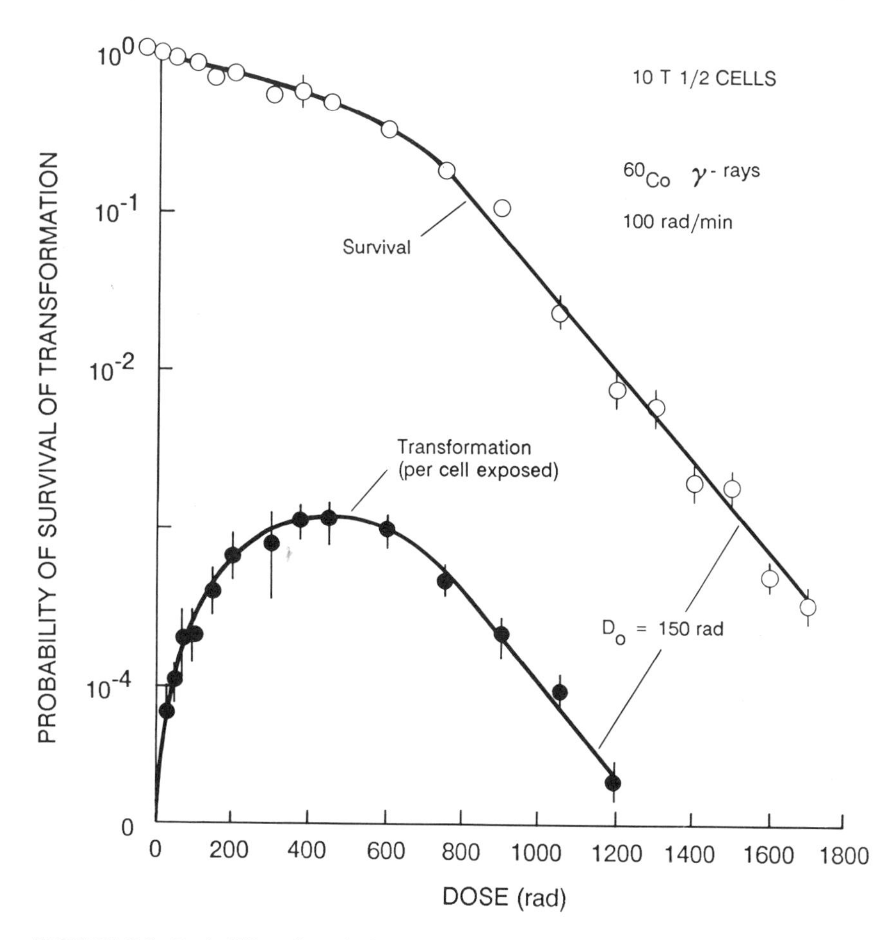

FIGURE 3-1 Probability of survival (top) and transformation per irradiated cell (bottom) as a function of dose (Ha80).

to a fat pad in a suitably prepared animal. Cell survival and transformation incidence can be determined in this way (Cl86a, Cl86b). Experiments using different initial cell densities or reseeded/diluted cell cultures have indicated that the malignant transformation of cells arises from very few carcinogen-treated cells (Ke85a, Cl86b). These results have led to the notion that the first event in carcinogenesis is a high frequency event as discussed earlier.

The dose-response relationship for the induction of radiogenic transformation reflects a balance between an increase with dose in the proportion of cells that are transformed and a decrease in cell survival. This is illustrated in Figure 3-1 (Ha80). For gamma rays and other low-LET radiations, the cell survival curve is characterized by a broad initial shoulder region before it becomes steeper and approaches an exponential function of dose at higher doses (Figure 3-1) (Ha80). Transformation incidence, as expressed by frequency per surviving cell, increases with dose up to a few Gray, and reaches a plateau at higher doses. While the transformation data are often plotted in terms of frequency per surviving cell, they can also be expressed as frequency per initial cell at risk when applying these in vitro data to whole organisms. This approach is also illustrated in Figure 3-1 where the dose-response transformation curve rises at low doses, reaches a maximum, and falls at higher doses to eventually parallel the cell-killing curve. The curve represents a balance between transformation and cell killing and indicates that cells destined to become transformed have a survival response similar to that of untransformed normal cells. The peak of the dose-response curve for transformation frequency per initial cell at risk often reaches higher values for densely ionizing radiations, such as neutrons and alpha particles than for x rays or gamma rays.

Dose Rate and Dose Fractionation

For low-LET radiations, the consensus is that cell survival is enhanced by a decrease in the dose rate or separation of the dose into a number of fractions. Effects on the yield of transformants, however, are more complex. It has been reported that for low-LET radiations, splitting or fractionating the dose or reducing the dose rate can either enhance (Bo74, Ha81, Li79) or decrease (Hi84) the transformation frequencies in a variety of in vitro transformation models. More recent studies suggest that the proliferative status of the cells may account for some of the observed variation (Lu85). Using C3H10T1/2 cells, Hill et al. (Hi85) have compared dose-response transformation curves for gamma rays and for fission spectrum neutrons delivered in both a single exposure or in multiple small fractions. Although fractionation was observed to result in a sparing effect on transformation by gamma rays, it increased the rate of transformation by fission spectrum neutrons (Ha79, Hi85). Since enhanced transformation was observed after exposure to multiple low doses or a continuous low dose rate, compared to high-dose-rate fission spectrum neutrons, the relative biological effectiveness (RBE) of neutrons relative to that of gamma rays was larger at low-dose rates than at high-dose rates. As outlined in chapter 1, these observations have important practical implications for the selection of an appropriate RBE for neutrons.

Linear Energy Transfer (LET)

Comparisons of various high- and low-LET ionizing radiations for their abilities to induce oncogenic transformation in several cell systems have been reported. In general, high-LET radiations are far more cytotoxic and oncogenic than low-LET radiations such as x rays or gamma rays. Furthermore, the RBE for oncogenic transformation and cytotoxicity increases with increasing LET of the radiation. Hence, if the transformation frequencies for each type of high-LET particle are plotted against the corresponding survival values, the curves obtained cannot be superimposed. This suggests that there is a real difference in the RBE between cell killing and transformation (He88, Ya85) and also indicates that there is a significant frequency of transformation at doses of high-LET radiations that have very little effect on cell survival.

Figure 3-2 (Ha87a) shows survival and transformation data for gamma rays and high-LET helium-3 ions. The cell survival curve for gamma rays has a broad initial shoulder, while that for helium-3 ions is an exponential function of dose. For high-LET particles, the transformation frequency peaks at a much lower dose than for gamma rays and reaches a value that is higher by a factor of about 5 than is the case for gamma rays (Ha87a).

Neutrons are also highly effective at inducing transformation. Figure 3-3 shows the variation of RBE with neutron energy over a wide range,

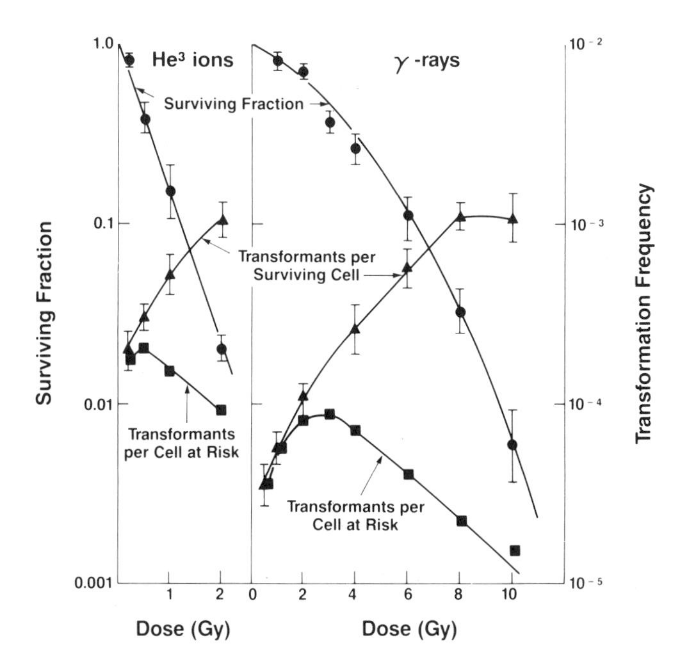

FIGURE 3-2 Cell survival curves and dose response relationships for oncogenic transformation for C3H10T1/2 cells irradiated with either gamma rays or high-LET helium-3 ions. Transformation frequencies are expressed in two ways; per surviving cell and per cell initially at risk (Ha87a).

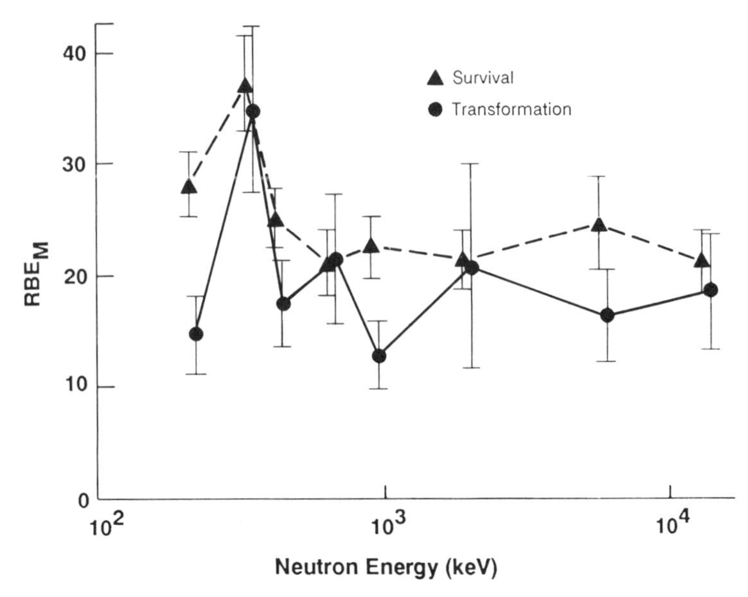

FIGURE 3-3 RBE_m for cell curvinal and for oncogenic transformation as a function of neutron energy and C3H10T1/2 cells irradiated with monoenergetic neutrons (Mi89).

which is similar to that received by individuals during the bombing of Hiroshima (Mi89). Energies of about 350 kiloelectron volts (keV) are most effective for both cell lethality and transformation. There is evidence that the effectiveness of neutrons increases with a decrease in the dose rate. As a consequence of this, RBE values are higher for a fractionated or a low-dose-rate exposure, than for a single, brief exposure, as mentioned above. It has been suggested that the misrepair of sublethal radiation damage in fission neutron-irradiated cells may account for the increased RBE values (Hi85).

Alpha Particles

The transforming ability of alpha particles also has been studied extensively with in vitro transformation systems. Robertson et al. (Ro83) showed that the RBE for transformation by plutonium-238 alpha particles in Balb/3T3 cells was substantially higher than that for cell lethality. It was also demonstrated that potentially lethal damage was repaired in x-irradiated 3T3 cells and was not repaired in alpha-particle irradiated cells, resulting in a high RBE value for oncogenic transformation in alpha-irradiated plateau-phase cultures.

Similar findings have also been reported by Hall and Hei who used the C3H10T1/2 cell system (Ha85). At equivalent doses, alpha particles were substantially more cytotoxic than gamma rays and were more efficient in inducing oncogenic transformation. The calculated RBE value for alpha particles ranged from 2.3 to 9 over the range of doses studied, with the highest RBE value at the lowest dose. Recent results have suggested the absence of a dose-rate effect with alpha particles (Hi87).

Previous studies by Lloyd et al. (Ll79) showed that at a dose corresponding to a surviving fraction of 37%, about 14 particles traversed the nucleus for each cell killed. The fact that on the average 13 particles may traverse a cell nucleus without killing the cell may explain the high efficiency with which high-LET particles induce transformed loci.

Agents That Modify Radiation Transformation

Many different classes of agents have been shown to modify radiation-induced transformation in vitro (Ke84a). The tumor promoting agent 12-O-tetradecanoyl phorbol acetate (TPA) has been studied in many laboratories for its ability to enhance radiation-induced transformation. It is of particular interest that promoting agents such as TPA can change the shape of the dose-response curve for radiation-induced transformation, making it linear (Figure 3-4) (Ke78). This alteration of the dose-response relationship also occurs in promotion by TPA of radiation carcinogenesis in vivo (Figure 3-5) (Fr84). While promotion can greatly enhance radiation transformation, other agents can suppress radiation transformation or the enhancement by TPA (Ke88). An example of the suppressive effect of the protease inhibitor antipain on radiation transformation and the TPA enhancement of radiation transformation is shown in Figure 3-6. Other examples of agents which suppress radiation transformation are selenium (Figure 3-7), which is thought to exert its inhibitory action by inducing glutathione peroxidases, and 5-aminobenzamide, which is an inhibitor of poly-ADP-ribose synthetase.

The frequency of transformation resulting from a given dose of radiation can also be modulated by the level of thyroid hormone in the serum. With high levels of T_3 hormone (corresponding to hyperthyroid conditions) the transformation incidence resulting from 3 Gray of x rays is increased, while with low levels of T_3 hormone, (corresponding to hypothyroid conditions), the transformation incidence is not detectable above the spontaneous level. The suppressing effects of some of these agents are illustrated in Figure 3-7 (Ha87a).

GENETICS OF CANCER

As noted above, much evidence supports the concept that mutation is involved in the etiology of cancer. Recent research has identified critical

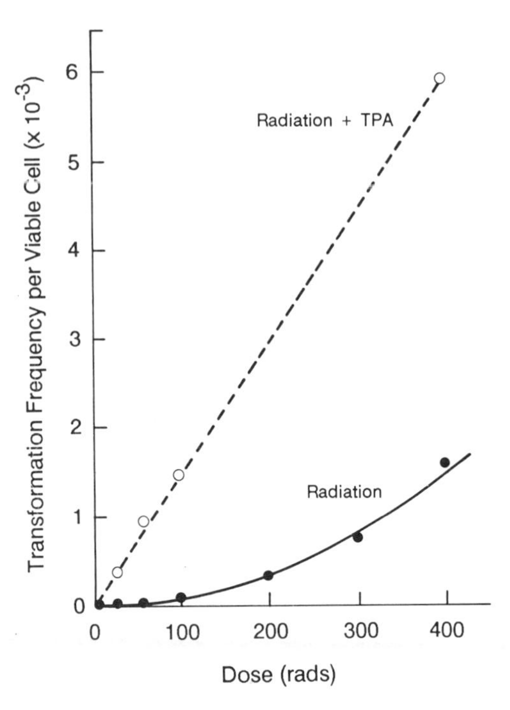

FIGURE 3-4 Dose-response curve for the induction of radiation transformation, with or without enhancement by TPA. Note how a promoter changes a linear quadratic response to a linear one (Ke78).

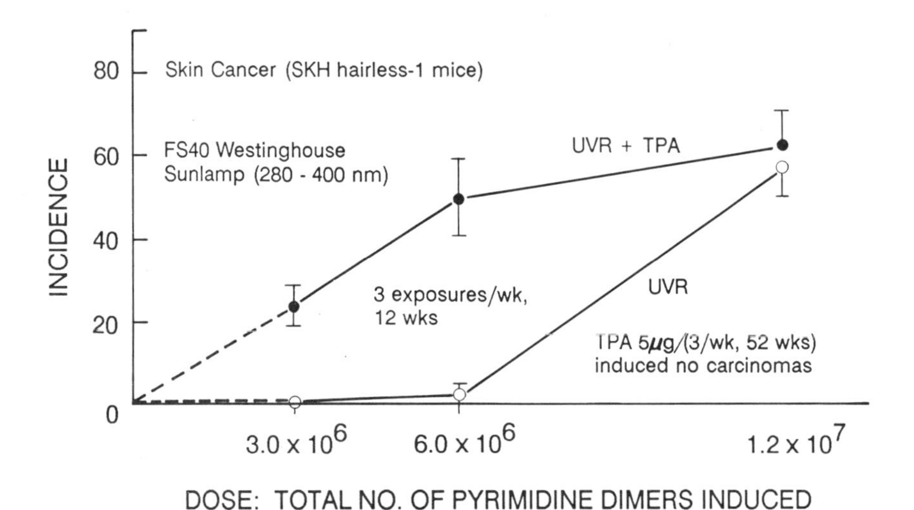

FIGURE 3-5 U.V. light-induced skin cancer, with and without promotion by TPA (Fr84).

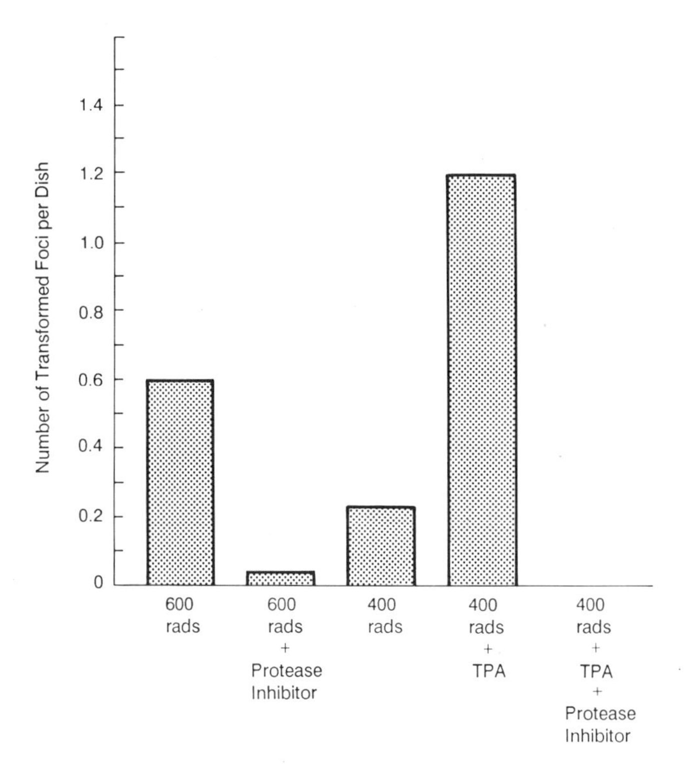

FIGURE 3-6 Suppressive effect of a protease inhibitor (antipain) on radiation transformation in vitro, both with and without promotion by TPA (Ke88).

genes that are thought to be the sites of oncogenic somatic mutations. Over the past decade, research on the mechanisms of carcinogenesis has focused on such genes, of which two broad classes are now known to exist: (1) protooncogenes and (2) tumor-suppressor genes, or antioncogenes (Kn85).

Protooncogenes

Protooncogenes, which may give rise to oncogenes, seem to be important in the origin of at least some forms of human cancer. The list of such genes has grown apace with new means for identifying them. Alterations of the *ras* protooncogene have now been observed in several different types of radiation-induced tumors, including murine lymphomas (Gu84a,b), plutonium-induced malignancies (Fr86b), and radiation-induced rat skin tumors (Sa87, Ga88, Ga86). Radiation has also been shown to activate other oncogenes presumed to be involved in carcinogenesis, including c-*myc* (Sa87, Ga86, Ga88) and oncogenes that are not members of the *ras*

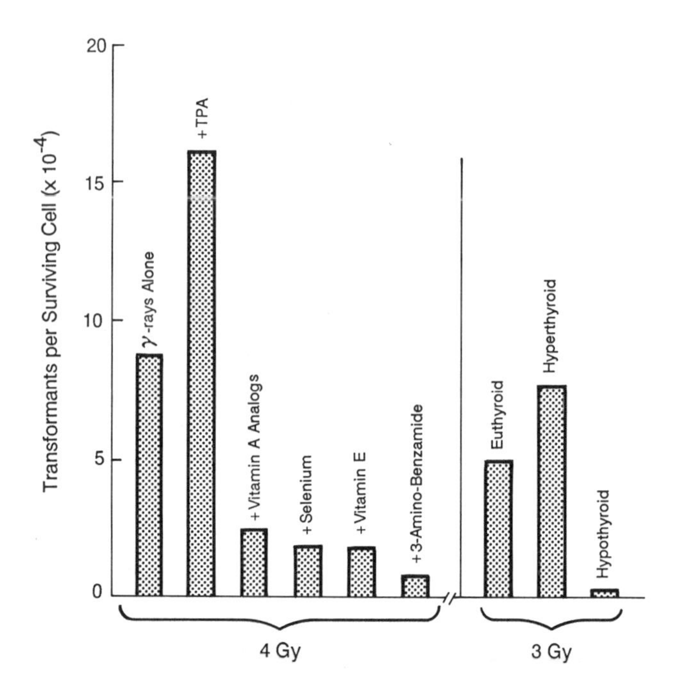

FIGURE 3-7 Effects of vitamin A analogues, selenium, vitamin E, 3-amino-benzamide, and TPA, at 4 Gy and T_3 (thyroid hormone) at 3 Gy on radiation transformation (Ha87a).

gene family but which cause transformation in the NIH 3T3 cell transfection assay system (Bo87, Ja88). The activation of *myc* has been shown to occur by amplification, translocation, and internal rearrangements.

Although there is evidence for some specificity in the pattern of oncogene alterations that is produced by a given carcinogen, it is still not possible on the basis of an oncogene "signature" to determine the cause of a given tumor, that is, whether the tumor was caused by radiation or some other carcinogen.

The stage at which a given oncogene is activated in the carcinogenic process also remains to be determined. While in some instances activation may occur as a late step in carcinogenesis (Su83, Su84, Ru84), evidence implies that in other instances it may occur early (Ba87, Ba87b). It is noteworthy that protooncogene loci are involved in the specific chromosomal changes that are associated with certain types of cancer (Ha87a, Ro84). This implies that such alterations of protooncogene structure or function play a causal role in the occurrence of those types of cancer. It is not known, however, whether the changes are early or late events in the origin of the neoplasms (Li80a, Fi81).

Some oncogene alterations clearly represent steps in tumor progression. An example is the amplification of the *myc* family of oncogenes in neuroblastomas and in small-cell carcinomas of the lung (Br84, Na86). This amplification is often cytogenetically evident in the form of double minute chromosomes consisting of repeated chromosomal pieces, including the oncogene in question. In these instances amplification signifies an advanced stage of disease and carries a poor prognosis.

A role for oncogenes in the earliest stage of oncogenic transformation could be better supported if individuals who carried such mutations in their germ lines were found. This has not been found as yet in humans, but susceptible mice have been produced experimentally by transgenically introducing an activated oncogene into the germ line. Mice with a strong predisposition for the development of lymphoma or mammary cancer have resulted from the introduction of a c-*myc* gene, fused with an immunoglobulin enhancer, or with the strong long terminal repeat (LTR) promoter of the mammary tumor virus, respectively (Ad85, St84). The tumors are clonally distinct, however, indicating that at least one somatic event occurred subsequently in their development. This finding parallels results of in vitro experiments showing a requirement for the activation of at least two different oncogenes in the transformation of normal rat embryo cells (La83a,b).

Tumor-Suppressor Genes (Antioncogenes)

The second class of cancer genes that has been identified was discovered through studies of individuals with inherited predispositions for specific cancers. For many cancers including carcinomas of colon, breast, lung, stomach, ovary, uterus, kidney and bladder, glioma, melanoma, leukemias, and lymphomas there is a subgroup of persons at higher than normal risk by virtue of the fact that they have inherited a specific mutation. This type of predisposition is transmitted in a Mendelian dominant fashion, although the different underlying mutations vary in their penetrances. Well-known examples of such predisposing conditions are familial polyposis coli (chromosome 5, Wilms' tumor (chromosome 11), and the hereditary form of retinoblastoma (chromosome 13). The latter tumor has been the prototype in research on this group of genes (Kn85).

About 40% of the individuals with retinoblastoma carry germ-line mutations that predispose them to the disease. The offspring of such persons have a 50% risk of developing the tumor. About 30% of the individuals with retinoblastoma have bilateral disease; all of the latter carry the germ-line mutation. A small fraction of cases (3-5%) bear a constitutional deletion in chromosome 13, a finding that has facilitated the search for the responsible gene. Genetic linkage studies have shown that the heritable cases without a deletion involve a mutation at the same site.

Although carriers of the mutation develop a mean of three to four tumors, the inherited mutation alone is not sufficient for the production of the cancer; another event is necessary. The second event that is necessary is the loss or mutation of the normal allele on the other chromosome 13 by nondisjunction, deletion, genetic recombination, or local mutation (Ca82, Kn85). The result in all cases is the same: the tumor cell contains no normal copy of the retinoblastoma gene. Hence, although inheritance of the predisposition is dominant, oncogenesis at the cellular level is recessive. Therefore, the normal allele can be viewed as protective, thus, the designation tumor-suppressor gene, or antioncogene.

Patients with retinoblastoma have a high risk of developing osteosarcoma of the orbit following radiation therapy. They also have a lesser predisposition to osteosarcoma in the absence of irradiation. In either case, the genetic change in the tumor cells is the loss of the two normal alleles of the retinoblastoma gene; thus, this gene is a tumor-suppressor gene for osteosarcoma (Ha85) as well as for retinoblastoma. The probability of mutation or loss of the normal gene in persons born with one mutant gene in the germ line is apparently increased by radiation, as would be expected.

The retinoblastoma gene has recently been cloned, an accomplishment that will greatly facilitate investigation of the relevant oncogenic mechanism, the identification of those at risk, and the study of the physiology of the gene in normal development (Fr86a, Fu87b, Le87a, Le87b). It has already been shown that the messenger RNA (mRNA) of the gene is absent or defective in virtually every case of retinoblastoma, whether it was inherited or not. In the nonhereditary cases, the two normal genes are lost or mutated as the result of two somatic events, the second events being of the same kinds as those observed in heritable cases (see above). The only difference between the two forms of tumor is that the first event is present in the germ line in one form and occurs after conception in the other.

The idea that recessive genes may suppress the oncogenic process is not new. Previous experiments with somatic cell hybrids have shown that the neoplastic character of most tumor cells can be suppressed by fusing the cells with normal cell partners (St76). On the other hand, it is clear that oncogenes are frequently abnormal in structure and/or function in many tumors. It is probable, therefore, that protooncogenes and tumor-suppressor genes are both important in carcinogenesis. Whether either or both are necessary in every case of cancer remains to be determined.

Recessive Breakage and Repair Disorders

These disorders, which include xeroderma pigmentosum, ataxia telangiectasia, Fanconi's anemia, and Bloom's syndrome, are recessively inherited conditions that predispose the chromosomes of an individual to breakage and/or defective repair of DNA damage (Han86). They do not involve cancer genes of the types discussed above but can be viewed as conditions that increase the probability of a cancer-producing mutation.

Thus, in xeroderma pigmentosum a defect in excision repair permits an increased rate of mutations at all genetic loci in cells exposed to sunlight. Ataxia telangiectasia predisposes the chromosome to breakage, especially in lymphocytes; the underlying molecular defect is not known, but it is thought to involve a defect in DNA repair. Patients with the syndrome are especially predisposed to lymphoid neoplasia, and their cells are highly sensitive to ionizing radiation. Chromosome breakage and rearrangement are regular features of Fanconi's anemia, which predisposes an individual to acute myelomonocytic leukemia; the underlying molecular defect for this is not known. Finally, Bloom's syndrome is associated with high rates of mutation and of sister chromatid, and even homologous chromosome, exchanges. The molecular defect apparently involves a ligase that is important in the repair of DNA damage (Ch87, Wi87). The syndrome predisposes an individual to several kinds of neoplasia, perhaps by facilitating mutation, somatic recombination, and the expression of recessive oncogenes.

Genetic Polymorphism for Metabolism of Carcinogens

In contrast to the aforementioned DNA repair disorders, in which the response to an environmental agent is altered, there are cases in which the response may be normal but the amount of radiant energy imparted is increased. Thus, albinos are sensitive to ultraviolet light because they absorb more of it, not because they have a defective DNA repair mechanism. Such a genetic predisposition is also known for many chemical carcinogens (Ca82, Ko82, Ay84, Go86). Hence, to the extent that the effects of a given chemical may promote the carcinogenic effects of radiation, traits affecting the metabolism of the chemical may alter susceptibility to radiation carcinogenesis.

Hereditary Fragile Sites

Another kind of inherited mutation that may predispose an individual to cancer is the hereditarily fragile genetic site. About 18 such sites are known. Fragility for a specific site can be elicited in vitro, and the fragility is transmitted in a Mendelian dominant fashion (He84). Although several of the sites have been found to be situated at or near break points that are known to be involved in various cancer-associated translocations (Le84), cancer does not appear to be common in families with such abnormalities.

The importance of these mutations in carcinogenesis thus remains to be determined.

EFFECTS OF AGE, SEX, SMOKING, AND OTHER SUSCEPTIBILITY FACTORS

As discussed in the preceding section, the carcinogenic process includes the successive stages of initiation and promotion. The latter phase, promotion, appears to be particularly susceptible to modulation, with cigarette smoking being a conspicuous example of a modulating factor. Susceptibility to the carcinogenic effects of radiation can thus be affected by a number of factors, such as genetic constitution, sex, age at initiation, physiological state, smoking habits, drugs, and various other physical and chemical agents (UN82). The mechanisms through which these factors influence susceptibility are, however, not well understood. Moreover, they depend on the particular type of cancer, the tissue at risk, and the specific modifying factor under consideration. Therefore, the Committee elected to discuss the factors affecting carcinogenesis at specific organ sites in Chapters 4 and 5.

Some general conclusions can be drawn from the observations reported in Chapter 4. Cancer rates are highly age dependent and, in general, increase rapidly in old age. The expression of radiogenic cancers varies with age in a similar way, so that the age-dependent increase in the excess risk of radiogenic cancer is conveniently expressed in terms of relative risk; that is, the increased risk tends to be proportional to the baseline risk in the same age interval. In some cases, however, such as breast cancer, the change in the baseline cancer rate with age is more complicated and possibly related to variations in hormonal status with age. Susceptibility to radiation-induced breast cancer may be similarly complicated, as outlined in Chapter 5, and there is some indication that protective factors for breast cancer in nonirradiated women, such as early age at the birth of the first child, may also be relevant for radiation-induced breast cancer.

The situation is less clear for the risk factors for lung cancer. The BEIR IV Committee found that smoking and prolonged exposure to inhaled alpha-particle emitters interacted in a multiplicative fashion, or nearly so, with the result that the increased risk of radiogenic lung cancer in those of a given smoking status was proportional to the baseline risk for the same smoking status (NRC88); however, this may not be the case for acute exposures to x rays and gamma rays. It is commonly believed that the data on lung cancer and smoking among the atomic-bomb survivors support an additive risk model, in which there is no interaction between radiation and tobacco use. Nevertheless, the BEIR IV Committee's analyses of these data indicated that the pattern of observed risk is also compatible with a multiplicative interaction. Currently, available data are ambiguous, as indicated in Chapter 5, and further studies are needed to explore the role of cigarette smoking as a risk factor for radiation-induced cancer.

For lung cancer and most other non-sex-specific solid cancers, it is unclear how a person's sex affects the risk of radiogenic cancer. In general, baseline rates for such cancers in males exceed those in females, possibly because of increased exposure to carcinogens and promoters in occupational activities and life-style factors, such as increased smoking and use of alcohol. While sex specific excess rates of cancer can generally be modeled adequately as being proportional to the corresponding sex-specific baseline rates, in many cases an additive excess risk model fits the data equally well; that is, the number of radiation-induced cancers per unit dose is nearly the same in both sexes. This means that the relative-risk coefficient for females compared with that for males is, to a good approximation, inversely proportional to the ratio of the sex-specific baseline rates (NRC88). For this reason, as outlined in Chapter 4 and in Annex 4D, the Committee tested a number of risk models that include sex as a modifying factor for the risk of radiogenic cancer.

REFERENCES

Ad85 Adams, J. M., A. W. Harris, C. A. Pinkert, L. M. Corcoran, W. S. Alexander, S. Cory, R. D. Palmiter, and R. L. Brinster. 1985. The c-myc oncogene driven by immunoglobin enhancers induces lymphoid malignancy in transgenic mice. Nature 318:533-538.

Ar54 Armitage, P., and R. Doll. 1954. The age distribution of cancer and a multistage theory of carcinogenesis. Br. J. Cancer 8:1-2.

Ay84 Ayesh, R., J. R. Idle, J. C. Ritchie, M. J. Crothers, and M. R. Hetzel. 1984. Metabolic oxidation phenotypes as markers for susceptibility to lung cancer. Nature 312:169-170.

Ba84 Balmain, A., M. Ramsden, G. T. Bowden, and J. Smith. 1984. Activation of mouse cellular Harvey-ras gene in chemically induced benign skin papillomas. Nature 307:658-660.

Ba87 Balmain, A., K. Brown, R. Bremner, M. Quintanilla, and M. Archer. 1987. The action of chemical carcinogens and oncogenic retroviruses in mouse skin tumor induction. Pp. 501-506 in Radiation Research Proceedings of the 8th International Congress of Radiation Research, vol. 2, Edinburgh, July 1987, E. M. Fielden, J. F. Fowler, J. H. Henry, and D. Scott, eds. Philadelphia: Taylor and Francis.

Ba87b Barbacid, M. 1987. ras genes. Annu. Rev. Biochem. 56:779-827.

Be75 Berenblum, I. 1975. Sequential aspects of chemical carcinogenesis: Skin. Pp. 323-344 in Cancer: A Comprehensive Treatise, Vol. 1, F. F. Becker. New York: Plenum.

Bi82 Birnboim, H. C. 1982. DNA strand breakage in human leukocytes exposed to a tumor promoter, phorbol myristate acetate. Science 215:1247-1249.

Bo14 Boveri, T. H. 1914. Zur Frage der Entstehung maligner Tumoren (On the problem of the origin of malignant tumors). Jena, German Democratic Republic: Fisher.

Bo74 Borek, C., and E. J. Hall. 1974. Effect of split doses of x rays on neoplastic transformation of single cells. Nature 252:499-501.

Bo74a Boutwell, R. K. 1974. The function and mechanism of promoters of carcinogenesis. CRC Crit. Rev. Toxicol. 2:419-443.

Bo87 Borek, C., A. Ong, and H. Mason. 1987. Distinctive transforming genes in x-ray transformed mamalian cells. Proc. Natl. Acad. Sci. USA 84:794-798.

Br77 Braun, A. 1977. The story of cancer: On its nature, causes and control. Reading, Mass.: Wesley.

Br84 Brodeur, G. M., R. C. Seeger, M. Schwab, H. E. Barmus, and J. M. Bishop. 1984. Amplification of N-myc in untreated human neuroblastomas correlates with advanced disease stage. Science 224:1121-1124.

Ca82 Cartwright, R. A., R. W. Glasham, H. J. Rogers, R. A. Ahmad, D. Barham-Hall, E. Higgins, and M. A. Kahn. 1982. The role of N-acetyltransferase phenotypes in bladder carcinogenesis: A pharmacogenetics epidemiological approach to bladder cancer. Lancet ii:842-846.

Ca83 Cavenee, W. K., T. P. Druja, R. A. Phillips, W. F. Benedict, R. Godbout, B. L. Gallie, A. L. Murphree, L. C. Strong, and R. L. White. 1983. Expression of recessive alleles by chromosomal mechanisms in retinoblastoma. Nature 205:779-784.

Ch87 Chan, J. Y. H., F. F. Becker, J. German, and J. H. Ray. 1987. Altered DNA ligase I activity in Bloom's syndrome cells. Nature 325:357-359.

Cl86b Clifton, K. H. 1986. Cancer risk per clonogenic cell in vivo: Speculation on the relationship of both cancer incidence and latency to target cell number. Proceedings of the 14th International Cancer Congress, Budapest.

Cl86a Clifton, K. H., M. A. Tanner, and M. N. Gould. 1986. Assessment of radiogenic cancer initiation frequency per clonogenic rat mammary cell in vivo. Cancer Res. 46:2390-2395.

Co83 Copeland, E. S., ed. 1983. A National Institutes of Health Workshop Report. Free radicals in promotion—a chemical pathology study section workshop. Cancer Res. 43:5631-5637.

Em81 Emwerit, I., and P. A. Cerutti. 1981. Tumor promoter phorbo1-12-myristate-13-acetate induces chromosomal damage via indirect action. Nature 293:144-146.

Fa77 Fabre, F., and H. Roman. 1977. Genetic evidence for inducibility of recombination competence in yeast. Proc. Nat. Acad. Sci. USA 74:1667-1671.

Fa80 Fahmy, M. J., and O. G. Fahmy. 1980. Intervening DNA insertions and the alteration of gene expression by carcinogens. Cancer Res. 40:3374-3382.

Fi81 Fialkow, P. J., P. J. Martin, V. Najfeld, G. K. Penfold, R. J. Jacobson, and J. A. Hansen. 1981. Evidence for a multistep pathogenesis of chronic myelogenous leukemia. Blood 58:158-163.

Fi85 Fisher, S. M., and L. M. Adams. 1985. Suppression of tumor-promoter induced chemiluminescence in mouse epidermal cells by several inhibitors of arachinoic acid metabolism. Cancer Res. 45:3130-3136.

Fr84 Fry, R. J. M., and R. D. Ley. 1984. Ultraviolet radiation carcinogenesis. Pp. 73-96 in Mechanisms of Tumor Promotion, Vol. II, Tumor Promotion and Skin Carcinogenesis, T. J. Slaga, ed. Boca Raton, FL: CRC Press.

Fr86a Friend, S. H., R. Bernards, S. Rogelj, R. A. Weinberg, J. M. Rappaport, D. M. Albert, and T. P. Dryja. 1986. A human DNA segment with properties of the gene that predisposes to retinoblastoma and osteosarcoma. Nature 323:643-646.

Fr86b Frazier, M. E., R. A. Lindberg, D. M. Mueller, A. Gee, and T. M. Seed. 1986. Oncogene involvement in plutonium-induced carcinogenesis. Int. J. Rad. Biol. 49:542-543.

Fu87 Fujiki, H., and T. Sugimura. 1987. New classes of tumor promoters: Teleocidin, aplysiatoxin and palytoxin. Adv. Cancer Res. 49:223-264.

Fu87b Fung, Y.-K. T., A. L. Murphree, A. t'Ang, J. Qian, S. H. Hinrichs, and W. F. Benedict. 1987. Structural evidence for the authenticity of the human retinoblastoma gene. Science 236:1657-1661.

Ga86 Garte, S. J., M. J. Sawey, and F. J. Burns. 1986. Oncogenes activated in radiation-induced rat skin tumors. Pp. 389-397 in Radiation Carcinogenesis and DNA alterations, F. J. Burns, A. C. Upton, and G. Silini, eds. New York: Plenum.

Ga88 Garte, S. J., M. J. Sawey, F. J. Burns, M. Felber, and T. Ashkenazi-Kimmel. 1982. Multiple oncogene activation in a radiation carcinogenesis model. In Anticarcinogenesis and Radiation Protection, P. A. Cerutti, O. F. Nygaard, and M. G. Simic, eds. New York: Plenum Press.

Go81 Goldstein, B. O., G. Witz, M. Amoruso, D. S. Stone, and W. Troll. 1981. Morphonuclear leukocyte superoxide anion radical (O_2) production by tumor promoters. Cancer Lett. 11:257-262.

Go86 Gonzalez, F. J., A. K. Jaiswal, and D. W. Nebert. 1986. P-450 genes: Evolution, regulation, and relationship to human cancer and pharmacogenetics. Cold Spring Harbor Symp. Quant. Biol. 51:879-890.

Gu84a Guerrero, I., P. Calzava, A. Mayer, and A. Pellicer. 1984. A molecular approach to leukemogenesis; mouse lymphomas contain an activated c-ras oncogene. Proc. Natl. Acad. Sci. 181:202-205.

Gu84b Guerrero, I., A. Villasante, V. Corces, and A. Pellicer. 1984. Activation of a c-K-ras oncogene by somatic mutation in mouse lymphomas induced by gamma radiation. Science 225:1159-1162.

Hah86 Hahn, P., L. N. Kapp, W. F. Morgan, and R. B. Painter. 1986. Chromosomal changes without DNA overproduction in hydroxyurea-treated mammalian cells: Implications for gene amplification. Cancer Res. 46:4607-4612.

Ha81 Hall, E. J., and R. C. Miller. 1981. The how and why of *in vitro* oncogenic transformation. Radiat. Res. 87:208-223.

Ha85 Hall, E. J., and T. K. Hei. 1985. Oncogenic transformation *in vitro* by radiations of varying LET. Radiat. Protect. Dosimetry 13:149-151.

Ha87a Hall, E. J., and T. K. Hei. 1987. Oncogenic transformation by radiation and chemicals. In Proceedings of the VIIth International Congress of Radiation Research, E. M. Fielden, J. F. Fowler, J. H. Hendry, and D. Scott, eds. London: Taylor and Francis.

Ha87b Haluska, F. G., Y. Tsujimoto, and C. M. Croce. 1987. Oncogene activation by chromosome translocation in human malignancy. Annu. Rev. Genet. 21:321-345.

Ha79 Han, A., and M. M. Elkind. 1979. Transformation of mouse C3H10T1/2 cells by single and fractionated doses of x rays and fission-spectrum neutrons. Cancer Res. 39:123-130.

Ha80 Han, A., C. K. Hill, and M. M. Elkind. 1980. Repair of cell killing and neoplastic transformation at reduced dose rates of ^{60}Co gamma rays. Cancer Res. 40:3328-3332.

Han86 Hanawalt, P. C., and A. Sarasin. 1986. Cancer-prone hereditary diseases with DNA processing abnormalities. Trends Genet. 2:124-129.

Han85 Hansen, M. F., A. Koufos, B. L. Gallie, R. A. Phillips, O. Fodstad, A. Brogger, T. Gedde-Dahl, and W. K. Cavenee. 1985. Osteosarcoma and retinoblastoma: a shared chromosomal mechanism revealing recessive predisposition. Proc. Natl. Acad. Sci. USA 82:1-5.

He83 Hennings, H., R. Shores, M. L. Wenk, E. F. Spangler, R. Tarone, and S. H. Yuspa. 1983. Malignant conversion of mouse skin tumors is increased by tumor initiators and unaffected by tumor promoters. Nature 304:67-69.

He84 Hecht, F., and G. R. Sutherland. 1984. Fragile sites and cancer breakpoints. Cancer Genet. Cytogenet. 12:179-181.

He88 Hei, T. K., E. J. Hall, and M. Zaider. 1988. Oncogenic transformation by charged particles of defined LET. Carcinogenesis.

Hi87 Hieber, L., G. Ponsel, H. Roos, S. Senn, E. Fromke, and A. N. Kellerer. 1987. Absence of dose-rate effect in the transformation of C_3H10 1/2 cells by alpha particles. Int. J. Rad. Biol. 52:859-869.

Hi84 Hill, C. K., A. Han, F. Buonaguro, and M. M. Elkind. 1984. Multifractionation of ^{60}Co gamma-rays reduces neoplastic transformation *in vitro*. Carcinogenesis 5:193-197.

Hi85 Hill, C. K., B. A. Carnes, A. Han, and M. M. Elkind. 1985. Neoplastic transformation is enhanced by multiple low doses of fission spectrum neutrons. Radiat. Res. 101:404-410.

Ja86 Jaffe, D. R., and G. T. Bowden. 1986. Ionizing radiation as an initiator in the mouse two-stage model of skin tumor formation. Radiat. Res. 106:156-165.

Ja87 Jaffe, D. R., J. F. Williamson, G. T. Bowden. 1987. Ionizing radiation enhances malignant progression of mouse skin tumors. Carcinogenesis 8:1753-1755.

Ja88 Jaffe, D. R., and G. T. Bowden. 1988. Enhanced malignant progression of mouse skin tumors by ionizing radiation and activation of oncogenes in radiation induced tumors. In Radiation Research: Proceedings of the 8th International Congress of Radiation Research, Vol. 1, Edinburgh, July 1987, E. M. Fielden, J. F. Fowler, J. H. Henry, and D. Scott, eds. Philadelphia: Taylor and Francis.

Ke78 Kennedy, A. R., S. Mondal, C. Heidelberger, and J. B. Little. 1978. Enhancement of x ray transformation by 12-O-tetradecanoyl phorbol 13 acetate in a cloned line C3H mouse embryo cells. Cancer Res. 38:439-443.

Ke84a Kennedy, A. R. 1984. Promotion and other interactions between agents in the induction of transformation in vitro in fibroblasts. Pp. 13-55 in Mechanisms of Tumor Promotion, Vol. III, Tumor Promotion and Carcinogenesis in Vitro, T. J. Slaga, ed. Boca Raton, Fla.: CRC Press.

Ke84b Kennedy, A. R., and J. B. Little. 1984. Evidence that a second event in x-ray induced oncogenic transformation *in vitro* occurs during cellular proliferation. Rad. Res. 99:228-248.

Ke85a Kennedy, A. R. 1985. Evidence that the first step leading to carcinogen-induced malignant transformation is a high frequency, common event. Pp. 455-364 in Carcinogenesis: A Comprehensive Survey, Vol. 9, Mammalian Cell Transformation: Mechanisms of Carcinogenesis and Assays for Carcinogenes, J. C. Barrett and R. W. Tennant, eds. New York: Raven.

Ke85b Kennedy, A. R. 1985. The conditions for the modification of radiation transformation in vitro by a tumor promoter and protease inhibitors. Carcinogenesis 6:1441-1446.

Ke86 Kennedy, A. R. 1986. Role of free radicals in the initiation and promotion of radiation-induced and chemical carcinogen induced cell transformation. Pp. 201-209 in Oxygen and Sulfur Radicals in Chemistry and Medicine, A. Breccia, M. A. J. Rodgers, and G. Semerano, eds. Bologna, Italy: Edizioni Scientifiche, Lo Scarabeo.

Ke88 Kennedy, A. R., and P. C. Billings. 1988. Anticarcinogenic actions of protease inhibitors. In Proceedings of the 2nd International Conference on Anticarcinogenesis and Radiation Protection, P. Cerutti, O. F. Nygaard, and M. Simic, eds. New York: Plenum.

Ki78 Kinsella, A., and M. Radman. 1978. Tumor promoter induces sister chromatid exchanges: Relevance to mechanisms of carcinogenesis. Proc. Natl. Acad. Sci. USA 75:6149-6153.

Kn85 Knudson, A. G. 1985. Hereditary cancer, oncogenes, and antioncogenes. Cancer Res. 45:1437-1443.

Ko82 Kouri, R. E., C. E. McKinney, D. J. Slomiany, D. R. Snodgrass, N. P. Wray, and T. L. McLemore. 1982. Positive correlation between high aryl hydrocarbon hydroxylase activity and primary lung cancer as analyzed in cryopreserved lymphocytes. Cancer Res. 42:5030-5037.

La83a Land, H., L. F. Parada, and R. A. Weinberg. 1983. Cellular oncogenes and multistep carcinogenesis. Science 222:771-778.

La83b Land, H., L. F. Parada, and R. A. Weinberg. 1983. Tumorigenic conversion of primary embryo fibroblasts requires at least two cooperating oncogenes. Nature 304:596-602.

Le84 LeBeau, M. M., and J. D. Rowley. 1984. Heritable fragile sites in cancer. Nature 308:607-608.

Le87a Lee, W. H., R. Bookstein, F. Hong, L. J. Young, J. Y. Shew, and E. Y.-H. P. Lee. 1987. Human retinoblastoma susceptibility gene: Cloning, identification, and sequence. Science 235:1394-1399.

Le87b Lee, W. H., J. Y. Shew, F. D. Hong, T. W. Sery, L. A. Donoso, L. J. Young, R. Bookstein, and E. Y.-H. P. Lee. 1987. The retinoblastoma susceptibility gene encodes a nuclear phosphoprotein associated with DNA binding activity. Nature 329:642-645.

Li77 Little, J. B., and J. R. Williams. 1977. Effects of ionizing radiation on mammalian cells. Pp. 127-155 in Handbook of Physiology, S. R. Geiger, H. L. Falk, S. D. Murphy, and P. H. K. Lee, eds. Bethesda, Md.: American Physiological Society.

Li78 Little, J. B., R. B. McGandy, and A. R. Kennedy. 1978. Interactions between polonium ^{210}alpha radiation, benzo(a)pyrene, and 0.9% NaCl solutions instillations in the induction of experimental lung cancer. Cancer Res. 38:1929-1935.

Li79 Little, J. B. 1979. Quantitative studies of radiation transformation with the A31-11 mouse Balb/3T3 cell line. Cancer Res. 39:1474-1480.

Li80a Lisker, R., L. Casas, O. Mutchinick, F. Perez-Chavez, and J. Labardini. 1980. Late-appearing Philadelphia chromosome in two patients with chronic myelogenous leukemia. Blood 56:812-814.

Li80b Little, J. W., S. H. Edmiston, L. Z. Pacelli, and D. W. Mount. 1980. Cleavage of the Escherichia coli lex A protein by the rec A portease. Proc. Natl. Acad. Sci. USA 77:3225-3229.

Ll79 Lloyd, E. L., M. A. Gemmell, C. B. Henning, D. S. Gemmell, and B. J. Zabransky. 1979. Transformation of mammalian cells by alpha particles. Int. J. Radiat. Biol. 36:467-478.

Lo44 Lorenz, E. 1944. Radioactivity and lung cancer: A critical review of lung cancer in the mines of Schneeberg and Joachimstal. J. Natl. Cancer Inst. 5:1-15.

Lu71 Lundin, F. E., J. K. Wagoner, Jr., and V. E. Archer. 1971. Radon daughter exposure and respiratory cancer: Quantitative and temporal aspects. NIOSH-NIEHS Joint Monograph No. 1. Washington, D.C.: U.S. Public Health Service.

Lu85 Lurie, A. G., and A. R. Kennedy. 1985. Single, split, and fractionated dose x-irradiation-induced malignant transformation in A31-11 mouse Balb-3T3 cells. Cancer Lett. 29:169-176.

Ma76 Maruyama, K., R. Natori, and Y. Nonomura. 1976. Down's syndrome and related abnormalities in an area of high background radiation in coastal Kerala. Nature 262:60-61.

Ma83 Marx, J. H. 1983. Do tumor promoters affect DNA after all? Science 219:158-159.

Me77 Meyn, M. S., T. Rossman, and W. Troll. 1983. A protease inhibitor blocks SOS functions in Escherichia coli; antipain prevents repressor in-activiation, ultraviolet mutagenesis and filamentous growth. Proc. Natl. Acad. Sci. USA 74:1152-1156.

Mi89 Miller, R. C., D. J. Brenner, C. R. Geard, K. Komatsu, S. A. Marino, and E. J. Hall. Neutron-Energy-Dependent Oncogenic Transformation of C3H/10T1/2 Mouse Cells. Radiat. Res. 117:114-127.

Mo64 Mole, R. H. 1964. Cancer production by chronic exposure to penetrating gamma irradiation. Natl. Cancer Inst. Monogr. 14:217-290.

Mo81 Moolgavkar, S. H., and A. G. Knudson. 1981. Mutation and cancer: A model for human carcinogenesis. J. Natl. Cancer Inst. 66:1037-1052.

Na79 Nagasawa, H., and J. Little. 1979. Effect of tumor promoters, protease inhibitors, and repair processes on X ray-induced sister chromatid exchanges in mouse cells. Proc. Natl. Acad. Sci. USA 76:1943-1947.

Na86 Nau, M. M., B. J. Brooks, D. N. Carney, A. F. Gazdar, J. F. Battey, E. A. Sausville, and J. D. Minna. 1986. Human small-cell lung cancers show amplification and expression of the N-myc gene. Proc. Natl. Acad. Sci. USA 83:1092-1096.

NRC88 National Academy of Sciences, National Research Council. Committee on the Biological Effects of Ionizing Radiations (BEIR IV). 1988. Health Risks of Radon and Other Internally Deposited Alpha Emitters. Washington, D.C.: National Academy Press.

Pa81 Parry, J. M., E. M. Parry, and J. C. Barrett. 1981. Tumor promoters induce mitotic aneuploidy in yeast. Nature 294:263-265.

Pe85 Pelling, J. C., and T. J. Slaga. 1985. Cellular mechanisms for tumor promotion and enhancement. Pp. 369-393 in Carcinogenesis, Vol. 8, M. J. Mass et al., ed. New York: Raven.

Ro83 Robertson, J. B., A. Koehler, J. George, and J. B. Little. 1983. Oncogenic transformation of mouse Balb/3T3 cells by plutonium-238 alpha particles. Radiat. Res. 96:261-274.

Ro84 Rowley, J. B. 1984. Biological implications of consistent chromosome rearrangements in leukemia and lymphoma. Cancer Res. 44:3159-3168.

Ru84 Rubin, H. 1984. Mutations and oncogenes—cause or effect. Nature 309:518.

Ry71 Ryser, H. J. P. 1971. Chemical carcinogenesis. N. Engl. J. Med. 285:721-734.

Sa84 Samet, J. M., D. M. Kutvirt, R. J. Waxweiler, and C. R. Key. 1984. Uranium mining and lung cancer in Navajo men. N. Engl. J. Med. 310:1581-1484.

Sa87 Sawey, M. J., A. T. Hood, F. J. Burns, and S. J. Garte. 1987. Activation of *myc* and *ras* oncogenes in primary rat tumors induced by ionizing radiation. Mol. Cell. Biol. 7:932-935.

Sc85 Scott, R. E., and P. B. Maercklein. 1985. An initiator of carcinogenesis selectively and stably inhibits stem cell differentiation: A concept that initiation of carcinogenesis involves multiple phases. Proc. Natl. Acad. Sci. 82:2995-2999.

Sh82 Shami, S., L. Thibideau, A. R. Kennedy, and J. B. Little. 1982. Proliferative and morphological changes in the pulmonary epithelium of the Syrian golden hamster lung during carcinogenesis initiated by ^{210}Po alpha radiation. Cancer Res. 42:1405-1411.

Sl83 Slaga, T. J., V. Solanki, and M. Logani. 1983. Studies on the mechanism of action of antitumor promoting agents: suggestive evidence for the involvement of free radicals in promotion. Pp. 471-485 in Radioprotectors and Anticarcinogens, O. F. Nygaard and M. G. Simic, eds. New York: Academic Press.

Sl83 Slaga, T. J., ed. 1980. Carcinogenesis: A Comprehensive Survey, Vol. 5. Modifiers of Chemical Carcinogenesis. New York: Raven Press.

Sp85 Sporn, M. B., and A. B. Roberts. 1985. Autocrine growth factors and cancer. Nature 313:745-747.

St76 Stanbridge, E. J. 1976. Suppression of malignancy in human cells. Nature 260:17-20.

Ste84 Stewart, T., P. K. Pattengale, and P. Leder. 1984. Spontaneous mammary adenocarcinomas in transgenic mice that carry and express MTV/myc fusion genes. Cell 38:627-637.

Su73 Suss, R., V. Kinzel, and J. D. Scribner. 1984. Cancer–experiments and concepts. New York: Springer-Verlag.

Su83 Sukumar, S., V. Notario, D. Martin-Zanca, and M. R. Barbacid. 1983. Induction of mammary carcinomas in rats by nitroso-methylurea involves malignant activation of H-ras-l locus by single point mutations. Nature 306:658-661.

Su84 Sukumar, S., S. Pulciani, J. Doniger, J. A. DiPaolo, C. Evans, B. Zbar, and M. Barbacid. 1984. Science 223:1197-1199.

Tr82 Trosko, J. E., L. P. Yotti, S. T. Warren, G. Tsushimoto, and C. C. Chang. Inhibition of cell-cell communication by tumor promoters. 1982. Pp. 565-585 in Carcinogenesis, Vol. 7, E. Hecker et al., eds.

UN82 United Nations Scientific Committee on the Effects of Atomic Radiation (UNSCEAR). 1982. Ionizing Radiation: Sources and Biological Effects. Report A/37/45. Thirty Seventh Session, Supplement No. 45. New York: United Nations.

Wa85 Wattenberg, L. W. 1985. Chemoprevention of cancer. Cancer Res. 45:1-8.

Wa88 Watanabe, H., M. A. Tanner, F. E. Domann, M. N. Gould, and K. H. Clifton. In press. Inhibition of carcinoma formation and of vascular invasion in grafts of radiation-initiated thyroid clonogens by unirradiated thyroid cells. Carcinogenesis.

We87 Weinberg, C. R., K. G. Brown, and D. G. Hoel. 1987. Altitude radiation, and mortality from cancer and heart disease. Radiat. Res. 112:381-390.

Wi76 Witkin, E. M. 1976. Ultraviolet mutagenesis and inducible DNA repair in escherichia coli. Bacteriol. Rev. 40:869-907.

Wi84 Wintersberger, U. 1984. The selective advantage of cancer cells; a consequence of genome mobilization in the course of the induction of DNA repair processes? (Model studies of yeast). Pp. 311-323 in Advances in Enzyme Regulation, Vol. 22, G. Weber ed. New York: Pergamon.

Wi87 Willis, A. E., and T. Lindahl. 1987. DNA ligase I deficiency in Bloom's syndrome. Nature 325:355-357.

Ya85 Yang, T. C. H., L. M. Craise, M. T. Mei, and C. A. Tobias. 1985. Neoplastic cell transformation by heavy charged particles. Radiat. Res. 104: S177-S178.

4

Risks of Cancer—All Sites

INTRODUCTION

This report seeks to present the best description that can be provided at this time of the risk of cancer resulting from a specified dose of ionizing radiation. However, this description is bound to be inexact since the etiology of radiation-induced cancer is complex and incompletely understood. The risk depends on the particular kind of cancer; on the age and sex of the person exposed; on the magnitude of the dose to a particular organ; on the quality of the radiation; on the nature of the exposure, whether brief or chronic; on the presence of factors such as exposure to other carcinogens and promotors that may interact with the radiation; and on individual characteristics that cannot be specified but which may help to explain why some persons do and others do not develop cancers when similarly exposed.

Although scientists understand some of the intra-cellular processes that are initiated or stimulated by radiation and which may eventually result in a cancer, the level of understanding is insufficient at present to enable prediction of the exact outcome in irradiated cells. Estimates of the risk of cancer, therefore, must rely largely on observations of the numbers of cancers of different kinds that arise in irradiated groups. Since nearly 20% of all deaths in the United States result from cancer, the estimated number of cancers attributable to low-level radiation is only a small fraction of the total number that occur. Furthermore, the cancers that result from radiation have no special features by which they can be distinguished from those produced by other causes. Thus the probability that cancer will result from a small dose can be estimated only by extrapolation from the increased rates of cancer that have been observed after larger doses, based on assumptions about the dose-incidence relationship at low doses.

In this report it is estimated that if 100,000 persons of all ages received a whole body dose of 0.1 Gy (10 rad) of gamma radiation in a single brief exposure, about 800 extra cancer deaths would be expected to occur *during their remaining lifetimes* in addition to the nearly 20,000 cancer deaths that would occur in the absence of the radiation. Because the extra cancer deaths would be indistinguishable from those that occurred naturally, even to obtain a measure of how many extra deaths occurred is a difficult statistical estimation problem. Like all such problems, the answers obtained are subject to statistical errors which can be exacerbated by a limited sample size. The largest series of humans exposed to radiation for whom estimates of individual doses are available consists of the populations of Hiroshima and Nagasaki who were exposed to atomic bomb detonations in 1945. There were 75,991 A-bomb survivors in the two cities for whom dose estimates are available and who have been traced through 1985 to learn the health effects of exposure (Sh87). But 34,272 of those survivors were so far from the hypocenters that their radiation doses were negligible—less than 0.005 Gy (0.5 rad)—and thus they serve as a comparison, or "control" group, leaving 41,719 whose doses are estimated at 0.005 Gy or more. Of these, 3,435 died from some form of cancer between 1950 and 1985. This cohort is not only the largest available, but it has been followed through 1985, that is, for forty years after irradiation, and is the most important source of data for analysis in this report. Even so, there are large statistical uncertainties as to the number of cancer deaths that were induced by radiation and (relatively) even larger uncertainties in the number of radiation-related cancers of particular kinds. The Committee has taken special care to quantify these uncertainties to the extent possible. Nevertheless, the limitations of the data bases on which the Committee's risk estimates are based have conditioned the kinds of estimates that can be developed.

Heretofore, cancer risk estimates for low-LET radiations have been made by BEIR committees on the basis of constant additive risk and constant relative risk models (NRC80), an approach followed also by UNSCEAR in its latest report (UN88). That is, after a minimum latent period, risks were assumed to be relatively independent of time after exposure. The continued follow up of the A-bomb survivors and persons in the ankylosing spondylitis study indicates that temporal variations in risk are too important to be ignored. Consequently, it is necessary to model, not only how the risk increases with dose, but also how it varies as a function of time for persons exposed at various ages. This puts a heavy burden on available data.

Only the A-bomb survivor cohort contains persons of all ages at exposure. Those survivors who were young when exposed are just now entering the age range at which cancer becomes an appreciable cause of death in the general population. Consequently, the number of excess cancer deaths that have occurred among them to date is small, and estimates of how the radiation-induced excess changes over time for those exposed as children introduce a large uncertainty into any attempt to project lifetime risks for the population as a whole. Moreover, the estimated risk is largest for this age group, so that final results are sensitive to the way in which the risk from childhood exposures is accounted for in the risk model.

Although the number of excess cases has increased as exposed groups have been followed for longer periods, the data are not strong when stratified into different dose, age, and time categories. Even though modern statistical methodologies facilitate the analysis of highly stratified data, the fact remains that the number of cases in a given dose, age, and time interval is small and often zero. In situations such as this, one cannot differentiate between various competing risk models because of large statistical uncertainties. This problem is particularly acute when using models which take into account time dependence, age at exposure, etc. and applying them to cancers at a specific site. Because of these limitations, it was not possible for the committee to provide risk estimates for cancers at all of the specific sites of interest. Rather, attention was focused on estimating the risk for leukemia, breast cancer, thyroid cancer, and cancers of the respiratory and digestive systems, where the numbers of excess cases are substantial. To obtain an estimate of the total risk of mortality from all cancers, the committee also modeled cancers other than those listed above as a group.

While this approach limits the application of these results for calculating the probability of causation of cancers at specific sites, the Committee judges it is preferable to aggregating data over age and time on the basis of simple risk models that do not adequately reflect the observational data. In this respect, the report differs from that of the United Nations Scientific Committee on the Effects of Radiation (UN88), which presented two lifetime risk estimates from fatal cancer at each of 10 individual organ sites, one estimate based on a simple additive risk model and the other based on a simple multiplicative risk model.

MODEL FITTING

Methods

The Committee's estimates of cancer risks rely most heavily on data from the Life Span Study (LSS) of the Japanese atomic bomb survivors at Hiroshima and Nagasaki, although other studies also were used for estimation of incidence or mortality risks for specific sites. The cohorts

TABLE 4-1 Major Characteristics of the Data Sets Used for Model Fitting

Study Population	Reference	Incidence or Mortality	Cancer Sites	Total Cases	Total Person Years
Atomic bomb survivors	Sh87	Mortality	All	5,936	2,185,335
	To87	Incidence	Breast	376	940,000
Ankylosing spondylitis patients	Da87	Mortality	Leukemia	36	104,000
			All except leukemia and colon	563	104,000
Canadian fluoroscopy patients	Mi89	Mortality	Breast	482	867,541
Mass. fluoroscopy	Hr89	Mortality	Breast	74	30,932
N.Y. postpartum mastitis	Sh86	Incidence	Breast	115	45,000
Israel tinea capitis	Ro84	Incidence	Thyroid	55	712,000
Rochester thymus	Sh85	Incidence	Thyroid	28	138,000

from which these various data sets derive are described in Annex 4A to this chapter. Table 4-1 provides a summary of the various data sets that the committee used in developing its risk estimates. All of the data sets were provided in grouped form, consisting of the numbers of cases at each cancer site, the number of person-years, and mean dose. These data were stratified by sex and time-related variables, e.g., age at exposure.

The Japanese LSS data consisted of 8714 records, stratified by sex, city, ten exposure groups (based on the kerma at a survivors' location using DS86), and five-year intervals of attained age, age at exposure, and time since exposure. Most analyses used a reduced data set of 3399 records obtained by collapsing over attained age. As outlined in Annex 4B, where the new dosimetry system (DS86) for A-bomb survivors is discussed, survivors exposures are stratified into ten groups and organ doses calculated by multiplying the neutron and gamma kermas for each stratum by city-specific and age-specific body transmission factors.

As the estimate of the neutron component under DS86 is quite small and not very different between the two cities, there is virtually no prospect for estimating the RBE for neutrons from the available data. The committee's analyses are based on an assumed RBE of 20. This is a comparatively large value for high dose rate neutrons relative to high dose and dose rate gamma ray exposures, but is necessarily prudent in view of the degraded neutron spectrum at the survivors locations (see Annex 4B) and the potential low bias in the DS86 estimates of neutron kerma (Ro87). The analysis of the sensitivity of the results to this assumption in Annex 4D shows that the estimated risks for A-bomb survivors change insignificantly for a neutron RBE of 10 vis à vis 20.

Under DS86, the dose response exhibited by A-bomb survivors levels off at high exposure levels. Therefore, to avoid errors in dose estimation at high doses, the records with organ dose equivalents greater than 4 Sv (based on RBE = 20) were eliminated from all analyses. The effect of excluding the observations at dose equivalents greater than 4 Sv is discussed in Annex 4D. Records of cancer mortality at attained ages greater than 75 years were omitted because of the lesser reliability of death certificate information in such cases, as outlined in Annex 4F. Except for breast and thyroid cancers, the committee did not find cancer from tumor registries of sufficient quality to justify model fitting and estimating the incidence of radiogenic cancer. However, the effects of radiation on cancer incidence can be estimated from mortality data (Ho89).

Mortality among A-bomb survivors due to leukemia, cancer of the respiratory tract, cancer of the digestive tract, breast cancer, and as a group, all "other" cancers was analyzed in detail for the lifetime risk projections described below. In making this selection, the committee fitted models for ten sites or groups of sites, with the number of cancer deaths ranging from 2034 to 34. Clearly the larger groups produced more stable estimates of the model parameters. In developing estimates of lifetime risks, it was necessary for the Committee to weigh the consequences of model misspecification in using a single model for all non-leukemia cancers (since some of the sites clearly behaved quite differently across time) against the larger random errors if each of the subsite models were used. If one were not extrapolating in time, these two options would probably give quite similar answers, since larger relative variability of the estimates for the rarer sites would be offset by their lower overall risks. However, it was noticed that the lifetime risk estimates for some sites which had strong time-related modifiers seemed to be unreasonably large, and the reason was inferred to be the instability of the model in regions where the data were too sparse. Faced with this trade-off between precision and possible bias, the Committee opted for a compromise, treating only cancers of the respiratory tract, breast, digestive tract, and thyroid separately.

The only other cohort study that provided data on all cancers was the ankylosing spondylitis series (ASS). Its data set was similarly structured, with two important differences. First, no dose information at the level of the individual was available, so the cohort was fitted as a single exposed group and risk coefficients were derived by dividing the excess estimates by the estimated mean dose, e.g., 1.92 Gy for whole body, 3.83 for bone marrow (Le88). Second, since there were no unexposed comparison subjects, national rates were used to derive an expected number of events in each cell of the cross tabulation. A total of 250 strata by sex and 2 1/2 year intervals of age at exposure and time after exposure were used in these analyses. Because the numbers of cases of cancer were relatively small, and because the risk of colon cancer may be related to ankylosing spondylitis itself, analyses were restricted to leukemia and, as a group, all other cancers except colon cancer.

Statistical Methods

The program AMFIT, described in Annex 4C, was used to fit various exposure-time-response models to these data sets. This program fits a general form of "Poisson regression" model, in which the observed number of events in each cell of the cross-tabulation is treated as a Poisson variate with parameters given by the predicted number of events under the model, the product of the person-years in that cell times the fitted rate. The specific models used can be formally expressed as follows. Let γ_0 denote the age-specific background risk of death due to a specific cancer for an individual at a given age. This background risk will also depend upon the individual's sex and birth cohort (that is year of birth). For a given radiation dose equivalent d in sievert (Sv) we write the individual's age-specific cancer risk $\gamma(d)$ as

$$\gamma(d) = \gamma_0[1 + f(d)g(\beta)]. \tag{4-1}$$

Let $f(d)$ represent a function of the dose d which in the committee's models is always a linear or linear-quadratic function, i.e., $f(d) = \alpha_1 d$ or $f(d) = \alpha_2 d + \alpha_3 d^2$. In general, the excess risk function, $g(\beta)$ will depend upon a number of parameters, for example, sex, attained age, age-at-exposure, and time-since-exposure. One can also write the age-specific risk as an additive risk model

$$\gamma(d) = \gamma_0 + f(d)g(\beta). \tag{4-2}$$

These models give similar results (see Annex 4D) as expected since the function $g(\beta)$ is allowed to depend on age, time, etc. This would not be the case if $g(\beta)$ were restricted to having a constant value other than for sex and age at exposure.

The models were fitted using maximum likelihood, i.e., the values of the unknown parameters which maximize the probability of the observed number of cases (the "likelihood function") are taken as the best estimates, and, where applicable, confidence limits and significance tests are derived from standard large-sample statistical theory.

It was expected that the form of the background term might vary considerably between populations at risk and is not of particular interest in terms of radiation risk. The committee chose not to model it, but rather to estimate the baseline rate nonparametrically by allowing for a large number of multiplicative rate parameters as is often done when fitting hazard models to ungrouped data (Co72, Ka80). Annex 4D provides some comparisons of the results with parametric and stratified background rates. Parametric models for breast cancer are described in Annex 4E.

To summarize, each model considered can be described in terms of the "point" estimates of the various parameters, their respective standard errors and significance tests, and an overall "deviance" for the model as a whole (see Annex 4D). Because of the extreme sparseness of the data, comparison of deviance to its degrees of freedom should not be used as a test of fit of the model. However, differences in deviance between nested alternative models (pairs of models for which all terms in one model are included in the other) have an asymptotic chi squared distribution with degrees of freedom equal to the difference in the degrees of freedom between the models being compared. Therefore, this test can be used to assess the improvement in fit as a result of adding terms to the dose response function. This test was used repeatedly by the committee to minimize potential over-specification of the risk models. Annex 4D provides some comparisons of the many alternative models that were considered.

Approximate confidence limits on parameter estimates can be constructed in the usual way by adding and subtracting the standard error times 1.65 (for 90% confidence) or 1.96 (for 95% confidence). However, in cases where the committee had reason to believe that the use of a normal distribution to estimate confidence limits is not valid, it reports "likelihood based" limits found by iteratively searching for the parameter values which led to a corresponding increase in the deviance (Co74).

The Committee's Preferred Risk Models

The committee's models for each site are discussed in the respective sections on site specific cancers in Chapter 5. Only a brief summary and the equations for dose response are presented here.

Leukemia (ICD 204-207): The final model for leukemia is a relative risk model with terms for dose, dose squared, age at exposure, time after exposure, and interaction effects. A minimum latency of 2 years is assumed. There is a distinct difference between the risks exhibited by individuals exposed before age 20 and those exposed later in life. Within these two groups, there does not appear to be any effect of age at exposure but simply a different time pattern within each group. A simple step function with two steps fit both groups rather well. As indicated in Chapter 5, splines can be used to smooth these transitions when desired (e.g., in the calculation of probability of causation).

The leukemia model mathematically is as follows (see the general equation 4.1):

$$f(d) = \alpha_2 d + \alpha_3 d^2$$
$$g(\beta) = \begin{cases} \exp[\beta_1 I(T \leq 15) + \beta_2 I(15 < T \leq 25)] \text{ if } E \leq 20 \\ \exp[\beta_3 I(T \leq 25) + \beta_4 I(25 < T \leq 30)] \text{ if } E > 20, \end{cases} \quad (4\text{-}3)$$

where the indicator function $I(\mathrm{T} \leq 15)$ is defined as 1 if $T \leq 15$ and 0 if $T > 15$, T is years after exposure, and E is age at exposure. The estimated parameter values and their standard errors, in parentheses, are:

$$\alpha_2 = 0.243(0.291), \alpha_3 = 0.271(0.314),$$
$$\beta_1 = 4.885(1.349), \beta_2 = 2.380(1.311), \beta_3 = 2.367(1.121),$$
$$\beta_4 = 1.638(1.321).$$

The standard errors for the dose effect coefficients were estimated by means of the likelihood method mentioned above and are both imprecise and highly skewed (see Annex 4F). The Monte Carlo analysis of the statistical uncertainty in the risk estimates for leukemia, described below in the section on uncertainty in point estimates, provides a better measure of the precision.

Cancers other than leukemia: In fitting the data for cancers other than breast cancer and leukemia, a 10-year minimum latency was assumed; this was done simply by excluding all the observations (cases and person-years) less than 10 years after exposure. As for leukemia, similar fits could be obtained with either additive or relative risk models, but with different modifying effects (see Annex 4D). As was the case for leukemia, relative risk models were more parsimonious or required weaker modifiers.

The committee subdivided solid tumors into cancers of the respiratory tract, breast, digestive tract, and other sites as described in the 8th revision of the International Classification of Diseases (ICD) (ICD67).

Respiratory cancer (ICD 160-163): The committee's preferred model is as follows:

$$f(d) = \alpha_1 d$$
$$g(\beta) = \exp[\beta_1 \ln(T/20) + \beta_2 I(S)], \quad (4\text{-}4)$$

where T = years after exposure and $I(S) = 1$ if female, 0 if male with $\alpha_1 = 0.636(0.291)$, $\beta_1 = -1.437(0.910)$, $\beta_2 = 0.711(0.610)$.

Under the committee's model, the relative risk for this site decreases with time after exposure. The coefficient for time after exposure, -1.437, means that the relative risk will decrease by a factor of about 5 over the period of 10 to 30 years post-exposure. The committee notes that few data are available, as yet, on respiratory cancer among those exposed as children. Finally, the relative risk is 2 times higher for females (owing to their much lower baseline rates) than for males, although the observed excess risks are similar.

The fit of a constant relative risk model to the data on respiratory cancer is not statistically different from that for the committee's preferred model. When testing departures from a constant relative risk model, the addition of a parameter for time after exposure resulted in the greatest improvement in describing the data. This finding is consistent with the decreasing relative risk observed in the Ankylosing Spondylitis study (Da87) which influenced the committee's choice of parameters. While the inclusion of a parameter for sex did not improve the model's fit to the data significantly, there was some improvement, and the committee felt that it was appropriate to include a parameter for sex. Although it had been used in other risk models for respiratory cancer, there was no improvement whatever when a term for age-at-exposure was added to the regression model. When in fact such a term was estimated, its value was sufficiently close to zero as to have no influence on the estimated risk.

Breast cancer (ICD 174): The breast cancer models are based on a parallel analysis of several cohorts. The important modifying factors found were age at exposure and time after exposure. The dependence of risk on age at exposure is complex, doubtless being heavily influenced by the woman's hormonal and reproductive status at that time. Lacking any data on these biological variables, the committee found that the best fit was obtained with the use of an indicator variable for age-at-exposure less than 16, together with additional indicator or trend variables depending on the data set. Both incidence and mortality models were developed. Although these differ, the highest risks are seen in women under 15-20 years of age at exposure. Risks are very low in women exposed at ages greater than 40. This suggests that risks decrease with age at exposure. Finally, risks decrease with time after exposure in all age groups. These issues are discussed in some detail in Annex 4E and the section on breast cancer, in Chapter 5.

The model for breast cancer age specific mortality (female only) is

$$f(d) = \alpha_1 d$$
$$g(\beta) = \begin{cases} \exp[\beta_1 + \beta_2 \ln(T/20) + \beta_3 \ln^2(T/20)] & \text{if } E \leq 15 \\ \exp[\beta_2 \ln(T/20) + \beta_3 \ln^2(T/20) + \beta_4(E-15)] & \text{if } E > 15, \end{cases} \quad (4\text{-}5)$$

where E is age at exposure and T is years after exposure with $\alpha_1 =$

1.220(0.610), $\beta_1 = 1.385(0.554)$, $\beta_2 = -0.104\ (0.804)$, $\beta_3 = -2.212\ (1.376)$, $\beta_4 = -0.0628\ (0.0321)$.

Digestive cancer (ICD 150-159): The most significant aspect of the LSS data is the greatly increased risk (factor of 7) for those exposed under the age of 30. Although the committee has no explanation for this observation, the LSS data strongly support this effect. There is no evidence of a significant change in the relative risk with time after exposure.

The committee's preferred model is:

$$f(d) = \alpha_1 d$$
$$g(\beta) = \exp[\beta_1 I(S) + \sigma_E] \tag{4-6}$$

where $I(S)$ equals 1 for females and 0 for males and

$$\sigma_E = \begin{cases} 0 \text{ if } E \leq 25 \\ \beta_2(E-25) \text{ if } 25 < E \leq 35 \\ 10\beta_2 \text{ if } E > 35 \end{cases}$$

with E = age at exposure. The estimated parameter values are $\alpha_1 = 0.809(9.327)$, $\beta_1 = 0.553(0.462)$, $\beta_2 = -0.198(0.0628)$.

Other cancers (ICD 140-209 less those listed above): This group of miscellaneous cancers contributes significantly to the total radiation-induced cancer burden. Finer subdivision of the group did not, however, provide sufficient cases for modeling individual substituent sites. When attempted, the models were quite unstable, resulting in risk estimates for which there was little confidence. The general group of "other cancers" was reasonably fit by a simple model with only a negative linear effect by age-at-exposure at ages greater than 10. There was no evidence of either an effect by sex or by time after exposure.

The preferred model is

$$f(d) = \alpha_1 d$$
$$g(\beta) = 1 \text{ if } E \leq 10 \text{ and } \exp\,[\beta_1(E-10)] \text{ if } E > 10, \tag{4-7}$$

where E = age at exposure and $\alpha_1 = 1.220(0.519)$, $\beta_1 = -0.0464(0.0234)$.

Nonleukemia: For risk estimation, the committee simply chose to sum the risks of the components of the nonleukemia cancer group (i.e. respiratory cancer, digestive cancer, etc.). Alternatively, modeling the risk for all nonleukemia cancers directly yielded models which are linear in dose with additional variables for sex and time. These models provided a significantly poorer fit than other reasonable models and also project greater estimated risks (see Annex 4D).

Analysis of the ankylosing spondylitis study (ASS) data for all cancers other than leukemia and colon gave a somewhat different picture. Here the fit was significantly improved by the addition of linear and quadratic terms for time after exposure, so that the risk essentially decreases to zero after about 20 years post-exposure. Part of the difference between the LSS and ASS data may be due to differences in the proportions of cancers of different sites. The most common cancers in the ASS series are lung cancer and breast cancer, the frequency of which declined with time after exposure in both data sets. On the other hand, cancers of the digestive system were very common in the LSS and showed no variation with time after exposure.

RISK ASSESSMENT

Point Estimates of Lifetime Risk

Methods: The committee used standard lifetable methods as outlined in Chapter 1. Vital Statistics of the United States 1980 was used as the source of baseline data on cancer mortality (PHS84). The fitted risk models described above were applied to a stationary population having United States death rates for 1979-81 (NCHS85) and lifetime risks calculated for the following patterns of exposure.

- Instantaneous exposure causing a dose equivalent to all body organs of 0.1Sv (10 rad of low-LET radiation), varying the age at exposure by 10-year intervals and taking the population-weighted average of the resulting estimates, weighted by the probability of surviving to a specified age in an exposed stationary population.
- Continuous lifetime exposure causing a dose equivalent in all body organs of 1 mSv (0.1 rad of low-LET radiation) per year.
- Continuous exposure from age 18 to age 65 causing a dose equivalent to all body organs of 10 mSv (1 rad of low-LET radiation) per year.

Application to low dose rates: Since the risk models were derived primarily from data on acute exposures (a single instantaneous exposure in the case of the LSS data, or fractionated but still high dose rate exposures in the case of most of the medical exposures), the application of these models to continuous low dose-rate exposures requires consideration of the dose rate effectiveness factor (DREF), as discussed in Chapter 1. For linear-quadratic models, there is an implicit dose-rate effect, since the quadratic contribution vanishes at low doses and, presumably, low dose-rates leaving only the linear term which is generally taken to reflect one-hit kinetics. The magnitude of this reduction is expressed by the DREF values. For the leukemia data, a linear extrapolation indicates that the lifetime risks per unit bone marrow dose may be half as large for continuous low dose rate as for instantaneous high dose rate exposures. For most other cancers in the

TABLE 4-2 Excess Cancer Mortality Estimates and Their Statistical Uncertainty—Lifetime Risks per 100,000 Exposed Persons[a]

	Male			Female		
	Total	Nonleukemia[b]	Leukemia[c]	Total	Nonleukemia	Leukemia
Single exposure to 0.1 Sv (10 rem)	770	660	110	810	730	80
90% confidence limits[d]	540–1,240	420–1,040	50–280	630–1,160	550–1,020	30–190
Normal expectation	20,510	19,750	760	16,150	15,540	610
% of normal	3.7	3.3	15	5	4.7	14
Total years of life lost	12,000			14,500		
Average years of life lost per excess death	16			18		
Continuous lifetime exposure[e] to 1 mSv/y (0.1 rem/y)	520	450	70	600	540	60
90% confidence limits[d]	410–980	320–830	20–260	500–930	430–800	20–200
Normal expectation	20,560	19,760	790	17,520	16,850	660
% of normal	2.5	2.3	8.9	3.4	3.2	8.6
Total years of life lost	8,100			10,500		
Average years of life lost per excess death	16			18		

Continuous exposure[e] to 0.01 Sv/y (1 rem/y) from age 18 until age 65	2,880	2,480	400	3,070	2,760	310
90% confidence limits[d]	2,150–5,460	1,670–4,560	130–1,160	2,510–4,580	2,120–4,190	110–910
Normal expectation	20,910	20,140	780	17,710	17,050	650
% of normal	14	12	52	17	16	48
Total years of life lost	42,200			51,600		
Average years of life lost per excess death	15			17		

[a] Based on an equal dose to all organs and the committee's preferred risk models—estimates rounded to nearest 10.
[b] Sum of respiratory, breast, digestive, and other cancers.
[c] Estimates for leukemia contain an implicit dose rate reduction factor.
[d] Additional sources of uncertainty are discussed in Annex 4F.
[e] A dose rate reduction factor has not been applied to the risk estimates for solid cancers.

LSS, the quadratic contribution is nearly zero, and the estimated DREFs are near unity. Nevertheless, the committee judged that some account should be taken of dose rate effects and in Chapter 1 suggests a range of dose rate reduction factors that may be applicable. It must be emphasized, however, that such reductions should be applied only to the non-leukemia risks, as the leukemia risks already contain an implicit DREF owing to the use of the linear-quadratic model. For this reason, the tables which follow report excess risks for leukemia and all other cancers separately even though the quadratic term for leukemia is numerically negligible at 0.1 Sv. Faced with a similar situation, the BEIR III Committee chose to estimate a DREF from the leukemia data and apply it to the nonleukemia data as a fixed constant. After considerable discussion, this committee concluded that it could not justify assuming the same dose-response model for all cancer sites and, therefore, fitted separate dose-response models, with no DREF.

The method of lifetime excess risk estimation used in this report differs slightly from that used in BEIR III (NRC80) and UNSCEAR (UN77,UN88) reports. In this report, separate lifetime risks are estimated for exposed and unexposed populations, and the excess risk is simply the difference between the two lifetime risk estimates. Competing risks due to other radiogenic cancers are included in the population decrement. In the other reports, the differences in age-specific rates between exposed and unexposed populations were multiplied by the survival probabilities for an *exposed population* and summed. Because an exposed population will have smaller survival probabilities, the method used here produces lower excess risk estimates, which more correctly reflect the difference in the lifetime risk of cancer mortality. Vaeth and Pierce (Va89) have shown that the ratio of the two estimates is approximately the lifetime probability of not dying of cancer, or in this case, about 0.8.

Results: Table 4-2 summarizes the estimates of lifetime risks for leukemia and all other cancers resulting from two continuous exposure situations (lifetime and ages 18-65) and a population-weighted instantaneous exposure to persons of all ages. These results were obtained using the committee's preferred relative risk models for each site and a lifetable analysis that accounts for all competing risks including those due to radiation-induced cancer. Stratification of these results by age at exposure and by cancer site, for the case of instantaneous exposure, is provided in Table 4-3. Results from alternative risk models are considered in Annex 4D to this chapter.

Table 4-4 provides a comparison of the risk projections under the preferred relative risk models from this report and the relative and absolute risk models in the BEIR III report. Overall, the risk estimates in this report are consistently higher than in the BEIR III report. This is due, in part,

TABLE 4-3 Cancer Excess Mortality by Age at Exposure and Site for 100,000 Persons of Each Age Exposed to 0.1 Sv (10 rem)

MALES

Age at Exposure	Total	Leukemia	Nonleukemia	Respiratory	Digestive	Other
5	1,276	111	1,165	17	361	787
15	1,144	109	1,035	54	369	612
25	921	36	885	124	389	372
35	566	62	504	243	28	233
45	600	108	492	353	22	117
55	616	166	450	393	15	42
65	481	191	290	272	11	7
75	258	165	93	90	5	—
85	110	96	14	17	—	—
Average[a]	770	110	660	190	170	300

FEMALES

Age at Exposure	Total	Leukemia	Nonleukemia	Breast	Respiratory	Digestive	Other
5	1,532	75	1,457	129	48	655	625
15	1,566	72	1,494	295	70	653	476
25	1,178	29	1,149	52	125	679	293
35	557	46	511	43	208	73	187
45	541	73	468	20	277	71	100
55	505	117	388	6	273	64	45
65	386	146	240	—	172	52	16
75	227	127	100	—	72	26	3
85	90	73	17	—	15	4	—
Average	810	80	730	70	150	290	220

[a] Averages are weighted for the age distribution in a stationary population having U.S. mortality rates and have been rounded to the nearest 10. See also footnotes to Table 4-2. 90% confidence interval for these risk estimates are listed in Annex 4D, Table 4D-4.

to this Committee's use of a linear dose response model for cancers other than leukemia rather than a linear quadratic one with an implicit DREF of nearly 2.5, as was the case in the BEIR III Committee's report. However, there are several other reasons for the differences between the two sets of results. These include the new dosimetry for the LSS data (Annex 4B), the additional years of follow-up, and the changes in the structure of the fitted models. In their work on the comparison of T65D and DS86 risk estimates using linear dose response models, Preston and Pierce (Pr88) concluded that while the changes in leukemia risk estimates were largely attributable to changes in dose estimates, the other two factors were more important for solid cancers; so that only 35-40% of the increase in their risk estimates was due to the use of the DS86 dose estimates.

TABLE 4-4 Comparison of Lifetime Excess Cancer Risk Estimates from the BEIR III and BEIR V Reports

	Continuous Lifetime Exposure, 1 mGy/y (deaths per 100,000)		Instantaneous Exposure, 0.1 Gy (deaths per 100,000)	
	Males	Females	Males	Females
Leukemia				
BEIR III[a]	15.9	12.1	27.4	18.6
BEIR V	70	60	110	80
Ratio BEIR V/ BEIR III	4.4	5.0	4.0	4.3
Nonleukemia				
BEIR III				
Additive risk model	24.6	42.4	42.1	65.2
Relative risk model	92.9	118.5	192	213
BEIR V	450	540	660	730
Ratio BEIR V/ BEIR III	4.8–18.3	4.6–12.7	3.4–15.7	3.4–11.2

[a] Based on Table V-16, page 203, and Table V-19, page 206 ($\overline{LQ-L}$ model for nonleukemia) (NAS80).

The major differences between the two sets of estimates in Table 4-4 are for the BEIR III Committee's additive risk models. It is the opinion of this committee that the assumption of a constant additive excess risk is no longer tenable in the face of the data now available and that the risk estimates from this model provided in the BEIR III report are therefore too low. The estimates presented in this report are also higher than those based on a simple additive risk model in the latest UNSCEAR report (UN88) but are not quite as high as those based on the simple multiplicative risk model in that report.

UNCERTAINTY IN POINT ESTIMATES OF LIFETIME RISK

The total uncertainty in the Committee's risk models is discussed in Annex 4F. In this section, the discussion is largely limited to the statistical uncertainty in the risk estimates made with the Committee's preferred models. Lifetime risk projections are subject to three types of uncertainty. The first is simply random error owing to sampling variation in the fitted coefficients of the final models; this is thought to be the largest component of uncertainty and is expressed in terms of confidence intervals on the fitted model parameters and the estimated lifetime risks. Second, there is

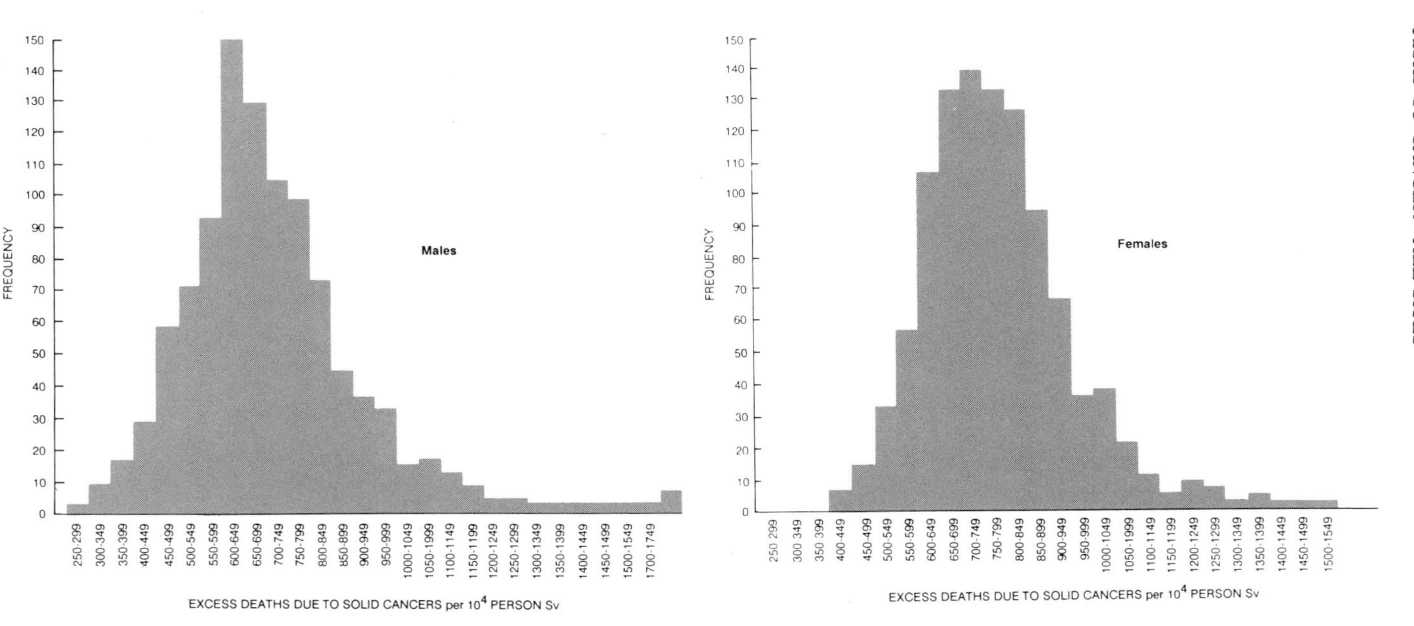

FIGURE 4-1 Excess mortality due to solid cancers per 10^4 person Sv (million person rem). Results of 1,000 Monte Carlo simulations and lifetable analyses of the excess mortality from all solid cancers following an acute total body dose of 0.1 Sv. The populations at risk are 100,000 males and 100,000 females. The Committee's preferred models yield a point estimate for males of 660 excess deaths; for females, 730. In 50% of the trials, the excess mortality for males was between 590 and 820 deaths; for females, between 670 and 860 deaths.

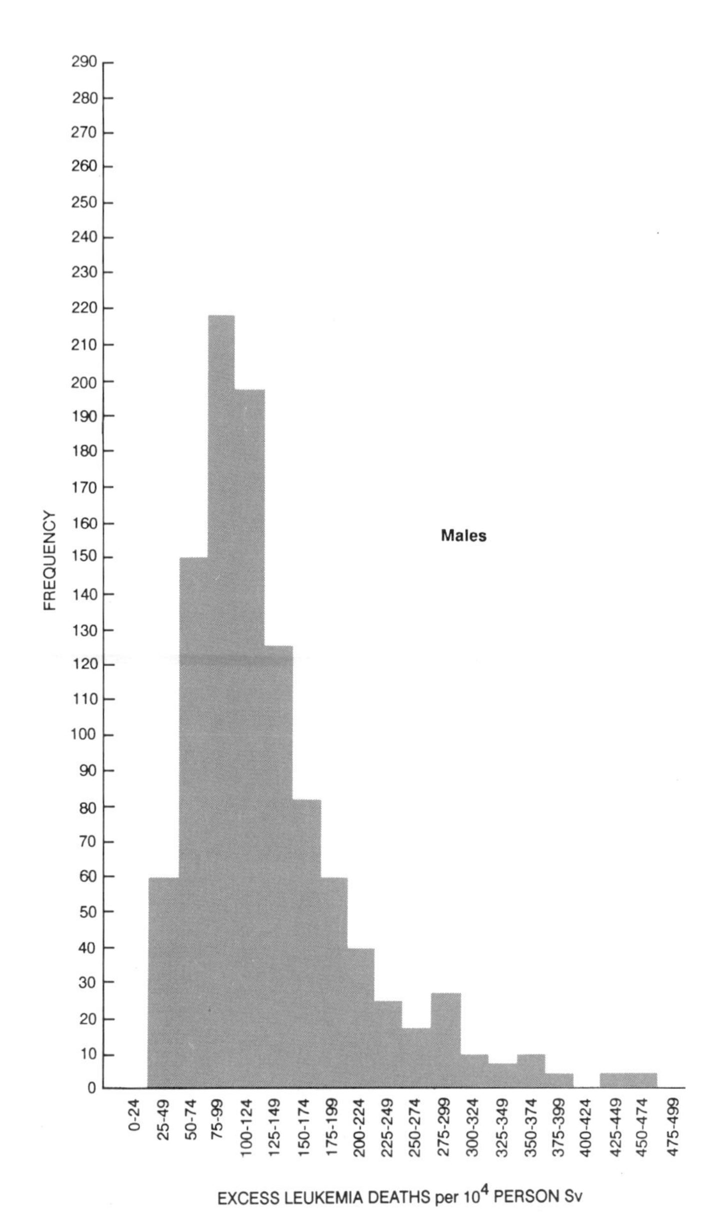

FIGURE 4-2 Excess leukemia fatalities per 10^4 person Sv (million person rem). Results of 1,000 Monte Carlo simulations and lifetable analyses of the excess mortality from leukemia following an acute total body dose of 0.1 Sv. The populations at risk are 100,000 males and 100,000 females. The point estimate for males is 111 excess deaths; for females, 82. In 50 percent of the trials the excess mortality for males was between 60 and 135 deaths; for females, between 55 and 115 deaths.

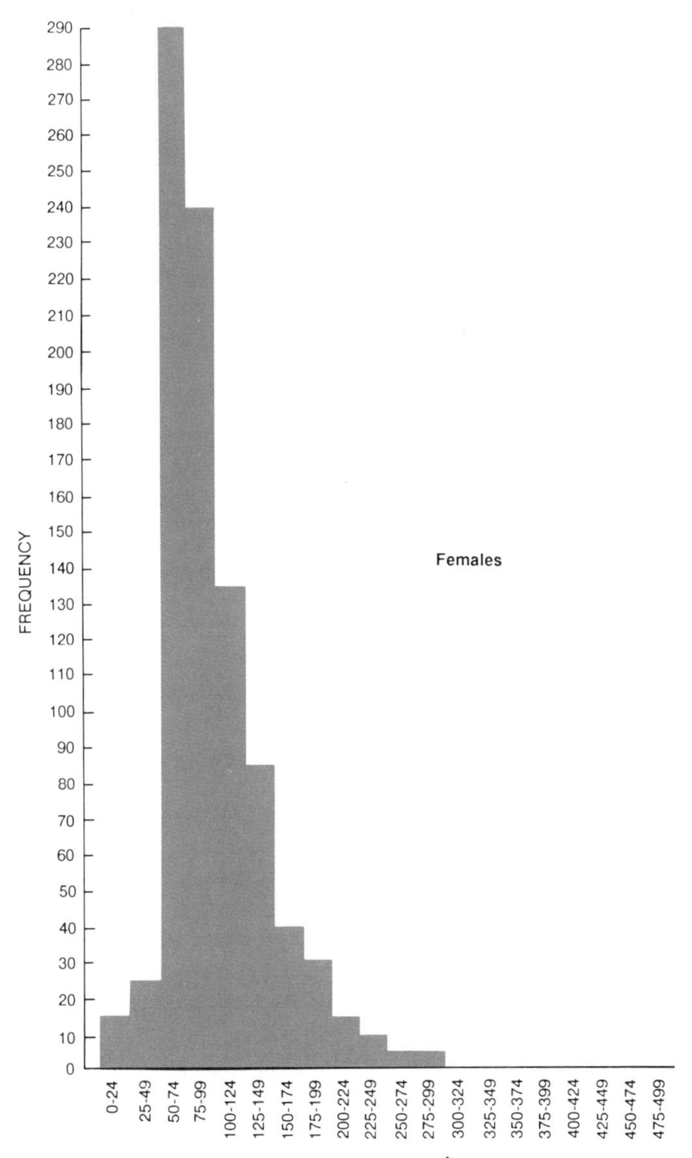
290
280
270
260
250
240
230
220
210
200
190
180
170
160
150
140
130
120
110
100
90
80
70
60
50
40
30
20
10
0
FREQUENCY
Females
0-24
25-49
50-74
75-99
100-124
125-149
150-174
175-199
200-224
225-249
250-274
275-299
300-324
325-349
350-374
375-399
400-424
425-449
450-474
475-499
EXCESS LEUKEMIA DEATHS per 10^4 PERSON Sv

uncertainty as to the correct form of the exposure-time-response model, since the true model could be misspecified in a number of ways. It is more difficult to assess this component of the uncertainty, but a sense of its importance can be obtained by considering the range of lifetime risks resulting from alternative well-fitting models as discussed in Annex 4D and 4F. In addition, there are various potential biases in the data themselves; while these cannot be quantified precisely, they are discussed in Annex 4F along with the Committee's judgment concerning their magnitude.

Since the lifetime risk is a complex function of the parameters of the fitted models, it is not a simple matter to translate the standard errors in risk coefficients into uncertainties in lifetime risk. This overall uncertainty depends not just on the uncertainty in the coefficient of dose, but also on the uncertainty in the coefficients of the modifying factors and their correlations. Furthermore, the distributions of the estimates of the coefficients are often quite skewed, leading to skewness in the resulting distribution of lifetime risks. For these reasons, the Committee undertook an uncertainty analysis by means of Monte Carlo simulation. In this approach, parameter vectors for each cancer site were randomly sampled from multivariate normal distributions with means and covariant matrices given by their maximum likelihood estimates. Any components that showed marked skewness were adjusted by multiplying the deviations of the sampled value from their means by the ratio of the likelihood-based to asymptotic confidence intervals for the corresponding 90% upper or lower tail. Lifetable calculations of risk were repeated for each randomly selected set of parameters, and in this way a distribution of lifetime risk estimates was produced.

Figure 4-1 presents results for each sex based on 1,000 Monte Carlo simulations and lifetable analyses of the excess mortality risk for all solid cancers following a 0.1 Sv acute total body dose to a stationary population. Figure 4-2 presents the same results for leukemia. These histograms give a good idea of the statistical uncertainty in the Committee's risk models.

Table 4-2 summarizes the resulting 90% confidence limits due to statistical uncertainty on the lifetime risk estimates for each of three exposure patterns. The intervals are wide indicating sparseness of data. For the most part, risk estimates derived from the alternative models described in Annex 4D are within these confidence intervals. Not included in Table 4-2 are several additional sources of uncertainty external to model parameters that are discussed in Annex 4F. The effect of these external sources of uncertainty on the risk estimates is not as well quantified as the uncertainty due to sampling variation shown in Figures 4-1 and 4-2; however, they probably contribute comparable uncertainty. The Committee's analysis in Annex 4F indicates these external factors increase the confidence intervals due to sampling variation in Table 4-2 by about a factor of 1.4.

Finally, it must be recognized that derivation of risk estimates for low doses and dose rates through the use of any type of risk model involves assumptions that remain to be validated. At low doses, a model dependent interpolation is involved between the spontaneous incidence and the incidence at the lowest doses for which data are available. Since the committee's preferred risk models are a linear function of dose, little uncertainty should be introduced on this account, but departure from linearity cannot be excluded at low doses below the range of observation. Such departures could be in the direction of either an increased or decreased risk. Moreover, epidemiologic data cannot rigorously exclude the existence of a threshold in the millisievert dose range. Thus the possibility that there may be no risks from exposures comparable to external natural background radiation cannot be ruled out. At such low doses and dose rates, it must be acknowledged that the lower limit of the range of uncertainty in the risk estimates extends to zero.

REFERENCES

Co72 Cox, D. R. 1972. Regresson models and lifetables (with discussion). J. R. Stat. Soc. B 34:187-200.

Co74 Cox, D. R., and D. V. Hinkley. 1974. Theoretical Statistics. London: Chapman and Hall.

Da87 Darby, S. C., R. Doll, S. K. Gill, and P. G. Smith. 1987. Long-term mortality after a single treatment course with X-rays in patients treated for ankylosing spondylitis. Br. J. Cancer 55:179-190.

Ho89 Hoel, D. G., and G. E. Dinse, 1989. Using mortality data to estimate radiation effects on breast cancer incidence. In press. Environ. Health Perspectives.

Hr89 Hrubec, Z., J. Boice, R. Monson, and M. Rosenstein. 1989. Breast cancer after multiple chest fluoroscopies: Second follow-up of Massachusetts women with tuberculosis. Cancer Res. 49:229-234.

ICD67 Eighth Revision International Classification of Diseases, Vol. 1. Public Health Service Publication No. 1639, Washington, D.C. Government Printing Office.

Ka80 Kalbfleisch J., and R. Prentice 1980. The Statistical Analysis of Failure Time Data, New York: John Wiley & Sons.

Le88 Lewis, C. A., P. G. Smith, I. M. Stratton, S. C. Darby, and R. Doll. 1988. Estimated Radiation Doses to Different Organs Among Patients Treated for Ankylosing Spondylitis with a Single Course of X-rays. Br. J. Radiol. 61:212-220.

Mi89 Miller, A., P. Dinner, G. Howe, G. Sherman, J. Lindsay, M. Yaffe, H. Risch, and D. Preston. 1989. Breast cancer mortality following irradiation in a cohort of Canadian Tuberculosis Patients. New England Journal Medicine (in press).

NCHS85 U.S. Decennial Life Tables for 1979-1981, Vol 1 No. 1, 1985. DHHS publication (PHS) 85-1150-1, Hyattsville, Md.: National Center for Health Statistics.

NRC80 National Research Council. 1980. Committee on the Biological Effects of Ionizing Radiations. The Effects on Populations of Exposure to Low Levels of Ionizing Radiation (BEIR III). Washington, D.C.: National Academy Press. Pp. 524.

PHS84 Vital Statistics of the United States 1980, Public Health Service, Hyattsville, Md.: National Center for Health Statistics.

Pr87 Preston, D. L., H. Kato, K. J. Kopecky, and S. Fujita. 1987. Life Span Study Report 10. Part l. Cancer mortality among a-bomb survivors in Hiroshima and Nagasaki, 1950-82. Radiat. Res. 111:151-178.

Pr88 Preston, D., and D. Pierce. 1988. The effect of changes in dosimetry on cancer mortality risk estimates in atomic bomb survivors. Radiat. Res. 114:437-466.

Ro84 Ron, E., and B. Modan 1984. Thyroid and other neoplasms following childhood scalp irradiation Pp. 139-151 in Radiation Cancergenesic: Epidemiology and Biological Significance. J. D. Boice and J. F. Fraumeni, Jr., eds. New York: Raven Press.

Sh85 Shore, R. E., E. Woodard N. Hildreth, P. Dvoretsky, L. Hempelmann, and B. Pasternack. 1985. Thyroid tumors following thymus irradiation. J. Natl. Cancer Inst. 74:1177-1184.

Sh87 Shimizu, Y., H. Kato, W. J. Schull, D. L. Preston, S. Fujita, and D. A. Pierce. 1987. Life Span Study Report 11, Part 1. Comparison of Risk Coefficients for Site-Specific Cancer Mortality Based on the DS86 and T65DR Shielded Kerma and Organ Doses. Technical Report RERF TR 12-87. Hiroshima Radiation Effects Research Foundation.

Sh86 Shore, R., N. Hildreth, E. Woodward, P. Dvoretsky, L. Hempelmann, and B. Pasternack. 1986. Breast cancer among women given x-ray therapy for acute postpartum mastitis. J. Natl. Cancer Inst. 77:689-696.

To87 Tokunaga, M., C. Land, T. Yamamoto, M. Asano, S. Tokuoka, H. Ezaki, and I. Nishimori. 1987. Incidence of female breast cancer among atomic bomb survivors, Hiroshima and Nagasaki, 1950-1985 Radiat. Res. 112:243-272.

UN77 United Nations Scientific Committee on the Effects of Atomic Radiation (UNSCEAR). 1977. Sources and Effects of Ionizing Radiation. Report E. 77. IX. 1. New York: United Nations. Pp. 725.

UN88 United Nations Scientific Committee on the Effects of Atomic Radiation. 1988. Sources, Effects, and Risks of Ionizing Radiations. U.N. Publication E.88.IX.7.647. New York: United Nations. Pp. 647.

Va89 Vaeth, M., and D. Pierce. 1989. Calculating excess lifetime risk in relative risk models. Environ. Health Persect. (in press).

ANNEX 4A: SUMMARY OF MAJOR EPIDEMIOLOGIC STUDIES USED IN BEIR V

The Life Span Study of A-Bomb Survivors

Cohort Source and Exposure

A mortality study (Sh87) of 120,321 individuals resident in Hiroshima or Nagasaki in 1950 make up the cohort. Among these there are 91,228 individuals who were exposed at the time of the bombing. This cohort continues to be followed up with deaths routinely determined through the Japanese household registries where ascertainment is essentially complete. Mortality data for the cohort has been completed for the period 1950-1985.

As discussed in Annex 4B, new dose estimates are now available for the A-bomb survivors of Hiroshima and Nagasaki. The main difference between the old and new dosimetry is that the estimated level of neutron kerma has been decreased by approximately an order of magnitude in Hiroshima and by a factor of two in Nagasaki. The result is that the neutrons are no longer a significant component of the dose in either of the two cities.

Mean organ doses have been calculated for twelve organs. For most high dose survivors, these doses are determined on an individual basis which includes a consideration of local shielding and orientation. The number of survivors in the life span study with new dose estimates, stratified by the kerma at the location where they were exposed, is as follows:

Kerma (Gray)	0	0.01-0.05	0.06-0.09	0.10-0.99	1.00-1.99	2.00+
Cohort size	34,272	19,192	4,129	15,346	1,946	1,106

Follow-up

The subcohort of approximately 76,000 subjects for which there are new dose estimates represents over two million person-years-at-risk. A total of 5,936 cancer deaths have been observed in the subcohort through 1985 The number of deaths due to cancer at sites showing a statistically significant excess are listed below.

Number of Cancer Deaths (Sh87)

leukemia	202	colon	232	ovary	82
esophagus	176	multiple myeloma	36	bladder	133
stomach	2007	female breast	155	lung	638

Incidence data are also being gathered and studied, the most prominent being data on breast cancer (To87).

Strengths and Limitations

This is the most important single cohort for estimating cancer risk from gamma radiation. The population is large and there is a wide range of doses. With these data it is possible to make determinations of dose-response and the effects of modifying factors such as age and time on the major cancer sites. The data are, however, limited at the high doses by the uncertainty in the dose estimates for highly exposed individuals. With this in mind, analyses in this report are carried out using only individuals with estimated doses to internal organs of less than 4 Gy.

The cohort of Japanese survivors is not a normal Japanese population, apart from their radiation exposures. Many young adult males were not present at the time of the bombing, but away in military service. It must be presumed that those who were still in the cities included persons whose physical condition barred them from active service. Children of both sexes and the elderly perished, in consequence of the bombing, at a greater rate than did young adults. While exact location and shielding situations played an important role in determining who survived and who did not, the possibility must be allowed for that the survivors were, in some sense, hardier than those who did not.

It has been hypothesized by Stewart (St84, St85, St88) that increased deaths, due to infections from suppressed immune function, resulted in a dose related survival-of-the-fittest plus permanent bone marrow damage at higher doses. The dose response for noncancer deaths does, in fact, have a U shaped behavior, as described by Stewart. However, there does not seem to be evidence of infectious disease; instead the lower mortality rates in the moderately exposed individuals result from lower rates of death from a variety of causes (Da85). It does not appear, at this point, that these differences in mortality contribute in any substantial way to cancer mortality risk estimates based upon data from this cohort.

Ankylosing Spondylitis

Source of Cohort and Exposure

The cohort consists of 14,106 patients treated with radiotherapy to the spine for ankylosing spondylitis in 87 centers in the United Kingdom between 1935 and 1954. Of this cohort, 7,431 individuals contributed an average of only 3.5 years of follow-up before they received a second course of radiotherapy and were then excluded from the study. Because the radiotherapy treatment was aimed at the spine, a large fraction of the body received substantial doses of radiation.

Dosimetry

Individual dose estimates are not available for the whole cohort, but radiotherapy records have been extracted for a random sample of 1 in 15 and Monte Carlo methods used to estimate individual organ doses for 30 organs or regions of the body and 12 bone marrow sites (Le88). Comparison of the mean marrow dose with earlier estimates based on phantom dosimetry are in good agreement.

Follow-up

The mortality of the cohort has been monitored using searches in the National Health Service central registry for death certificates. Mortality has been reported to the end of 1982, at which point 727 cancer deaths and 104,146 person-years of follow-up had been observed (Da88). Results have been reported for a number of sites, but colon cancer has been excluded because of its suspected association with ankylosing spondylitis.

Strengths and Limitations

This is a large irradiated series with a substantial number of organs, including the bone marrow, receiving fairly high doses. The underlying population is likely to be genetically similar to that of the U.S. but the applicability to a general population of the results from such patients, who have a condition that affects several causes of mortality, remains an issue. Comparisons of the cohort to date have mainly been made with general population rates, though it should be noted that a follow-up of ankylosing spondylitis patients not treated with radiotherapy has indicated that the comparison to the general population is not likely to be biased by the presence of the disease (Sm77). Doses were largely unfractionated, and no individual doses for all cohort members are available. Only cancer mortality, and not cancer incidence data are available for the cohort.

Study of Women Treated for Cancer of the Cervix

Cohort Source and Exposure

The cohort consists of approximately 150,000 women treated for cancer of the uterine cervix who were either registered in one of 19 population-based cancer registries or treated at one of 20 clinics in a number of countries. A substantial proportion of these women (approximately 70%) were treated with radium implants or external radiotherapy, which resulted in substantial radiation doses to a number of organs close to the cervix and moderate doses to organs located more distantly in the body.

Dosimetry

The original radiotherapy treatment records of the 4,188 women in the cohort who subsequently developed a second primary cancer were used to estimate individual organ doses for the organs of interest. Similar estimates were made for a control series consisting of 6,880 women who did not develop a second primary.

Follow-up

Follow-up of the cohort was carried out using the population-based cancer registries to identify second primaries occurring in the cohort. As indicated above, a total of 4,188 such cases have been identified, and their prior radiation experience has been compared to that of the 6,880 age-matched controls in order to estimate the relative risk for the various second cancers. The results of this analysis have been reported (Bo88).

Strengths and Limitations

This is a very large follow-up study with a substantial number of cancers for a number of organs of interest. Substantial doses were received by a number of organs, and moderate doses by a number of others, and these doses have been estimated with a good deal of care on an individual patient basis. The choice of a case-control analysis in order to make such dose estimates computationally feasible, however, means that absolute risk estimates can be made only by imputation. The most serious limitation of this study arises from the fact that the subjects had all developed cancer of the cervix, with its many associated risk factors, particularly those relating to socio-economic status. Although an internal control group has been used in the analysis, extrapolation of the results to the general population must be made with some caution.

Canadian Fluoroscopy Study

This cohort consists of 31,710 women, first treated for tuberculosis in Canadian sanatoria between 1930 and 1952. A substantial proportion of these women were exposed to multiple fluoroscopies in conjunction with artificial pneumothorax treatment for tuberculosis, and 8,380 (26.4%) received breast tissue doses of 10 rads or more. The maximum dose received was over 2000 rads. That part of the cohort which was treated in the province of Nova Scotia was generally treated in the anterior-posterior (AP) position, in contrast to the more usual PA orientation in the rest of Canada, and this sub-cohort was therefore exposed to particularly high doses to the breast. A similar number of men have also been included in this cohort, but to date, no analyses have been reported for the males.

Dosimetry

Individual breast tissue doses have been estimated for the 31,710 women. These estimates are based on a count of the number of fluoroscopies recorded in the medical records, interviews with a number of physicians using fluoroscopy during the relevant time period, and on phantom measurements and Monte Carlo simulations (Sh78). Although counts are based on individual records, the dose per fluoroscopy is an average figure which is a function only of the province where most exposures were received (Nova Scotia vs. the others), and the year the exposure was received (after 1945 or before). Doses to other organs have not been reported for this cohort.

Follow-up

The cohort has been monitored for mortality between 1950 and 1980 using computerized record linkage to the Canadian National Mortality Data Base. By 1980, 482 breast cancer deaths and 867,541 women years of follow-up had been observed. Analyses of these results have been reported (Mi88). No cancer incidence data have yet been obtained for this cohort.

Strengths and Limitations

This cohort has reported the largest number of breast cancer deaths observed to date in a single cohort, and the exposure is highly fractionated, and in a North American population. However, these subjects all had tuberculosis, and although comparisons are made internally within the cohort, extrapolations to the general population may require caution. Only organs in the direct beam (notably breast and lung) are likely to have received doses leading to any measurable increase in risk, and the averaging involved in the dose estimation procedure will inevitably lead to some misclassification of dose. To date, only cancer mortality and not incidence is available for this cohort.

New York State Postpartum Mastitis Study

Cohort Source and Exposure

The cohort consists of 601 women treated with radiotherapy for postpartum acute mastitis in New York State during the 1940's and 1950's, together with 1,239 non-exposed women consisting of women with mastitis not treated by radiotherapy, and siblings of both groups of women with mastitis. Doses were received in a small number of series, with breast tissue dose ranging from 60 to about 1,400 rads. The age range at first exposure was limited, with few under age 20 or over age 40 at entry.

Dosimetry

Individual breast tissue doses have been estimated for all 601 women from the original radiotherapy records.

Follow-up

Follow-up to ascertain breast cancer incidence has been carried out using mailed questionnaires, and results for such incidence have been reported for up to 45 years of follow-up (Sh86). During this period 115 breast cancer cases were observed.

Strengths and Limitations

This is a fairly small cohort, with most exposure limited to the breasts. The exposure was largely unfractionated, and estimates of breast tissue dose are probably accurate. However, the interpretation of possible differences in response of breast tissue with an inflammatory condition and subject to the hormonal changes due to pregnancy compared to the response of breast tissue unaffected by these factors is not clear.

Massachusetts Fluoroscopy Study

Sample Source and Exposure

The cohort consists of 1,742 women first treated between 1930 and 1956 in two Massachusetts sanatoria, one of which treated only those under the age of 17. Of these women, 1,044 were subjected to regular fluoroscopy in conjunction with treatment by artificial pneumothorax, and consequently received substantial doses of low-LET radiation to the breast.

Dosimetry

Individual breast tissue doses have been estimated from the original patient records, by interviews with physicians conducting the treatment during the time period of interest, measurements on fluoroscopes of the relevant vintage, and by Monte Carlo simulations (Bo78, Bo81).

Follow-up

The vital status of 97% of the cohort through 1980 has been determined from hospital records, death certificates, and mailed questionnaires (Hr88). A total of 74 breast cancer cases have been observed in this cohort, with a total accumulation of 30,932 women-years at risk.

Advantages and Limitations

The exposure in this study was highly fractionated, and the population is a U.S. one. Dosimetry has been carefully reconstructed and complete follow-up carried out. The major disadvantage of this cohort is its size, which is small, thus limiting the interpretation of results within subgroups of the cohort. Extrapolation of the results from a cohort with tuberculosis to the general population, however, requires cautious interpretation.

References

Bo78 Boice, J. D. Jr., M. Rosenstein, and D. E. Trout. 1978. Estimation of breast cancer doses and breast cancer risk associated with repeated fluoroscopic chest examinations of women with tuberculosis. Radiat. Res. 73:373-390.

Bo81 Boice, J. D. Jr., R. R. Monson, and M. Rosenstein. 1981. Cancer mortality in women after repeated fluoroscopic examinations of the chest. J. Natl. Cancer Inst. 66:863-867.

Bo88 Boice, J. D. Jr., G. Engholm, R. A. Kleinerman et al. 1988. Radiation dose and second cancer risk in patients treated for cancer of the cervix. Radiat. Res 116:3-55.

Da85 Darby, S. C., R. Doll, and M. C. Pike. 1985. Detection of late effects of ionizing radiation: why deaths of a-bomb survivors are a valuable resource. Letters to the Editor. Int. J. Epidemiol. 14:637-638.

Da88 Darby, S. G , R. Doll, and P. G. Smith. 1988. Paper 9. Trends in long term mortality in ankylosing spondylitis treated with a single course of x-rays. Health effects of low dose ionising radiation. BNES, London.

DS86 US-Japan Joint Reassessment of Atomic Bomb Radiation Dosimetry in Hiroshima and Nagasaki, Dosimetry System 1986 (DS86), Final Report. Radiation Effects Research Foundation, Volume 1.

Hr88 Hrubec, Z., J. D. Boice Jr., R. R. Monson, and M. Rosenstein. 1988. Breast cancer after multiple chest fluoroscopies: second follow-up of Massachusetts women with tuberculosis. Cancer Res. (in press).

Le88 Lewis, C. A., P. G. Smith, I. M. Stratton, S. C. Darby, and R. Doll. 1988. Estimated radiation doses to different organs among patients treated for ankylosing spondylitis with a single course of x-rays. Br. J. Radiol. 61:212-220.

Mi88 Miller, A. B., G. R. Howe, G. J. Sherman, J. P. Lindsay, and M. J. Yaffe. 1988. Breast cancer risk in relation to low-LET radiation: the Canadian study of cancer following multiple fluoroscopies. N. Engl. J. Med. (submitted).

Sh78 Sherman, G. J., G. R. Howe, A. B. Miller, and M. Rosenstien. 1978. Organ dose per unit exposure resulting from fluoroscopy for artificial pneumothorax. Health Phys. 35:259-269.

Sh87 Shimizu, Y., H. Kato, W. Schull, D. Preston, S. Fujita, and D. Pierce. 1987. Comparison of risk coefficients for site-specific cancer mortality based on the DS86 and T65DR shielded kerma and organ doses. Technical Report TR 12-87. Radiation Effects Research Foundation.

Sh86 Shore, R. E., N. Hildreth, E. Woodard, P. Dvoretsky, L. Hempelmann, and B. Pasternack. 1986. Breast cancer among women given x-ray therapy for acute postpartum mastitis. JNOI 77(3):689-696.

Sm77 Smith, P. G., R. Doll, and E. P. Radford. 1977. Cancer mortality among patients with ankylosing spondylitis not given x-ray therapy. Br. J. Radiol. 50:728.

St84 Stewart, A. M., and G. W. Kneale. 1984. Non-cancer effects of exposure to a-bomb radiation. J. Epidemiol. Comm. Health 38:108-112.

St85 Stewart, A. M. 1985. Detection of late effects of ionizing radiation: why deaths of a-bomb survivors are so misleading Int. J. Epidemiol. 14:52-56.

St88 Stewart, A. M., and G. W. Kneale. 1988. A-bomb survivors as a source of cancer risk estimates: confirmation of suspected bias. Proceedings of the 14th Gray Conference. Radiat. Prot. Dosimetry. In Press.

To87 Tokunaga, M., C. E. Land, T. Yamamoto, M. Asano, S. Tokuoka, H. Ezaki, and I. Nishimora. 1987. Incidence of female breast cancer among atomic bomb survivors, Hiroshima and Nagasaki, 1950-1980. Radiat. Res. 112:243-272.

ANNEX 4B: CHANGES IN THE ESTIMATED DOSE FOR A-BOMB SURVIVORS

The New Dosimetry, DS86

The analyses of radiation effects among the Japanese A-bomb survivors in this report make use of new dose estimates developed in a five-year study by Japanese and American scientists. This binational study resulted in a new dosimetry system, designated DS86, which is documented in two recent Radiation Effects Research Foundation (RERF) reports (RERF87, RERF88). The reassessment of A-bomb dosimetry consisted of a careful review of information on the number of fissions that occurred in the A-bomb explosions and detailed calculations of neutron and gamma ray transport through weapons materials and the intervening air. This was followed by Monte Carlo calculations of the radiation field within Japanese houses, which also take into account the shielding provided by neighboring houses, and finally, the organ doses received by survivors having various shielding circumstances, location, orientation, and size.

The calculational program was supported by new measurements of gamma-ray kerma to roof tiles by means of thermal luminescence and a reevaluation of the measurements of neutron-induced radioactivity that were made after the bombings by Japanese scientists. The dose reassessment was reviewed by a National Research Council (NRC) panel which concluded that the new dose estimates are more accurate and more soundly based than those used previously, and that they should be used in the assessment of radiation risks (NRC87). Nevertheless, investigations to determine the precision of the estimated doses and to account for differences between measured and calculated thermal neutron fluences are continuing.

A Comparison of DS86 and T65D

Doses estimated with DS86 differ from the tentative 1965 dosimetry (T65D) system estimates (Au77, Mi68) used by RERF before 1987 and by previous BEIR Committees (NRC72, NRC80). Before outlining these differences, it is necessary to identify the various ways dose estimates for the A-bomb survivors have been specified, as this can be a source of confusion when comparing results obtained with the new and older dosimetries.

In RERF reports, particularly those on the Life Span Study, individual survivors are categorized in terms of the incident radiation, i.e., the kerma, at the location where a survivor was exposed. If a survivor was outside and not near buildings or other structures, the kerma at this location is the "free field tissue kerma in air" (FIA kerma), but more often survivors were in houses or otherwise shielded. In such cases, the kerma is smaller than the FIA kerma at the same location.

For risk estimation, the mean dose within a given organ is the governing dosimetric parameter. This organ dose is smaller than the kerma due to the self-shielding provided by the body itself. How much smaller depends on the location of a particular organ within the body and the orientation of the survivor in the radiation field. In this report, as in the BEIR III report, risk estimates are based on organ doses, not the kerma at a survivor's location. This is in contrast to RERF reports on the Life Span Study in which results are often reported in terms of kerma.

Because neutrons have a larger effect per unit dose than gamma rays, the quantity dose equivalent is used to express the organ dose due to both radiations in combination. As indicated in Chapter 1, organ dose equivalents are calculated by multiplying the organ dose due to neutrons by an appropriate value of the neutron RBE and adding this product to the organ dose due to gamma radiation. Therefore, the difference between the new and old dosimetries, in terms of organ dose equivalent, also depends on what RBE value is assigned to neutrons. This point is particularly important when considering organ dose equivalents under T65D for the Hiroshima survivors. Because of the dissimilarity between the atomic weapons used at Hiroshima and Nagasaki, it was assumed under the T65D dosimetry system that neutrons made a major contribution to the doses at Hiroshima but not at Nagasaki. The new dosimetry indicates that the neutron doses in both cities were quite small compared to the organ dose from gamma rays.

Although differences between the two dosimetries vary somewhat with distance, the following generalities hold. At Hiroshima, neutron FIA kerma is about a factor of ten smaller under DS86 than under T65D. Conversely, the gamma ray FIA kerma at Hiroshima is greater under DS86 than under T65D. At Nagasaki, the newly estimated gamma ray and neutron FIA kermas are somewhat smaller than for T65D. These results are illustrated in the first panels of Figure 4B-1, Hiroshima, and 4B-2, Nagasaki. The results shown for Hiroshima are for a distance from ground zero of 1,150 meters; those for Nagasaki for 1,275 meters. These are "average" ranges in that approximately one-half of the collective dose (person rad) was delivered within these distances in the respective cities.

Although the estimated FIA gamma kerma at Hiroshima is greater under DS86 than T65D, the gamma kerma to house shielded survivors is smaller since the shielding provided by a house was underestimated under T65D (Figure 4B-1) This is important since most of the survivors who received appreciable doses were shielded from blast and thermal effects. Under DS86, the gamma ray kerma incident on survivors at Nagasaki is about a factor of two less than under T65D (Figure 4B-2). Conversely, the amount of shielding provided by the body was overestimated under T65D, so that in spite of the smaller shielded kerma at Hiroshima, organ doses are

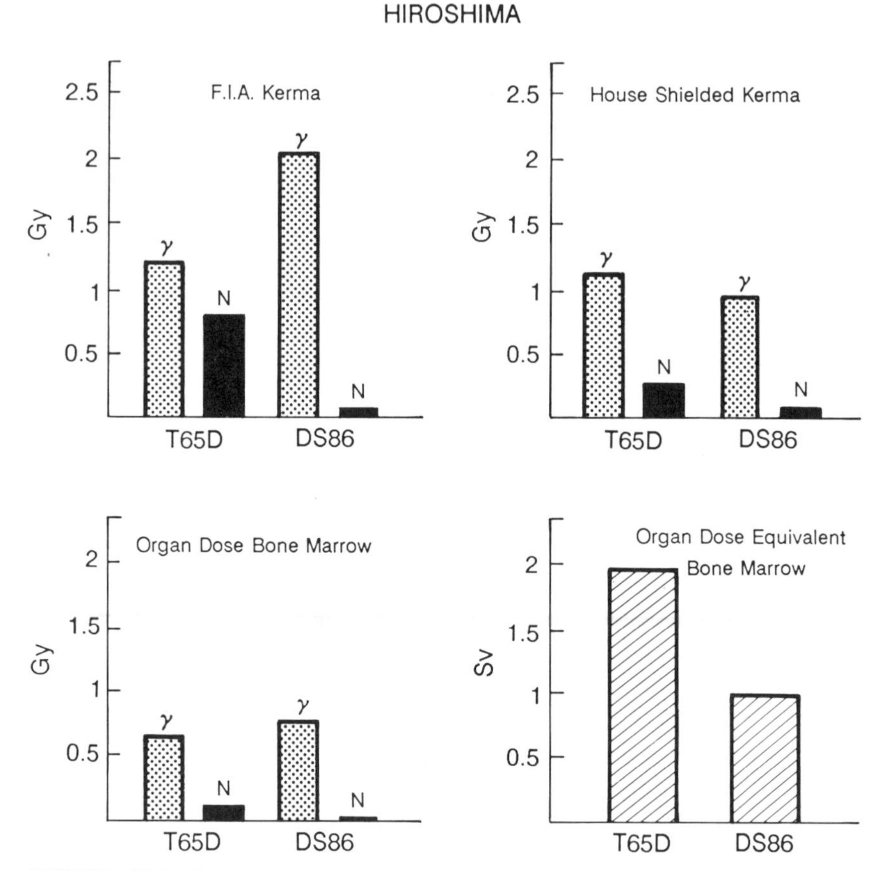

FIGURE 4B-1 Comparison of T65D and DS86 dose estimates for gamma rays and neutrons in Hiroshima.

slightly higher under DS86 than T65D (Figure 4B-1); at Nagasaki, organ doses are smaller under DS86 than T65D (Figure 4B-2).

At first glance, the near equality in organ rad under both the old and new dosimetries would indicate little net change with the introduction of the new dosimetry, DS86. This is not always thc casc. Where neutrons have been assigned a large RBE, such as in the BEIR III report (NRC80), they make a substantial contribution to the dose equivalent under T65D but not under DS86. For a neutron RBE of 20, the dose equivalents in bone marrow at Hiroshima becomes a factor of two smaller with the new dosimetry than with T65D (Figure 4B-1). Similar results are found for other internal organs. For survivors at Nagasaki, the estimated neutron doses under T65D are so small that RBE has little effect on the estimated dose equivalent (Figure 4B-2). It is important to note that, compared to T65D, organ dose equivalents at Nagasaki are somewhat smaller with the new dosimetry. Historically, risks have been lower per estimated unit dose or per unit dose equivalent in Nagasaki than in Hiroshima, a difference that was attributed to the neutrons in Hiroshima. Under DS86, observed risks per unit dose or per unit dose equivalent are still somewhat lower in Nagasaki than in Hiroshima, but the difference is small and not statistically significant. Moreover, neutron doses are so low in both cities that the

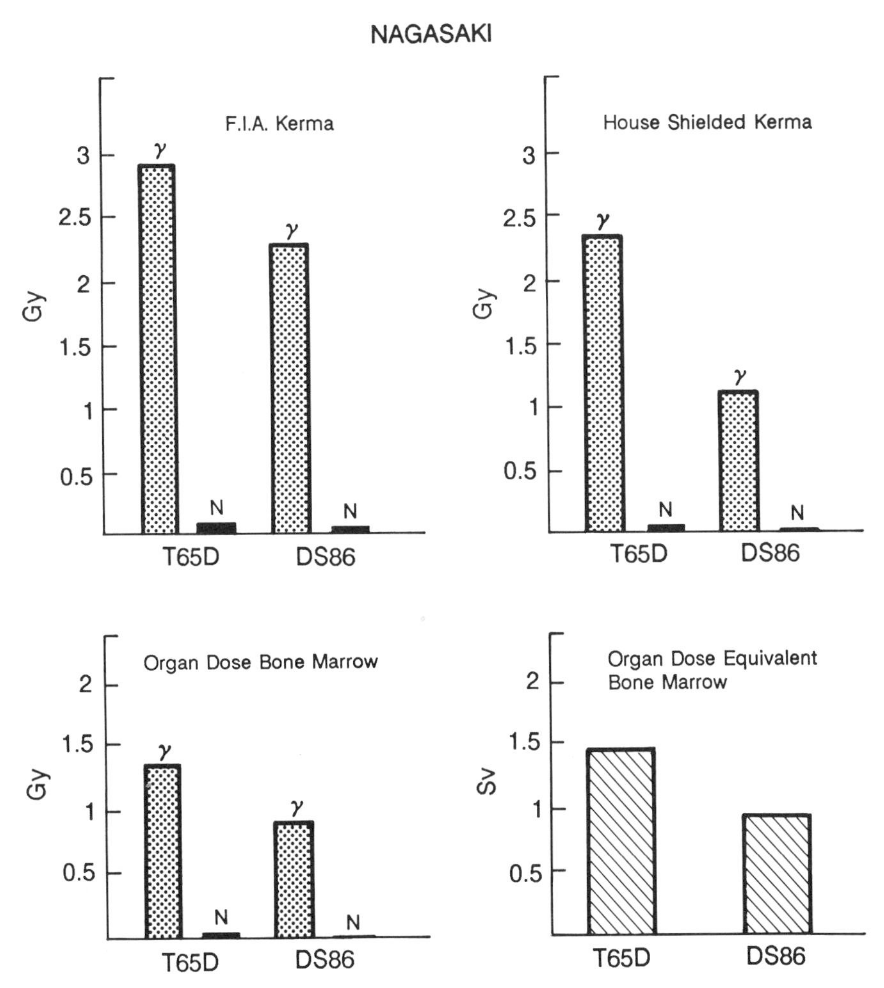

FIGURE 4B-2 Comparison of T65D and DS86 dose estimates for gamma rays and neutrons in Nagasaki.

A-bomb survivor data contain no information on the RBE of neutrons for human carcinogenesis (Pr87).

The Committee's Use of DS86

The A-bomb survivor data made available to the Committee by RERF pertain to the DS86 subcohort used to prepare Life Span Study Report 11 (Sh87). This subcohort is composed of 75,991 survivors for whom sufficient information was available in 1987 to calculate DS86 dose estimates. This subcohort is somewhat smaller than the exposed population covered by T65D dosimetry, because more data on shielding are required under DS86 protocols to compute a survivor's dose. Little information is lost by this restriction, since those excluded were mainly distal survivors whose shielding circumstances are poorly defined or unknown. The sex- and city-specific mortality data used by the Committee were stratified in terms of both the gamma and neutron kerma at the survivor's location in one of ten categories. The lower bounds of these categories are 0, 0.006, 0.05, 0.10, 0.20, 0.50, 1.0, 2.0, 3.0, and 4.0 Gy. The gamma and neutron kerma in each strata is a person-years weighted average for the survivors at risk in a specified age and city category.

Organ doses were calculated by RERF for each survivor but were not used directly. Instead, average age-specific and city-specific body transmission factors are being used to estimate organ doses. Organ-specific transmission factors averaged over survivors of all ages are listed in Table 4B-1. Although there is some variation in the transmission factors for neutrons, the high energy gamma radiation from the bombs resulted in an uncommon degree of uniformity in the dose to internal organs due to low-LET radiation (Table 4B-1). Application of t'.e transmission factors was straightforward. The stomach was used for the category *cancers of the digestive system* and as surrogate for all organs in the category *all solid cancers*. For the category *other cancers*, the transmission factor for the large intestine was used to estimate the neutron and gamma-ray dose.

As discussed in Chapter 4, an organ specific dose equivalent for each strata in the dose-response regressions was calculated using an RBE of 20 for neutrons. In this regard, it should be noted that the bomb neutron spectrum at distances where survivorship frequently occurred is considerably less energetic than an unattenuated spectrum of fission neutrons. Because of neutron scattering in bomb materials and well over a kilometer of air, a large fraction of the neutron kerma is below 1 MEV. For example at 1200 meters in Hiroshima, 50 percent of the incident kerma was between 0.1 and 1 MEV (Ka89). In such circumstances, the recoil protons in tissue have energies of a few hundred keV and are near the LET for maximum biological effectiveness (see Figure 3.3).

TABLE 4B-1 Averages of the Body Transmission Factors Under DS86[a] (Sh87)

Organ	Gamma	Neutron	n,γ[b]
Bone marrow	0.81	0.37	0.42
Stomach	0.75	0.28	0.40
Colon	0.74	0.19	0.41
Lung	0.80	0.33	0.37
Bladder	0.76	0.22	0.40
Liver	0.76	0.29	0.39
Pancreas	0.72	0.18	0.42
Breast	0.85	0.61	0.32
Ovary	0.74	0.16	0.39
Uterus	0.73	0.14	0.40
Testis	0.78	0.32	0.38
Thyroid	0.85	0.41	0.43

[a] The body transmission factor is the ratio of the organ dose in a male survivor to the kerma at his location. The values in the table are averages for 19,113 survivors and are largely independent of city and distance but do depend on age (body size).

[b] Gamma radiation following neutron capture within the body.

The Committee deliberated on whether risk estimates in terms of the kerma at a survivor's location would be a worthwhile addition to this report but decided against such an approach because the radiation field from the A-bombs is not representative of exposure situations that are often encountered in radiation protection practice. Because the gamma radiation from the bombs is so energetic, the degree of self-shielding provided by the body is small. Moreover, the A-bomb radiation had a substantial vertical component which leads to a rather atypical exposure geometry. Effective application of the Committee's risk estimates to other exposure situations are dependent therefore on a careful consideration of the dose distribution within the body and the resultant organ doses, as illustrated in Table 4B-1 for the A-bomb survivors.

References

Au77 Auxier, J. A. 1977. Ichiban. Technical Information Center, Energy Research and Development Administration, TID-27080, National Technical Information Service, U.S. Department of Commerce, Springfield, VA 22161.

Ka89 Kaul, D. C., and S. D. Egbert. 1989. Cumulative Fraction of Neutron Dose. Communication to RERF Office at NAS, SAIC, San Diego, Calif.

Mi68 Milton, R., and T. Shohoji. 1968. Tentative 1965 Radiation Dose Estimation for Atomic Bomb Survivors, Hiroshima and Nagasaki, 1968. ABCC TR 1-68. Hiroshima, Japan: ABCC.

NRC72 National Academy of Sciences. 1972. The Effects on Populations of Exposure to Low levels of Ionizing Radiation. Advisory Committee on the Biological Effects of Ionizing Radiation. Washington, D.C.

NRC80 National Academy of Sciences. 1980. The Effects of Populations of Exposure to Low levels of Ionizing Radiation. Advisory Committee on the Biological Effects of Ionizing Radiation. Washington, D.C.

NRC87 National Research Council. 1987. An Assessment of the New Dosimetry for A-bomb Survivors. Panel on the Reassessment of A-bomb Dosimetry. W. H. Ellett, ed. Washington, D.C.: National Academy Press,

Pr87 Preston, D., and D. Pierce. 1987. The Effect of Change in Dosimetry on Cancer Mortality Risk Estimates in the Atomic Bomb Survivors. TR 9-87. Hiroshima, Japan. RERF.

Sh87 Shimizu, Y., H. Kato, W. Schull, D. Preston, S. Fujita, and D. Pierce. 1987. Life Span Study Report 11, Part 1. Comparison of Risk Coefficients for Site-Specific Cancer Mortality Based on the DS86 and T 65R Shielded Kerma and Organ Doses TR 12-87. Hiroshima, Japan: RERF

RERF87 Radiation Effects Research Foundation. 1987. U.S.-Japan Joint Reassessment of Atomic Bomb Radiation Dosimetry in Hiroshima and Nagasaki—Final Report, Vol. 1. Hiroshima, Japan.

RERF88 Radiation Effects Research Foundation. 1988. U.S.-Japan Joint Reassessment of Atomic Bomb Radiation Dosimetry in Hiroshima and Nagasaki—Final Report, Vol. 2. Hiroshima, Japan.

ANNEX 4C: AMFIT

Parameter estimates for the relative and excess risk models used for risk projections in this report were obtained using AMFIT, a program for the analysis of cohort survival data which was written by Dale Preston and Donald Pierce. The detailed cross-tabulations of person-years and cases used as input to AMFIT were generally constructed using PYTAB, which was written by Dale Preston. Both programs were originally developed for analyses of mortality and incidence in the RERF Life Span Study. These programs have been used extensively in recent analyses of the RERF data, including the two most recent Life Span Study reports (Pr87, Sh87, Sh88), and the comparison of DS86 and T65D risk estimates (Pr88). The programs were also used by the BEIR IV Committee in their analyses of lung cancer risks among miners exposed to radon (NRC88).

AMFIT makes use of Poisson regression methods for the analysis of cohort survival data stratified on time and other factors (Fr83, Ho76, Ra86, Pi87, Br87). AMFIT computes maximum likelihood estimates of parameters in a general class of hazard function models, which includes both excess and relative risk (proportional hazards) models, using a Newton-Raphson algorithm which is equivalent, for fully parametric models, to the iteratively weighted least squares algorithm used for Poisson regression as in the systems, GLIM (Ba78) and PREG (Fr83, Fr85). Some of the simpler relative risk models available in AMFIT can be fit using GLIM or PREG, and these three programs produce identical estimates in such cases. The committee chose to use AMFIT in the development of risk projection models because of its ease of use and the broad range of models available in the program.

AMFIT can be used to fit relative risk models in which the background is a function of a large number (possibly several hundred) of stratum parameters. In order to avoid the inversion of a (potentially) large matrix in such cases, AMFIT uses a Gauss-Seidel iteration (Th88). Upon convergence for stratified models, the covariance matrix of the non-strata parameters is adjusted to take into account the stratum parameter estimates. The fitting of stratified Poisson regression models for grouped survival data in which time is one of the stratification variables is closely related to partial likelihood methods for ungrouped survival data (Co72, Ka80). The models used by the committee were generally fully parametric models, i.e., they did not contain stratum parameters.

For fully-specified parametric models, AMFIT can be used to produce residuals and other components for generalized regression diagnostics (Mc83, Pr81). These statistics were used in some of the goodness-of-fit evaluations carried out by the committee.

Current PC versions of AMFIT and PYTAB can be obtained from the Radiation Epidemiology Branch of the National Cancer Institute.

References

Ba78 Baker, R. J., and J. A. Nelder. 1978. The GLIM System: Release 3 Numerical Analysis Group, Oxford.

Br87 Breslow, N. J., and N. E. Day. 1987. Statistical Methods in Cancer Research Volume II—The Design and Analysis of Cohort Studies. IARC Scientific Publication 82. Lyon: International Agency for Research on Cancer.

Co72 Cox, D. R. 1972. Regression models and life tables (with discussion). J. R. Stat. Soc. Ser. B. 34:187-220.

Fr83 Frome, E. L. 1983. The analysis of rates using Poisson regression models, Biometrics 39:665-674.

Fr85 Frome, E. L., and H. Checkoway. 1985. Use of Poisson regression models in estimating incidence ratios and rates. Am. J. Epidemiol. 121:309-323.

Ho76 Holford, T. 1976. Life tables with concomitant information. Biometrics 32:587-597.

Ka80 Kalbfleisch, J. D., and R. A. Prentice. 1980. The analysis of failure time data. New York: Wiley.

Mc83 McCullagh, P., and J. A. Nelder. 1983. Generalized Linear Models. New York: Chapman and Hall.

NRC88 National Research Council, Committee on the Biological Effects of Ionizing Radiations (BEIR IV). 1988. Health Risks of Radon and Other Internally

Deposited Alpha Emmitters. Washington, D.C.: National Academy Press. 602 pp.
Pi87 Pierce, D. A., and D. L. Preston. 1987. Developments in cohort analysis with application 46th Session of the International Statistics Institute 31.2, 557-569.
Pr81 Pregibon, D. 1981. Logistic regression diagnostics. Ann. Stat. 9:705-724.
Pr87 Preston, D. L., H. Kato, K. Kopecky, and S. Fujita. 1987. Life span study Report 10 Part 1. Cancer mortality among A-bomb survivors in Hiroshima and Nagasaki, 1950-82.
Pr88 Preston, D. L., and D. A. Pierce. 1988. The effect of changes in dosimetry on the risk of cancer mortality among A-bomb survivors. Rad. Res. 114:151-178.
Ra86 Radford, E. P., D. L. Preston, and K. J. Kopecky. 1986. Methods for the study of delayed health effects of A-bomb radiation in cancer in atomic bomb survivors. GANN monograph on Cancer Research No. 32. I. Shigematsu and A. Kagan, eds. New York: Plenum Press.
Sh87 Shimizu, Y., H. Kato, W. J. Schull, D. L. Preston, S. Fujita, and D. A. Pierce. 1987. Life Span Study Report 11, Part 1, Comparison of risk coefficients for site specific cancer mortality based on the DW86 and T65 Dr shielded kerma and organ doses. RERF TR 12-87. Hiroshima: Radiation Effects Research Foundation.
Sh88 Shimizu Y, H. Kato, and W. J. Schull. 1988. Life Span Study Report 11, Part 2, Cancer mortality in the years 1950-1985, based on the recently revised doses. RERF TR 5-88. Hiroshima: Radiation Effects Research Foundation.
Th88 Thisted R. 1988. Elements of Statistical Computing: Numerical Computation. New York: Chapman and Hall.

ANNEX 4D: THE COMMITTEE'S ANALYSIS OF A-BOMB SURVIVOR DATA

Data Used

As outlined in Annex 4A, the RERF LSS data comprise the primary data set used by the committee for risk modeling. The data supplied to the committee by RERF covered follow-up through 1985 and were the same stratified data as used by RERF to prepare LSS Report 11. Two RERF reports have compared risk estimates under the new DS86 and old T65D dosimetries (Pr87, Sh87). As the aim of this report is to provide risk estimates based on the best available data, the committee confined itself to analyses using just the DS86 data. The primary data file used by the Committee contained a total of 3,399 strata, compartmentalized by cancer mortality at a specific site, person years at risk, age at exposure, time after exposure, dose, city, and sex.

The committee combined the cancer deaths into five categories; leukemia, breast, respiratory, digestive and "other" cancers. These broad categories were chosen to ensure adequate numbers for detailed modeling of modifying effects without combining cancers that showed distinctly different epidemiologic patterns. In addition, studies of the accuracy of death certificates by specific cause showed that for some sites errors in

TABLE 4D-1 Effects of Varying RBE on Relative Risk Models for Radiation-Induced Cancer

Dose Coefficients (Std. Dev.)			
Leukemia[a]			
	α_2	α_3	
RBE	(Per. Sv)	(Per. Sv)	Deviance
1	0.257(0.313)	0.310(0.349)	498.46
5	0.254(0.309)	0.301(0.341)	498.37
10	0.251(0.303)	0.290(0.331)	498.27
20	0.243(0.292)	0.271(0.314)	498.08
50	0.219(0.258)	0.225(0.276)	497.65
Nonleukemia[b]			
	α_1		
RBE	(Per. Sv)	Deviance	
1	1.158(0.381)	1,453.34	
5	1.113(0.366)	1,452.95	
10	1.061(0.349)	1,452.54	
20	0.969(0.320)	1,451.95	
50	0.763(0.253)	1,451.15	

[a]Linear, α_2, and quadratic, α_3, coefficients for dose response using the committee's preferred model, Equation 4.3; observations for organ dose greater than 4 Sv are excluded.

[b]Linear coefficient α_1, for dose response for all solid cancers using age at exposure and sex as risk factors with 10-year minimum latency.

certification were numerous; this was especially true for cancers of the liver and pancreas which were often assigned to stomach cancer on the death certificates. This provided an additional reason for modelling all cancers of the digestive system as a group.

The kerma categories were replaced with the corresponding organ doses, based on age-, city-, and organ-specific transfer coefficients and an RBE for neutrons of 20. Table 4D-1 describes the results of varying the RBE in relative risk models for nonleukemia cancers and leukemia. Although the slope of the dose-response curve decreased with increasing RBE, the fit of the model (as judged by the column "Deviance") was unaffected and there was no change in the estimate of any of the parameters for modifying variables.

The RERF data show a tendency toward decreased risk per Gy in the highest dose groups, which may reflect either cell-killing or overestimation of the doses in this group. The committee considered various ways of dealing with this problem, including adding terms to the dose-response part of the model and adjusting the highest doses downward. In the end, it was decided simply to exclude the two highest dose groups. Table

TABLE 4D-2 Effect of Excluding High-Dose Groups on Fitted Dose-Response Relationships

Exclusion (Sv)	Linear Dose Coefficient Per Sv (Std. Dev.)	Score for Adding Quadratic Term
Leukemia[a]		
None	0.575(0.503)	0.08
>5	0.763(0.625)	1.76
>4	0.482(0.550)	2.14
>3	0.254(0.520)	1.45
>2	0.050(0.556)	1.49
Nonleukemia[b]		
None	0.781(0.248)	−2.04
>5	0.823(0.274)	−1.88
>4	0.969(0.320)	−0.41
>3	0.980(0.331)	−0.31
>2	1.136(0.448)	0.66

[a] Linear fit using the risk modifiers in the preferred model with an RBE of 20.

[b] Linear fit using age at exposure and sex as risk modifier with a 10-year minimum latency and an RBE of 20.

4D-2 illustrates the results of this exclusion on fitted linear models for nonleukemia and leukemia. For both outcomes, the slope of the linear dose-response relation is highest when doses over 4 Gy (using an RBE of 20) are excluded. For nonleukemia cancers, there is no sign of a positive quadratic component at any restriction, but as shown in Table 4D-2 for leukemia, the evidence for a positive quadratic component is strongest upon restriction to under 4 Gy. With further restriction, the standard errors of all model parameters begin to increase to unacceptable levels. Hence it was decided to restrict all further analyses to the subgroup under 4 Gy.

Model Selection

While the BEIR III report used both additive and relative risk models, this committee prefers relative risk models. The relative risk models provide not only a more parsimonious description of the data but also have additional advantages. For example, relative risks are less affected by losses of cause of death assignments due to data arising from errors in certification by site, unless the errors are correlated with radiation dose. In contrast, absolute risks are strongly affected by losses due to erroneous certification. Investigation of the RERF autopsy data base shows that erroneous certifications are essentially unrelated to the dose estimates for A-bomb survivors.

One can show mathematically that the additive risk model and the relative risk model can be made equivalent if the variables used in the excess risk terms are also the ones used for estimating the background. Therefore, this committee does not make a distinction between additive risk and relative risk models. In BEIR III, however, the excess risk functions for the additive and relative risk models were either constant or approximately constant and as such, there needed to be a definite distinction between additive risk and relative risk. It is clear from the present analyses that such simple additive or relative risk models do not provide an adequate description of the data. Therefore, the committee choose to estimate risk with inclusion of several explanatory variables in the excess risk term. Functionally, the committee chose to use the relative risk formulation with a stratified or nonparametrically estimated background. The reason is simply that this avoids using the necessary but complicated functions to estimate the background.

Three modeling approaches are illustrated in Table 4D-3: additive risk with its necessarily modeled background; relative risk model with a modeled background; and the relative risk with the stratified background, which the committee chose to use. Three sets of parameters were used in these illustrated models. They all provide a fairly reasonable fit, although some of them are statistically superior, based on the values of deviance. The average risks for these various models do vary as one might suspect. However, they are reasonably close to one another, generally within a factor of 2 and, for the most part, are well within the statistical confidence intervals given for the committee's preferred models, which differentiate between cancer types.

Previous risk analyses (e.g., UNSCEAR), for the group of all non-luekemia cancers, have used a constant relative risk model with adjustments for sex and age-at-exposure. The second model, #5 in the relative risk-stratified background group in Table 4D-3, is essentially this model since the coefficient for time since exposure (0.0775) is effectively zero. This model, however, provides a significantly poorer fit than the other two models (#4, #6) as measured by deviance. Secondly, the risk estimates are considerably larger than for the other two models.

In Table 4D-3 we have included the risk estimates for acute exposure at age 5. These values can be quite large and tend to vary to a much greater degree than the all-age average. This is not surprising when it is realized that there are few data for survivors exposed at the early ages, because they are only now reaching the age at which cancer rates are measurable.

TABLE 4D-3 Three Illustrative Models for all Cancers Other Than Leukemia

Model No.	α_1	β_1	β_2	β_3	β_4	DEV	Risk/10^4 Person Sv: Males AVE	Males AGE 5	Females AVE	Females AGE 5
Relative Risk—Modeled Background										
1	0.4054	−2.451	0.8062		0.4134	1,723.6	426	801	529	1,080
2	0.3246		−0.1466	−0.6688	0.5709	1,731.1	557	1,483	817	2,362
3	0.4054	−1.570		−0.2855	0.5062	1,726.6	371	697	503	1,031
Relative Risk—Stratified Background										
4	0.3933	−2.463	0.9199		0.4770	1,449.7	448	877	592	1,259
5	0.3251		0.0775	−0.6688	0.5446	1,456.6	669	1,923	960	2,991
6	0.4322	−1.592		−0.2866	0.4671	1,443.3	393	1,052	513	1,052
Additive Risk—Modeled Background										
7	7.222	2.334	0.5466		0.2309	1,725.2	350	600	630	1,108
8	8.145		1.772	0.8784	0.1015	1,723.3	315	313	490	476
9	7.548	2.697		−0.0930	0.2829	1,727.0	302	453	566	863

where

$$f(d) = \alpha_1 d$$
$$g(\beta) = \exp[\beta_1 \ln(A/50) + \beta_2 \ln(T/20) + \beta_3 \ln(E/30) + \beta_4 I(S)]$$
$$I(S) = \begin{cases} 1 \text{ if female} \\ 0 \text{ if male} \end{cases}$$

with A = attained age, T = time since exposure, E = age at exposure.

TABLE 4D-4 Excess Risk Estimates and 90% Confidence Intervals with the Preferred Models (0.1 Sv Acute Exposure to 100,000 Males of Each Age)

Age at Exposure	Leukemia	Nonleukemia
5	111 (20–455)[a]	1,165 (673–1,956)
15	109 (21–450)	1,035 (642–1,775)
25	36 (8–87)	885 (534–1,442)
35	62 (21–134)	504 (272–947)
45	108 (43–223)	492 (257–883)
55	166 (59–338)	450 (217–815)
65	191 (65–369)	290 (137–572)
75	165 (56–316)	93 (38–233)
85	96 (33–183)	14 (5–44)

[a](5%, 95%) 200 replications.

Therefore estimates for the young are, in a sense, a model dependent extrapolation from the data for older ages.

The degree of precision in the projections for the cancer risk at young ages is illustrated further in Table 4D-4. In that table for leukemia, the estimated excess risk is 111 cases for exposure at age 5, with a 90% confidence interval extending from 20 to 455, i.e., the upper bound is about 4 times the point estimate. On the other hand, for ages 35, 45, etc., the upper bound of the 90% confidence limit is within about a factor of 2 of the point estimate. Confidence limits do not vary as much with age at exposure for nonleukemia mortality (Table 4D-4). Nevertheless, the risks for nonleukemia are relatively high and imprecise for early ages at exposure, so that considerably more experience will be needed before there are sufficient data to estimate more precisely the lifetime risks for those exposed at early ages.

Alternative Models

The committee considered a variety of models before selecting the preferred models described in Chapter 4. Some of these alternative models and their deviance are described in Table 4D-5 for the various types of cancer considered in the chapter. In each case, model 0 is the committee's preferred model described in Chapter 4. In general, the preferred models fit the data as well as the alternatives and have fewer terms. This was not the sole criterion for model selection. The committee paid particular attention to how risks were proportioned between various age groups. Lifetime risks following an acute exposure of 0.1 Sv under these models

TABLE 4D-5 Alternative Models

Leukemia

Model	α_2	α_3	β_1	β_2	β_3	β_4	β_5	Deviance
0	See Equation (4.3) in Chapter 4 for the preferred leukemia model							498.1
1	2.087	2.206	−1.921	−1.791	−0.442	−2.030		500.0
2	1.809	1.975	−2.531	−1.728	−0.688			503.3
3	1.890	2.062	−2.345	−1.772	−0.592		−0.753	502.8

where

$$f(d) = \alpha_2 d + \alpha_3 d^2$$
$$g(\beta) = \exp[\beta_1 \ln(T/20) + \beta_2 \ln^2(T/20) + \beta_3 \ln(E/30) + \beta_4 \ln(T/20) I(E \le 20) + \beta_5 \ln(T/20) I(E \le 15)]$$
$$I(E \le 20) = \begin{cases} 1 \text{ if } E \le 20 \\ 0 \text{ if } E > 20 \end{cases}$$
$$I(E \le 15) = \begin{cases} 1 \text{ if } E \le 15 \\ 0 \text{ if } E > 15 \end{cases}$$

Respiratory

Model	α_1	β_1	β_2	Deviance
0	0.635	−1.440	0.710	710.5
1	0.420		0.766	712.7
2	0.869	−1.453		711.8
3	0.615			714.2

where

$$f(d) = \alpha_1 d$$
$$g(\beta) = \exp[\beta_1 \ln(T/20) + \beta_2 I(S)]$$
$$I(S) = \begin{cases} 1 \text{ if female} \\ 0 \text{ if male} \end{cases}$$

Digestive

Model	α_1	β_1	β_2	β_3	β_4	Deviance
0	0.809		−0.198		0.553	1,191.3
1	0.009		0.264	4.455	0.336	1,186.1
2	1.027	−0.553	−0.219		0.519	1,190.7
3	0.107			2.106	0.412	1,187.9

where

$$f(d) = \alpha_1 d$$
$$g(\beta) = \exp\{\beta_1 \ln(T/20) + \beta_2 [(E - 25) I(25 \le E < 35) + 10\, I(E \ge 35)] + \beta_3 I(E < 30) + \beta_4 I(S)\}$$
$$I(25 \le E < 35) = \begin{cases} 1 \text{ if } 25 \le E < 35 \\ 0 \text{ otherwise} \end{cases}$$
$$I(E \ge 35) = \begin{cases} 1 \text{ if } E \ge 35 \\ 0 \text{ if } E < 35 \end{cases}$$
$$I(E < 30) = \begin{cases} 1 \text{ if } E < 30 \\ 0 \text{ if } E \ge 30 \end{cases}$$
$$I(S) = \begin{cases} 1 \text{ if female} \\ 0 \text{ if male} \end{cases}$$

Table 4D-5 *Continued*

Other Cancers						
Model	α_1	β_1	β_2	β_3	β_4	Deviance
0	1.220		−0.0464			1,124.2
1	0.824	−3.676	−0.0225	0.2843		1,117.0
2	1.295	−0.370	−0.0542			1,124.0
3	1.174		−0.0452		−1.481	1,122.3

where

$$f(d) = \alpha_1 d$$

$$g(\beta) = \exp[\beta_1 ln(T/30) + \beta_2(E - 10)I(E \geq 10) + \beta_3(E - 10)ln(T/30)I(E \geq 10) + \beta_4 ln(T/30)I(E < 20)]$$

$$I(E \geq 10) = \begin{cases} 1 \text{ if } E \geq 10 \\ 0 \text{ if } E < 10 \end{cases}$$

$$I(E < 20) = \begin{cases} 1 \text{ if } E < 20 \\ 0 \text{ if } E \geq 20 \end{cases}$$

TABLE 4D-6 Alternative Models—Lifetime Cancer Mortality Risk per 10,000 Person Sv Acute Dose Equivalent (10^6 person rem)

	Male			Female		
Age at Exposure	5	45	Avg.	5	45	Avg.
Leukemia						
Model 0[a]	111	108	111	75	73	82
1	66	75	66	42	51	48
2	41	65	57	27	45	42
3	44	69	57	29	47	43
Respiratory						
Model 0	17	353	188	48	277	150
1	249	246	207	226	207	171
2	65	492	276	26	186	100
3	370	379	316	146	141	113
Digestive						
Model 0	361	22	167	655	71	288
1	367	23	164	508	56	222
2	234	12	122	403	60	206
3	412	22	184	637	63	274

Table 4D-6 *Continued*

Other Cancers						
Model 0	787	117	300	625	100	222
1	64	642	310	46	602	253
2	639	85	241	509	86	184
3	219	121	165	185	109	131
Nonleukemia						
Model 0	1,165	492	655	1,457	468	730
1	680	912	681	920	886	717
2	939	590	638	1,078	356	563
3	1,000	522	655	1,105	339	592

[a]Model 0 is the committee's preferred model.

are shown in Table 4D-6 for ages of exposure 5 and 45 and averaged over all ages. Although the averaged risks generated by the various models are comparable, this is less true for risks at specified age of exposure.

References

Pr87 Preston, D. L., and D. A. Pierce. 1987. The effect of changes in dosimetry on cancer mortality risk estimates in the atomic bomb survivors. RERF TR 9-87. Radiation Effects Research Foundation.

Sh87 Shimizu, Y., H. Kato, W. J. Schull, D. L. Preston, S. Fujita, and D. A. Pierce. 1987. Life Span Study Report 11. Part 1. Comparison of risk coefficients for site-specific cancer mortality based on the DS86 and T65DR shielded kerma and organ doses. RERF TR 12-87. Radiation Effects Research Foundation.

ANNEX 4E: MODELING BREAST CANCER

Introduction

A general description of the Committee's final models for radiation induced breast cancer incidence and mortality risks was given in Chapter 4. This annex contains additional information on these models and on issues considered in their development. The topics to be considered herein include: summary information on the cohorts used; background rate models; relative versus absolute time-dependent risk models; cohort effects; the shape of the dose-response relationship; and effects due to age-at-exposure and time-after-exposure. This annex concludes with a summary of the parameter estimates in the Committee's preferred risk models.

Description of the Cohorts

The Committee's parallel analyses made use of mortality data from two cohorts: the Canadian TB Fluoroscopy Study (CAN-TB) (Mi89) and the subcohort of the RERF Life Span Study (LSS) for which DS86 doses are available (Sh87). Data from three cohorts were used in the incidence analyses. These cohorts included: a subset of women in the 1950 to 1980 LSS incidence series (To87) for whom DS86 dose estimates were available (LSS-I); data on women in the New York Acute Postpartum Mastitis study (NY-APM) (Sh86); and data on women in the Massachusetts TB Fluoroscopy (MASS-TB) cohort (Hr89). In all of the Committee's analyses, data on the first five-years of follow-up and, as described below, data on women with the highest exposures have been omitted. Tables 4E-1 and 4E-2 summarize the follow-up and exposure information for the mortality and incidence cohorts used in these analyses.

Background Rate Models

For the LSS, and CAN-TB cohorts there were enough deaths in the zero dose group to allow the use of internal estimates of the base line rate for breast cancer mortality. For each of these series the background rates were modelled as a log-linear spline of attained age with a single inflection point at age 50 and a log-linear trend in the age-specific rates with time (years since 1945). Table 4E-3 contains the parameter estimates for these models as estimated in the Committee's preferred mortality and incidence models.

Because the MASS-TB and NY-APM data did not include enough information on the evidence of breast cancer among unexposed women to allow internal estimation baseline rates, they were described using cohort-specific standardized incidence ratios (SIRs) relative to age- and time-specific breast cancer rates in Connecticut obtained from the SEER registry (NCI86). The estimated SIR for the MASS-TB series was 0.75 (90% confidence interval 0.59 – 0.94) while that for the NY-APM series was 1.6 (90% confidence interval 1.3 to 1.9). The difference between these SIRs was highly significant ($p < .001$).

In order to compare Connecticut and Japanese background incidence rates, a model of the form used for the LSS data was fitted to the Connecticut rates. Figure 4E-1 compares the fitted rates for several birth cohorts. The fitted age-specific breast-cancer *incidence* rates in Connecticut are 2.5 to 6 times the corresponding fitted rates in the LSS. The largest differences are seen in the earlier birth cohorts. Figure 4E-2 presents a similar comparison of the fitted background *mortality* rates for the LSS and CAN-TB cohorts.

TABLE 4E-1 Summary of Cohorts Used for BEIR V Breast Cancer Incidence Analysis

Cohort[a]	Subcohort	Person Years (1,000's)[b]	Cases	Mean Dose (Gy)[c]	Crude Rate per 1,000 Person Years
NY-APM	TOTAL	44.7	118		2.64
(~1940–1980)	Exposed women	13.7	56	2.04	4.08
	Irradiated breasts	8.7	49	3.21	5.60
	Unirradiated breasts	5.0	7		1.40
	Siblings of exposed	16.8	38		2.27
	Unexposed controls	9.4	15		1.59
	Siblings of controls	4.8	9		1.88
MASS-TB	TOTAL	36.5	65		1.78
(1930–1980)	Exposed	21.2	49	1.01	2.31
	Unexposed	15.3	16		1.04
RERF	TOTAL	940.3	367		0.39
(1950–1980)	Hiroshima	748.6	307	0.18	0.41
	Dose ≥ 0.005 Gy	379.8	170	0.35	0.45
	Dose < 0.005	368.8	137		0.37
	Nagasaki	191.7	60	0.16	0.31
	Dose ≥ 0.005 Gy	99.2	36	0.31	0.36
	Dose < 0.005	92.6	24		0.26
GRAND TOTAL		1,021.5	550		0.54

[a] In all three studies only women with at least five years of follow-up have been included. In the MASS-TB and RERF studies women with doses greater than 4 Gy have been excluded, while in the NY-APM cohort women with doses greater than 6.5 Gy have been omitted. For all three cohorts, only women with known doses have been included.

[b] In the NY-APM study both breasts did not receive the same dose. For this reason, time-at-risk computations in this study were originally done in terms of breast years. These values were then converted to person years (divided by two) for use in the analyses.

[c] Mean doses are weighted by person years.

Cohort Effects Under Relative Risk and Additive Risks

The excess relative risk for the incidence of breast cancer in the LSS was estimated to be about 50% greater than that in the two U.S. cohorts, but this difference was not statistically significant ($p = 0.4$). There was no evidence of differences in the relative risk between the NY-APM and MASS-TB cohorts. The additive excess incidence rates per unit dose in the LSS were about half of the average for the two U.S. cohorts. This difference was statistically significant ($p = 0.01$).

On the basis of the Committee's analyses of these data it was decided to use a relative risk model in which the excess relative risk was estimated using the pooled data from all three incidence series, with allowance for

TABLE 4E-2 Summary of Cohorts Used for BEIR V Breast Cancer Mortality Analyses

Cohort[a]	Subcohort	Person Years (1,000's)	Cases	Mean Dose (Sv)[b]	Crude Rate per 1,000 Person Years
CAN-TB	TOTAL	774.3	473		0.61
(1950–1980)	Nova Scotia				
	Dose ≥ 0.005 Gy	23.9	58	2.46	2.43
	Dose < 0.005	29.3	13		0.44
	Non-Nova Scotia				
	Dose ≥ 0.005 Gy	287.9	156	0.25	0.54
	Dose < 0.005	433.2	246		0.57
RERF	TOTAL	1,163.2	153		0.13
(1950–1985)	Hiroshima	804.4	112		0.14
	Dose ≥ 0.005 Sv	490.0	75	0.32	0.15
	Dose < 0.005	314.4	37		0.12
	Nagasaki	358.8	41		0.11
	Dose ≥ 0.005 Sv	163.4	21	0.22	0.13
	Dose < 0.005	195.4	20		0.10
GRAND TOTAL		1,937.5	626.0		0.32

[a] In both studies only women with at least five years of follow-up have been included. In the RERF cohort women with doses greater than 4 Sv have been excluded.
[b] Mean doses are weighted by person years.

TABLE 4E-3 BEIR V Breast Cancer Models—Log Rate Parameter Estimates for the Background Models

Incidence				
LSS			Connecticut	
Effect	Estimate	S.E.	Estimate	
Constant	0.97	0.18	2.46	
Log(age/50)	3.35	0.41	3.41	
Log(age/50) if (age ≧ 50)	−4.50	0.67	−2.51	
Years since 1945	0.038	0.006	0.020	
Mortality				
LSS			Canadian TB	
Effect	Estimate	S.E.	Estimate	S.E.
Constant	0.58	0.07	1.39	0.17
Log(age/50)	4.38	0.48	4.38	0.48
Log(age/50) if (age ≧ 50)	−4.71	0.82	−3.57	0.69
Years since 1945	−0.003	0.009	0.021	0.006

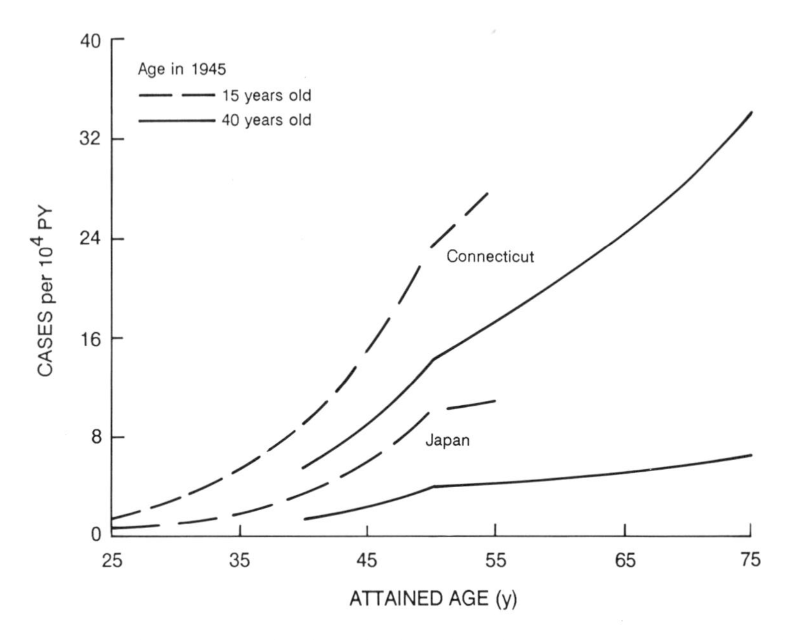

FIGURE 4E-1 Breast cancer incidence in the U.S. (Connecticut) and Japan by attained age for women who were 15 and 40 years old in 1945.

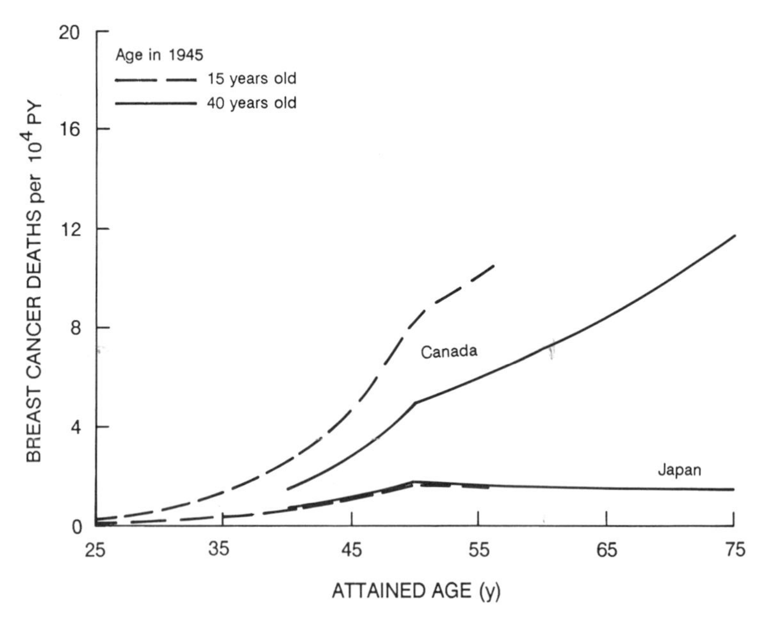

FIGURE 4E-2 Breast cancer mortality in the Japanese RERF Life Span and Canadian TB Studies by attained age for 15- and 40-year-old cohorts in 1945.

analysis of the breast cancer incidence in these cohorts (La80) had suggested that, while relative risk models provide a better fit within each cohort than do constant excess additive risk models, the additive excess risks averaged over the (then current) follow-up periods were roughly comparable for the different cohorts. In contrast, the Committee's analyses indicate that constant additive excess risk models do not adequately describe either the mortality or incidence data. In addition, the Committee's analyses suggest that if one allows the additive risks to depend on time, the excess risk of breast cancer seen in the LSS data is lower than the excess risks in the U.S. data while the relative risks are roughly comparable. Differences between the present findings and those of the earlier parallel analyses can be attributed to various factors, including additional follow-up for the U.S. cohorts, the introduction of the DS86 doses along with the consequent changes in the makeup of the LSS cohort, and the use of time-variable excess risk models.

For the case of breast cancer mortality, striking and highly significant ($p < 0.001$) differences were seen in both the estimated relative and additive excess risks within the Canadian cohort. In particular, the estimated risks per unit dose for the Nova Scotia women were about six times those for women in other provinces. It was suggested that this difference could be attributed to nonlinearities in the dose-response since the estimated doses for the women treated in Nova Scotia were much higher than those for women treated in other Canadian provinces. However, it was found that the differences in risk between Nova Scotia and the other provinces remain significant in a linear-quadratic dose-response model. This topic will be discussed further below.

The estimated excess relative risk per unit dose for women in the LSS was two to three times that for Canadian women from provinces other than Nova Scotia and about half that seen for Nova Scotia women. Neither of these differences were statistically significant ($p = 0.12$ for the LSS–non-Nova Scotia comparison and $p = 0.2$ for the LSS-Nova Scotia contrast). Since Japanese background rates are considerably lower than those in Canada, the LSS additive excess risks per unit dose were significantly less than those for Nova Scotia women ($p < .001$), but were not significantly lower than those for other Canadian women ($p > .5$).

The large, if not always statistically significant, differences in the magnitude of risk between and within the mortality cohorts complicate the choice of a preferred model for use in lifetime risk projections. The Committee's final choice was to estimate the level of risk per Gy using the pooled LSS and non-Nova Scotia CAN-TB data, but to use data on all women in both cohorts in describing temporal factors affecting the dose response. This choice was based upon an assumption that relative risks for breast cancer mortality and incidence should be roughly similar across cohorts and a judgment that the estimated relative risks for mortality due to breast cancer in the Nova Scotia subcohort of the CAN-TB series were larger than one might reasonably expect on the basis of estimated relative risks obtained from the incidence data.

In theory, the excess risk of breast cancer mortality or incidence following irradiation can be described equally well by suitably rich time-dependent relative or additive risk models. Indeed, for the mortality data, models of either type with similar numbers of parameters were found to fit the data equally well. However, for the incidence data, the relative risk models considered by the Committee had fewer parameters and, on the basis of deviance comparisons, fit better than did the additive risk models considered. Based upon these results and the fact that relative risks are less subject to bias as a result of incomplete (non-dose related) ascertainment, the Committee decided to use time-dependent relative risk models for their lifetime risk estimates for both breast cancer incidence and mortality.

Dose-Response Relationships

There is strong evidence for a flattening of the dose-response curve at high doses in all of the cohorts except the CAN-TB series, in which the curvature appears to be in the opposite direction, i.e., concave upward. It has been suggested that the flattening in the dose-response function at doses in excess of 4 Gy or so is the result of cell-killing effects. However it is unlikely that this curvature is solely a result of cell-killing since:

1. For the fluoroscopy cohorts (MASS-TB and CAN-TB) the doses were highly fractionated and it is unlikely that any single exposure involved doses which were high enough to cause appreciable cell-killing.
2. While it is likely that some survivors in the LSS received doses large enough to cause cell-killing, there is a large positive bias in the highest dose estimates as a result of the combination of: (a) random errors in the dosimetry; (b) the fact that only survivors are included in the cohort.

Since the emphasis in this report is on low dose effects, the committee decided to restrict the dose range in order to eliminate the need to consider the shape of the dose response at high doses.

Even when the women who received the highest doses are excluded, it is difficult to reach firm conclusions about the shape of the dose-response function at low doses. The incidence data provide weak evidence for a negative quadratic response ($p = 0.1$), while the Canadian mortality data indicate evidence for a positive quadratic component when the Nova Scotia data are included in the analyses. However, after allowing for this nonlinearity, a significant difference between the risk per unit dose in the two Canadian subcohorts remains. In contrast, if one allows for in the two Canadian subcohorts remains. In contrast, if one allows for this subcohort difference, the quadratic component of the dose response is not statistically significant ($p = 0.5$). Based upon these analyses the Committee's preferred models for breast cancer incidence and mortality are linear dose-response models.

Effects Due to Age at Exposure Effects

For both incidence and mortality there is a strong association between age-at-exposure and the subsequent risk of breast cancer following exposure to low-LET radiation. The general pattern is for the relative risks to decrease with increasing age at exposure. It is clear that relative risks for women who are over age 40 at exposure are quite small. There remains considerable uncertainty about the excess risk among women exposed under the age of 10, since these women are just now reaching ages at which baseline breast cancer incidence rates become appreciable.

In the incidence data, it was found that the estimated relative risks for women between 15 and 19 years old at the time of exposure in the NY-APM cohort were significantly lower than the risks for women initially exposed at the same ages in the other two cohorts ($p = 0.05$). Except for this effect, there were little variability and no significant differences in the relative risk estimates between the 0-9, 10-14, and 15-19 age-at-exposure categories. The estimated relative risk in the combined 0- to 19-year-old category (allowing for the reduced effect among the NY-APM 15-19 group) was significantly higher than that for the women in the 20-40 year age-at-exposure group. For women over age 40 at exposure, the excess relative risk estimate is about half of that for women who were between 20 and 40 when first exposed; however, this estimate is neither significantly lower than that for 20- to 40-year-olds nor significantly greater than 0. If one looks at the estimated excess relative risks for women under the age of 10 at exposure, a similarly ambiguous result is seen. As noted above, the point estimate of the excess relative risk for this group differs little from that for the non NY-APM 10- to 19-year-olds, but, because of the small number of cases (23) among women in this age-at-exposure group, their estimated relative risk is also not significantly greater than 0.

Although attempts were made to model the age-at-exposure effects on incidence as a log-linear trend or log-linear spline, it was found that these models did not fit the incidence data as well as step functions with discontinuities at age of exposure 20 and 40. Thus, in the committee's preferred model, the age-at-exposure effect on the excess relative risk is modelled as a step function with steps at ages 20 and 40.

In the case of breast cancer mortality, the highest estimated relative risks were seen among women aged 10 to 14 at exposure. The excess relative risk in this age group appeared to be significantly greater than that for women aged 15-19 at exposure. In contrast to the incidence data, there is as yet little evidence of any excess risk of breast cancer mortality among those exposed under the age of 10. However, the risk in this group was also not significantly lower than that seen among 10- to 14-year-olds. Because the total number of breast cancer deaths in the youngest age-at-exposure group is low (7), and because of the suggestion of an elevated risk in the incidence data, it was decided to pool the 0-9 and 10-14 age-at-exposure categories in the final model.

The excess relative risk of breast cancer mortality for women over age 40 at exposure was lower than that seen for women who were between 20 and 40 years of age when exposed ($p < 0.1$); in fact, the point estimate of the relative risk for this group was slightly, but not significantly, less than one. It was found that the variability in the excess relative risk as a function of age-at-exposure for women who were 15 or over at the time of exposure was best described by a decreasing log-linear trend in risk with age-at-exposure. As described in Chapter 4, the committee's final model for breast cancer mortality allows for this age-at-exposure trend together with an elevated risk for women who were under age 15 at exposure. The function is not constrained to be continuous at age 15.

The Effect of Time-After-Exposure

The Committee's analyses suggest that for both the incidence and mortality data the relative risks of breast cancer following exposure to low-LET radiation are not constant in time. The pattern that emerges from these analyses is that the relative risk for breast cancer incidence increases with time until about 15 years after exposure then begins to decrease. Similarly, the mortality data suggest that risks increase for about 20 years and then begin to decline. The decreases in the relative risk 15 to 20 years after exposure are of sufficient magnitude to result in predictions of decreases in the additive excess risks by the age of 50 for women who were exposed more than 20 years before.

For the case of breast cancer incidence, a log-quadratic model in log time-after-exposure was found to fit the data marginally better than a time-constant relative risk model. However, when the temporal pattern of risk was modeled as a log-linear spline in log time-after-exposure with a knot at 15 years after exposure, the fit was improved significantly ($p = 0.01$) relative to the time-constant model. The primary difference between the spline and quadratic models is that the spline yields a sharper peak and a less rapid decline in the risks following the peak than does a quadratic model. In order to assess the significance of the decrease in the excess relative risk after the peak, the committee considered a model in which and then remain constant thereafter. The unconstrained spline fit the data significantly better than this constrained model ($p = 0.02$). On the basis of these analyses the Committee's preferred breast cancer incidence model is a log-linear spline with a single knot at 15 years after exposure.

In the case of breast cancer mortality there is a suggestion ($p = 0.1$) of a temporal pattern similar to that seen in the incidence data. In particular, the risk appears to reach a peak at about 20 years after exposure. A log quadratic function of log time-since-exposure fit the data slightly better than a log-linear spline with a single inflection point knot. The Committee has chosen to use a log-quadratic model for the variation in the excess relative risk with time in its preferred risk model for breast cancer mortality.

TABLE 4E-4 BEIR V Breast Cancer Incidence Analysis—Preferred Model

Effect	Estimate	S.E.	Z	RR
Constant	−0.73	0.28	−2.61	1.48
Cohort effects				
NY-APM			[−0.80]	
MASS-TB			[−0.62]	
LSS			[1.13]	
Age-at-exposure effects				
<20	1.49	0.30	4.97	3.14
0–9			[−0.19]	
10–14			[0.31]	
15–19			[−0.17]	
NY-APM and 15–19	−2.26	1.65	−1.37	1.22
20–30			[−0.34]	
30–40			[0.34]	
40+	−0.90	1.06	−0.85	1.20
Time-since-exposure (T) effects				
Log(T/30)	−1.28	0.54	−2.37	
Log(T/15) if ($T < 15$)	6.67	3.92	1.70	

NOTES: RR is the relative risk at 1 Gy 30 years after exposure. For the constant term this is the risk for a woman exposed at age 20. For the other estimates RR is the relative risk in the corresponding subgroup.

In the fitted model the estimated excess relative risk at 1 Gy is a loglinear function of the parameters. Thus the estimated relative risk at dose d is

$$RR = 1 + d\exp(\mathbf{BZ}),$$

where **B** is the vector of parameter estimates and **Z** is a vector of covariates. The dose response is assumed to be linear.

Values in [ ]'s are the signed square roots of score statistics for a test of the null hypothesis that the corresponding parameter has no effect. These statistics are asymptotically distributed as standard normal deviates.

TABLE 4E-5 BEIR V Breast Cancer Mortality Analysis Preferred Model

Effect	Estimate	S.E.	Z	RR
Constant	−0.21	0.50	−0.41	1.81
Cohort effects				
CAN-TB Nova Scotia	1.14	0.42	2.72	3.54
CAN-TB Non-Nova Scotia			[−1.35]	
LSS			[1.35]	
Age-at-exposure (E) effects				
<15	1.39	0.55	2.50	4.25
(E − 15) if (E > 15)	−0.06	0.03	−1.95	
0–9			[−1.29]	
10–14			[1.28]	
15–19			[−0.34]	
20–30			[0.43]	
30–40			[0.75]	
40+			[−0.87]	
Time-since-exposure (T) effects				
Log(T/30)	−1.90	0.84	−2.28	
Log(T/30)**2	−2.22	1.38	−1.61	

NOTES: RR is the relative risk at 1 Gy 30 years after exposure. For the constant term this is the risk for a woman exposed at age 15. For the other estimates RR is the relative risk in the corresponding subgroup.

In the fitted model the estimated excess relative risk at 1 Gy is a loglinear function of the parameters. Thus the estimated relative risk at dose d is

$$RR = 1 + d\exp(\mathbf{BZ}),$$

where **B** is the vector of parameter estimates and **Z** is a vector of covariates. The dose response is assumed to be linear.

Values in [ ]'s are the signed square roots of score statistics for a test of the null hypothesis that the corresponding parameter has no effect. These statistics are asymptotically distributed as standard normal deviates.

Assuming that the risk of radiation-induced breast cancer does not appear until at least the age of 25, i.e., until the earliest ages at which naturally occurring breast cancer appears, and allowing a minimal latency period of five years for women over the age of 20 at exposure, the committee found no evidence that the temporal pattern of risk was affected by dose or age-at-exposure. It should be noted that although a 5-year minimum latency was used in the development of the preferred model, no excess breast cancer risk was observed within ten years of exposure. Therefore, in the calculations of lifetime risk for various patterns of exposure presented in Chapter 4, a 10-year minimum latency was assumed in life table calculations.

Final Models

The analyses which led to the Committee's preferred models have been discussed in the earlier sections of this annex. Tables 4E-4 and 4E-5 contain the estimates and standard errors for all of the parameters in the excess relative risk models used as a basis for the breast cancer risk estimates and lifetime risk projections presented in Chapter 4. These tables also include score test statistics for some of the other parameters considered in the modeling. For parameters included in the final models, Wald statistics (ratios of the parameter estimate to its standard error) are given (in the column labeled Z). The *p*-values reported in this annex were based upon likelihood ratio statistics which provide a better guide to the statistical significance of an effect.

References

Hr89 Hrubec, F., J. Boice, R. Monson, and M. Rosenstein. 1989. Breast cancer after multiple chest flouroscopies: Second follow-up of Massachusetts women with tuberculosis. Cancer Res. 40:229-234.

La80 Land, C. E., J. D. Boice, Jr., R. E. Shore, J. E. Norman, and M. Tokunaga. 1980. Breast cancer risk from low-dose exposure to ionizing radiation: Results of parallel analysis of three exposed populations of women. J. Natl. Cancer Inst. 65:353-365.

Mi89 Miller, A., P. Dinner, H. Risch, and D. Preston. 1989. Breast cancer mortality following irradiation in a cohort of Canadian tuberculosis patients. N. Engl. J. Med. (in press).

NCI86 National Cancer Institute. 1986. Forty-Five Years of Cancer Incidence in Connecticut: 1935-1979. NCI Monograph 70. Bethesda, Md.: National Cancer Institute.

Sh87 Shimizu, Y., H. Kato, W. J. Schull, D. L Preston, S. Fujita, and D. A. Pierce. 1987. Life Span Study Report 11, Part 1. Comparison of risk coefficients for site-specific cancer mortality based on the DS86 and T65DR shielded kerma and organ doses. RERF TR 12-87. Radiation Effects Research Foundation.

Sh86 Shore, R. E., N. Hildreth, E. Woodard, P. Dvoretsky, L. Hempelmann, and B. Pasternack. 1986. Breast cancer among women given X-ray therapy for acute postpartum mastitis. J. Natl. Cancer Inst. 77:689-696.

To87 Tokunaga, M., C. E. Land, T. Yamomoto, M. Asano, S. Tokuoka, H. Ezaki, and I. Nishimori. 1987. Incidence of female breast cancer among atomic bomb survivors, Hiroshima and Nagasaki, 1950-1980. Radiat. Res. 112:243-272.

ANNEX 4F: UNCERTAINTY, PROBABILITY OF CAUSATION, AND DIAGNOSTICS

Uncertainty

Estimates of radiation risks formulated on the basis of epidemiological data are far from precise. The data show, as expected, considerable sample variation due to the relatively small number of cases in a given category. Such statistical uncertainties are additional to those arising from other sources which are not readily evaluated. These include uncertainties inherent in dose estimates, in the selection of an appropriate risk model, and in the applicability of risk estimates measured in one populaton to other exposed groups.

Population Effects

A Japanese population is the most important source of data for this report, and for some types of cancer the only source. Since baseline (naturally occurring) cancer rates are different in the U.S. from those in Japan for many kinds of cancer, it is not clear whether cancer risks derived in one population are applicable to the other, and if so, whether relative or absolute risks should be used. The answer to this question may vary from cancer site to site; in fact, it may be that neither absolute nor relative risks can be extrapolated with assurance.

The general applicability of the experience of the Japanese A-bomb survivors is uncertain on additional grounds. Most human exposures to low-LET ionizing radiation are to x rays, while the A-bomb survivors received low-LET radiation in the form of high energy gamma rays. These are reported to be only about half as effective as ortho-voltage x rays (ICRU86). While that is not a conclusion of this Committee, which did not consider this question in detail, it could be argued that since the risk estimates that are presented in this report are derived chiefly (or exclusively) from the Japanese experience they should be doubled as they may be applied to medical, industrial, or other x-ray exposures.

Certification of Cause of Death

An additional source of uncertainty that affects the estimates of risk of death from specific cancers is the fact that specification of cause of death on death certificates (the source of data for almost all analyses of mortality) is not always accurate. The Committee has been provided with data by the RERF leading to the conclusion that great specificity as to cancer site cannot be justified on the basis of certificate-based data (e.g., cancer of the uterus is reasonably well reported, but not cancer of the uterine corpus). A further conclusion is that, at least in that body of data, the accuracy of diagnosis from death certificates declines rather sharply beyond age 75, to the point that little reliance can be placed on the data for specific sites. The Committee has refrained from basing analyses on data that it considers unreliable.

Sex Differences

Baseline cancer rates differ markedly between the sexes for most forms of cancer. The effect of radiation may, then, also be different for males and females. Sex was included specifically in all of the models that were fitted except for the group "other" cancers and for leukemia, where the effect was small and not statistically significant. Where sex is included in the models, uncertainties associated with sex differences are taken into account explicitly. Because sex does not appear in the final models for leukemia and "other" cancers, a residual uncertainty of 10% is assessed in the risk estimates for these cancers.

Time-Related Effects

It is difficult enough to determine the cancer risk over a lifetime; if one asks what is the risk at a particular time following exposure, the number of cases available for analysis becomes so small as to frustrate attempts at direct estimation of risks. This problem is avoided by estimating instead a mathematical function that describes the time-course, but that function is subject to uncertainties of two kinds: the proper functional form to use in the first place, and the values of the parameters that enter into it.

Age-Related Effects

How does radiation sensitivity vary with the age of the person exposed? Is it true that very young children are at greater risk than older persons? Is there some age after which sensitivity disappears and there is no risk? The Committee has addressed these questions explicitly in devising mathematical models for cancer risk as functions of kinds of cancer, sex, age at exposure, and time after exposure (latency). All of these factors were considered for each site for which models were fitted. For some cancers, not all of these factors were influential. For example, the leukemia model does not vary by sex, and the model for respiratory cancer does not depend upon age at exposure. An especially difficult problem is encountered at the very youngest and oldest ages; since there were few cases of breast cancer in women more than 55 years of age at the time of exposure, the risk of breast cancer in such women is poorly estimated. Similarly, since there is no follow-up information from the Life-Span Study until 5 years after exposure, the risk of death from leukemia after a latent period of 5 years or less are rather uncertain. It will be noticed, however, from the accompanying table of uncertainty that large geometric standard deviations usually apply to quite small estimates of risk, so that although the uncertainties may be large as proportions of the risk estimates, their absolute values are not large.

Shape of the Dose-Response Curve

Is the cancer risk from a given dose of radiation strictly proportional to the dose? Are larger doses more effective than linear extrapolation of low dose risks would imply? Are the effects of repeated doses, separated in time, the same as if the entire dose had been delivered at once? Are the effects of a given total dose received at very low dose rates the same as those from the same dose at high dose rates? Are there doses so small that they have no effect? Specifically, since the effects are measured in populations that have had rather large doses delivered at a very high dose rate, how shall we use that information to assess the effects of small doses, received at low dose rates? The latter problem is faced by those who must establish limitations for occupational and general population exposures. As is suggested in Chapter 1 of this report, it may be desirable to reduce the estimates derived here by a "Dose Rate Effectiveness Factor" (DREF) of about 2 for application to populations or persons exposed to small doses at low dose rates. On the other hand, as mentioned above, the estimates could be too small by a factor of about 2 for application to the consequences of x-ray exposures. It may be that these two factors (DREF and the relative biological effectiveness of gamma rays) could, in some cases, simply offset each other.

Procedures Employed

The approach taken here follows that used by the NIH Committee in its report on the Radioepidemiological Tables (NIH85). In brief, that approach is to assess the magnitude of the error that may be attributable to each independent component of an estimate and then to combine the individual estimates into an overall estimate. Some of the components of error, such as the statistical variability in the number of deaths in a population group, can be evaluated in a conventional way; others, however, like the uncertainty associated with the application of risks in a Japanese population to a U.S. population, cannot be evaluated objectively. Instead, we resort to a consensus of expert opinion as to the uncertainty, expressed in a number on a scale commensurate with ordinary statistical measures of variability.

Uncertainty is expressed as the "Geometric Standard Deviation," (or GSD), that is in *ratio* terms; by an uncertainty of 1.2 (20%) it is meant that the range of uncertainty of the estimate is from its value *divided* by 1.2 to the value *multiplied* by 1.2. If, for example, some excess relative risk is estimated to be 0.3 per Gy, with an uncertainty (exp σ) of 1.4 ($\sigma = 0.336$), we would mean that it is believed that the chance is 68% that the value lies in the range from 0.3 divided by 1.4 = 0.21 to 0.3 times 1.4 = 0.42. We call such an interval a "68% credibility interval." We use the term "credibility interval," instead of the commonly used statistical term "confidence interval" because the values are obtained, at least in part, by judgment, not calculation.

A basic assumption is that the error in the final estimate of risk is distributed lognormally, that is, that the logarithms of the errors are normally distributed. This assumption gains credibility from the fact that the logarithm of the total error is the sum of the logarithms of the individual components of error. There is a well-known mathematical result that the distribution of a sum of variables will be approximately normal, so the assumption is unlikely to be seriously wrong. In order to obtain an interval with any desired credibility coefficient, say 90%, the factor exp ($1.645 \times \sigma$) would be used. In the example above, σ was assumed to be 0.336, so a 90% interval would require division and multiplication by $\exp(1.645 \times 0.336) = 1.74$. The 90% interval on the estimated risk of 0.3 would be from 0.17 to 0.52.

The value of the error attributable to all of the independent sources is obtained by the usual method of calculation for the logarithmic errors. That is, if σ_T denotes the standard deviation of the logarithm of the total error, and σ_1, σ_2, etc. denote the standard deviations of the logarithms of the individual components, then

$$\sigma_T = \sqrt{[(\sigma_1)^2 + (\sigma_2)^2 + \ldots]}.$$

Models used in this report are, generally, of the form:

$$\text{Excess Relative Risk} = D \exp \{\beta_0 + \beta_1 X_1 + \beta_2 X_2 + \ldots\}.$$

where D represents the organ dose equivalent in sievert and the X's are covariates such as age at exposure, etc., and the β's are their respective coefficients.

If we denote the logarithm of the excess relative risk by $\ln(R)$, we have, then,

$$\ln(R) = \ln(D) + \beta_0 + \beta_1 X_1 + \beta_2 X_2 + \ldots$$

We suppose that the covariates are known without error, only their coefficients, which have been calculated from the available data, will have statistical error. Then the variance of $\ln(R)$, which we call V will be:

$$V = V(\beta_0) + 2X_1 \operatorname{Cov}(\beta_0, \beta_1) + \ldots$$

The maximum-likelihood fitting procedures employed supply the variance-covariance matrix applicable to the coefficients, and these values have been used to obtain the variance of V and its standard error.

Uncertainties External to the Parametric Model

Although it can be assumed that such factors as age and time of death are known without error, there can be no such assurance concerning the estimates of radiation dose. Dose estimates for medical exposures are based upon recorded parameters of the x-ray exposure; such estimates cannot be exact, but the uncertainty to be attributed to them is not known. Dose estimates for the Japanese A-bomb survivors are based upon statements by the survivors concerning their location at the time of the bombing, their shielding situation, and estimates of the air dose curves, the exact location of the hypocenters, shielding characteristics of building materials and, for doses to specific organs, the attenuation of external dose by tissues overlaying the organ of interest. For breast cancer, especially, the orientation of the survivor with respect to the direction of the bomb is of importance, but cannot be known with any precision. The magnitude of the uncertainty in the new DS86 dose estimates for A-bomb survivors is still being evaluated. Preliminary assessments indicate that bias in the risk estimates resulting from random errors in the dose estimates is about 10% when organ doses are limited to 4 Sv, as is the case here (Pi89). Further review of this issue, including the role of bias in the estimated neutron kerma, is required.

Although the magnitude of some of the sources of uncertainty (such as the effect of statistical variability on risk estimates) can be evaluated explicitly, others, like the error of "transportation" (application of risks determined in one population to another population) cannot be. In such cases we rely on consensus judgment; we judge what is the range within which it is believed that the variable lies with 95 percent "credibility." A "standard deviation" can be obtained by dividing the width of that range by 3.92. All of the standard deviations, both those actually calculated and those estimated as just explained, can be combined by the methods described above to obtain a combined measure of uncertainty which we call a "standard error" and used to obtain "credibility intervals" by the same procedure that would be used to obtain "confidence intervals" were the uncertainty measures really statistically determined standard errors.

The sources of uncertainty that can be evaluated in a straightforward way, using conventional statistical theory, are those that derive from sampling variability as it affects the fitting of specific models for the excess risk of particular cancers that result from radiation exposure. Such models have been fit for cancer mortality from leukemia, and for cancers of the respiratory system, the digestive system, the female breast and other sites. Most of the models have used the data on the Japanese A-bomb survivors, for whom 40-year follow-up data have been made available by the Radiation

Effects Research Foundation. As discussed in Annex 4E, several additional sources of epidemiologic data have been used for breast cancer.

Our task has been somewhat simplified by the fact that several of the factors that contribute to uncertainty, mentioned above, were considered explicitly in the model-fitting procedures, and their uncertainties are incorporated in the model uncertainties. These include age at exposure, time from exposure (latent period), sex, and the possible contribution of the square of dose in addition to radiation dose itself. Only for leukemia was the dose-squared factor significant. In any case, the statistical variability of the models includes the contributions from all of these factors. The most important element that is not accounted for in the models themselves is the *population* factor, that is, the applicability of risks determined in a Japanese population to populations of different ethnic composition, having different diets, industrial exposures, and, generally, different life styles. For cancer of the breast, however, data were available for mortality not only in Japanese but also in Canadian and U.S. women. Interestingly, for reasons that have not yet been elucidated, the only important differences were within the Canadian series, where it appeared that women in Nova Scotia had significantly different risks from those in other parts of Canada and from the other series. Apart from the Nova Scotia series there were no significant differences among the other series. We evaluate the population uncertainty at 20%, that is, the GSD corresponds to an uncertainty factor of 1.2.

Another source of uncertainty, which cannot be captured by usual statistical methods, is possible *mis-specification* of the model finally fitted to the data. Many variables (factors) were considered as candidates for inclusion in the final models; those selected were often the "best" in the statistical sense. Nevertheless, there can be no assurance that the models finally chosen were "correct" in that the factors included were just the right ones. The importance of possible mis-specification was evaluated by considering the variations in estimated risk for the fitted different models described in Annex 4D, weighting the risks from the various models by the reciprocals of their deviances. By this test, model mis-specification for males (1.16) was larger than for females (1.08). For children aged 5 at exposure, the mis-specification uncertainty is about 1.55 for both sexes.

Results

Uncertainties that result from the model fitting are displayed in Table 4F-1. Unlike the Monte Carlo generated estimates of uncertainties in lifetime risk given in Chapter 4 and Annex 4D, the uncertainties in Table 4F-1 are shown explicitly as functions of age at exposure, latency and sex when these factors are significant. This level of detail is not practical with

Monte Carlo techniques. It will be seen that the models for respiratory and digestive cancers do not show risk variation by age at exposure, so that the uncertainty factor for each sex varies only by time after exposure. For leukemia and the group "other cancers" there is significant variation by age and latency, but sex seems not to play an important role. The possible effect of sex on the uncertainty in these two cases is considered below.

In general, where data are relatively sparse, as is true for leukemia, the uncertainties are large, varying from nearly 2 to 8 for different ages and latencies. Uncertainties are usually not large for respiratory or digestive cancers or for breast cancer except for a short latency of 10 years.

Uncertainties not accounted for in the model themselves (referred to as non-model) derive from population differences (e.g., Japanese vs. Caucasians vs. Blacks) and uncertainty in the dosimetry estimates. The Committee's assessment of the magnitude of their contributions in terms of geometric standard deviations (GSD) are:

(A) Model mis-specification	
Males	:1.16
Females	:1.08
(B) Population differences	:1.20
(C) Dosimetry system	:1.10
(D) Sex (leukemia and "other" cancers)	:1.10

TOTAL GSD

All Except Leukemia and "Other"		"Other" Cancers and Leukemia	
Males	Females	Males	Females
1.29	1.25	1.31	1.27

Comparison with Table 4F-1 indicates that the uncertainties in the Committee's preferred model due to sampling variation are usually much larger than those due to the factors considered above. Where required, the non-model component shown above, can be added in quadrature to the model-based component shown in the tables using the methods outlined above.

Probability of Causation

In the Report of the National Institutes of Health Ad Hoc Working Group to Develop Epidemiological Tables (NIH85), the formula for the *PC* (probability of causation) is given as:

$$PC = R/(1 + R),$$

where R, really $R(D,X)$, is the *excess* relative risk that results from the dose D to a person with characteristics X. The *PC* is an estimate of the probability

TABLE 4F-1 Estimates of the Excess Relative Cancer Risk from 0.1-Sv Acute Dose and Their Geometric Standard Deviations (GSD) due to Sampling Variation

Cancer Type	Age at Exposure	Time After Exposure	Male GSD	Male Risk	Female GSD	Female Risk
Breast cancer mortality	5	15	—	—	1.90	0.418
		25	—	—	1.60	0.427
		35	—	—	1.57	0.230
		45	—	—	1.89	0.105
	15	15	—	—	1.90	0.418
		25	—	—	1.60	0.427
		35	—	—	1.57	0.230
		45	—	—	1.89	0.105
	25	15	—	—	1.77	0.056
		25	—	—	1.54	0.057
		35	—	—	1.60	0.031
		45	—	—	1.99	0.014
	35	15	—	—	1.90	0.030
		25	—	—	1.76	0.030
		35	—	—	1.85	0.016
	45	15	—	—	2.31	0.016
		25	—	—	2.25	0.016
	55	15	—	—	2.99	0.008
Breast cancer incidence	<20	15	—	—	1.45	0.52
		25	—	—	1.24	0.27
		35	—	—	1.30	0.18
		45	—	—	1.44	0.13
	20 to 39	15	—	—	1.35	0.12
		25	—	—	1.26	0.06
		35	—	—	1.40	0.04
		45	—	—	1.57	0.03
	≥40	15	—	—	2.90	0.05
		25	—	—	2.88	0.02
		35	—	—	2.99	0.02
Respiratory cancer mortality	All ages	15	1.59	0.096	1.47	0.196
		25	2.03	0.046	1.76	0.094
		35	2.63	0.028	2.27	0.058
		45	3.23	0.020	2.80	0.040
Digestive cancer mortality	All ages	All times	1.50	0.081	1.33	0.141
		>10 yr	1.50	0.081	1.33	0.141
			1.88	0.011	1.77	0.019
			1.88	0.011	1.77	0.019

(*continued*)

Table 4F-1 *Continued*

Cancer Type	Age at Exposure	Time After Exposure	Male or Female	
			GSD	Risk
Leukemia mortality	≤20	<15	2.80	3.63
		16 to 25	2.53	0.291
		≥26	3.32	0.027
	>21	≤25	1.83	0.287
		26 to 30	2.52	0.139
		≥31	3.32	0.027
Other cancer mortality	5	All times >10 yr	1.53	0.123
	15		1.40	0.097
	25		1.31	0.061
	35		1.45	0.038
	45		1.75	0.024
	55		2.17	0.015
	65		2.71	0.009

that a given radiation dose in the history of a patient was the cause, in some sense, of a subsequent malignant neoplasm that has actually occurred. The value of R in any given case can be obtained from the formulas provided in Chapter 4. Since the formulas for R for the malignancies other than leukemia are linear functions of the dose, D, Table 4F-1, can be used to obtain, not only the value of R but also its Geometric Standard Deviation, which leads immediately to an estimate of the associated uncertainty in the *PC*. These formulas do not, of course, take into account any lack of precision in the estimate of the radiation dose to the relevant organ; often this uncertainty will be comparable in magnitude to the uncertainty inherent in the models.

The data for breast cancer incidence in Table 4F-1 shows that the excess relative risk for breast cancer in a woman aged 20 through 39, 25 years after exposure, is 0.06 per 0.1 Gy (10 rads). The GSD (uncertainty) is 1.26. Assume that a woman who had an exposure that gave a dose of 2 rads to the breast at age 25 developed a breast cancer 25 years later, at age 50. Then the excess relative risk (R) would be $2/10 \times (0.06) = 0.012$. A 68% "credibility interval" for R would be from 0.01 to 0.015 (dividing and multiplying by the GSD, 1.26) and the *PC* would then be calculated as:

Lower limit	$:0.010 \div (1+0.0010) = 0.010$, or 1%
Best estimate	$:0.012 \div (1+0.012) = 0.0118$, or 1.2%
Upper limit	$:0.015 \div (1+0.015) = 0.0147$, or 1.5%

If a 90% or 95% credibility interval is desired, the GSD (1.26) must be raised to the power 1.64 or 1.96, respectively; the values turn out to be 1.46 and 1.57 and the intervals become:

90% :0.8% to 1.7%
95% :0.8% to 1.8%

Similar calculations can be made for any of the models presented in Chapter 4. Values for the GSD can be obtained by interpolation in Table 4F-1 with sufficient accuracy.

Diagnostic Examination of the Committee's Risk Models

Throughout its development of analytical models of cancer risk as a function of dose and other variables, the committee used a number of diagnostic tests to examine the degree of correspondence between a given model and the data on which it is based (Be80, Mc83). As noted in Chapter 4, decrements of deviance were used as a measure of the improvement in model "fit" gained by adding additional terms. This is, however, not a test of concordance between the data and the model as it is obvious that while the difference between the respective deviancies can perhaps discriminate with an acceptable fineness between two rival nested models, this procedure does not guarantee that either rival "fits" very well. It is important to know how well a given model fits, or describes, a given set of data, not just that it describes the data "better" than an alternative model.

There are several aspects to the issue of concordance. A "good fit" does not prove that the model is correct; it simply suggests that, at a chosen level of significance, the sample at hand does not provide any empirical evidence against the model in question. A "poor fit" suggests that there are problems with either the model or with the data. In either case, however, the issue of fit, if based solely on the the criterion of a goodness of fit statistic, such as Pearson's chi-squared χ^2, may lead to errors of inference simply because the assumptions required for the validity of the chi-squared approximation to the sampling distribution of the selected measure of concordance are not satisfied.

A measure of whether a model "fits" a given set of Poisson distributed data is the difference between observed and fitted values. The concordance, or goodness-of-fit, of a model of size k with a set of data of size n can be described by the "distance" $(\mathbf{y} - \hat{\mu})$ between the vector of observations, $\mathbf{y} = (y_1, y_2, ..., y_n)$ and the vector of expectations, $\hat{\mu} = (\hat{\mu}_1, \hat{\mu}_2, ..., \hat{\mu}_n)$. Here y_i is the *observed* number of cancer cases in the ith cell of the cross-classification of the data, $\hat{\mu}_i$ is the number of cases *expected* if the estimated model is correct and n is the number of cells, or records, in the cross-classification. The components, $(y_i - \hat{\mu}_i)$, $1 \leq i \leq n$, of $(\mathbf{y} - \hat{\mu}_i)$, or more properly, functions thereof, are described as the *residuals*.

There are two forms of residuals that are most commonly deployed for Poisson regression models. These are the deviance, d_i, and the Pearson chi-squared, χ_i; these are defined as, $d_i = \text{sgn}(y_i - \hat{\mu}_i)\ \{2[y_i \log(y_i/\hat{\mu}_i) -$

$(y_i - \hat{\mu}_i)]\}^{1/2}$ and $\chi_i = (y_i - \hat{\mu}_i)/\sqrt{\hat{\mu}_i}$. Note that d_i includes the ratio $y_i/\hat{\mu}_i$ and χ_i includes the ratio $(y_i/\sqrt{\hat{\mu}_i}$. Thus, it is obvious that as $\hat{\mu}_i$ approaches zero when $y_i > 0$, both d_i and χ_i become quite large.

On the null hypothesis, H_0, that the model is correct, the respective sums of squares of d_i and χ_i are, for "large" μ_i, distributed as chi-squared variates on $(n - k)$ degrees of freedom where n is the sample size and k is the size of the model; for a model with s strata and p free parameters, $k = s + p$. The sums of squared residuals are the aggregate statistics of goodness-of-fit:

Deviance, $D = \Sigma d_i^2 \sim \chi^2(n - k)$
Pearson chi-squared, $\chi^2 = \Sigma\chi_i^2 \sim \chi^2(n - k)$.

In general, for a model that "fits" the data and for which the set of $\hat{\mu}_i$ are acceptably "large", $\Sigma d_i^2 \simeq \Sigma\chi_i^2 = (n - k)$.

There is another form of residual that is quite useful to Poisson regression models of "sparse" data. This is the Freeman-Tukey residual, g_i, defined as (Bi75, Fr50, Fr83a, Fr83b, Ho85):

$$g_i = \sqrt{y_i} + \sqrt{y_i + 1} - \sqrt{4\hat{\mu}_i + 1}$$

g_i is a standardized residual, that is, it is distributed as $N(0, 1)$. Thus, Σg_i^2 is distributed approximately as chi-squared on $(n - k)$ degrees of freedom.

It is immediately evident that the g_i residuals will be "well-behaved" at both $y_i = 0$ and $\hat{\mu}_i \to 0$. This behavior is quite different from that of the d_i and χ_i residuals. That is, g_i is a more robust measure of discrepancy, $(y_i - \hat{\mu}_i)$, between the observed and expected numbers of cases when $\mu_i \to 0$, than are the d_i and χ_i residuals, in sparse data.

In using the aggregate statistics, Σd_i^2 or $\Sigma\chi_i^2$, as measures of overall fit, it is common practice to combine, or pool, the sparse observations in the cells of the cross-classification until $\hat{\mu}_i \geq 1$. This maneuver "adjusts" the number of degrees of freedom (*df*) by reducing the number, n, of cells. Then, Σd_i^2 and $\Sigma\chi_i^2$ are distributed asymptotically as chi-squared on $(n' - k)$ degrees of freedom where $n' < n$. But for the sum of squares of Freeman-Tukey residuals, Σg_i^2, an alternative and more satisfactory adjustment to the degrees of freedom, can be achieved (Fr83b, Ve81) by subtracting (Tukey correction) the sum, $c = \Sigma(1 - \hat{\mu}_i)^2$, from the usual measure of degrees of freedom to give the *adjusted degrees* of freedom:

$$df^* = (n - k) - c.$$

The sum is over all n^* cells for which $\mu_i < 1$. Then, Σg_i^2 is distributed as chi-squared on df^* degrees of freedom and the pth quantile, u_p, of the cognate chi-squared distribution can be obtained from the fact that $\sqrt{2\Sigma g_i^2}$

is distributed approximately Normally with expected value $\sqrt{2df^* - 1}$ and variance 1 (Br65):

$$u_p = \sqrt{2\Sigma g_i^2} - \sqrt{2df^* - 1}.$$

When the data are "sparse," then many of the $\hat{\mu}_i < 1$, and $\Sigma d_i^2 \neq \Sigma g^2 \neq \Sigma \chi_i^2 \neq (n - k)$. In particular, Σd_i^2 and $\Sigma \chi_i^2$ are inflated, as each is a function of μ_i^{-1}. In general, $\Sigma \chi_i^2$ is inflated more than $\Sigma {d_i}^2$ since χ_i is a stronger function of μ_i^{-1} than d_i. Moreover, d_i and χ_i are not defined for $\hat{\mu}_i = 0$, although g_i is. Thus, as Breslow (Br85) has cautioned, for sparse data neither the deviance, $\Sigma {d_i}^2$, nor Pearson chi-squared, χ_i^2 statistic, is suitable as an aggregate measure of the concordance of model and data. If data are not too sparse, Tukey has pointed out that Σg_i^2 is still a useful aggregate statistic of goodness-of-fit (Fr83, Ve81).

Tests of the Committee's Preferred Models

When stratified by dose, age and time, the LSS data for leukemia, digestive, respiratory and the group "other cancers" are very sparse. The proportions of records for which there are one or more cases, $y_i \geq 1$, are 0.061, 0.336, 0.155, and 0.259, respectively. This results, of course, in an excessive number of records for which $\hat{\mu}_i < 1$ for any model. Therefore, the committee found the analysis of Freeman-Tukey residuals to be the most useful measure of goodness of fit. It is well-known that, on occasion, aggregate statistics of concordance such as the deviance, Σd_i^2, may indicate that a model "fits the data" but examination of the set of component residuals, d_i, $1 \leq i \leq n$, may disclose that the model is, "grossly inconsistent with the data" (Ro86). However, on other occasions, especially if the data are sparse, the selected aggregate statistic of fit, say Σg_i^2, may lead to the opposite inference, indicating that a model does not fit the data when in fact, more sensitive tests that are based on the distributions of the g_i residuals rather than on the (sampling) distribution of their sum of squares may, as shown below, disclose a quite acceptable degree of concordance with the data.

Some of the results of analyses of the residuals of the respective BEIR V models of the LSS (DS86) data are presented in Table 4F-2. It will be noted that the deviance, d_i, and chi-squared, χ_i, residuals are greatly inflated; moreover, their respective distributions are decidedly *skewed*. However, the cognate Freeman-Tukey residuals, g_i, are much smaller and more "well-behaved," with rather symmetric distributions in each case, and with means more nearly equal to zero than is the case for the d_i and χ_i residuals, as shown in Figure 4F-1 for cancers at specified sites. The figure presents the superpositions of the histograms for two samples of random variates, both of sizes n, where n is the number of records in the

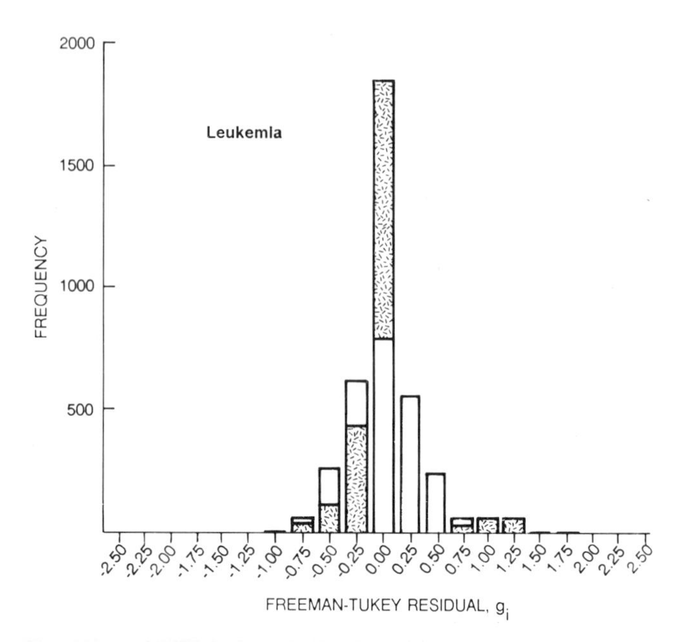

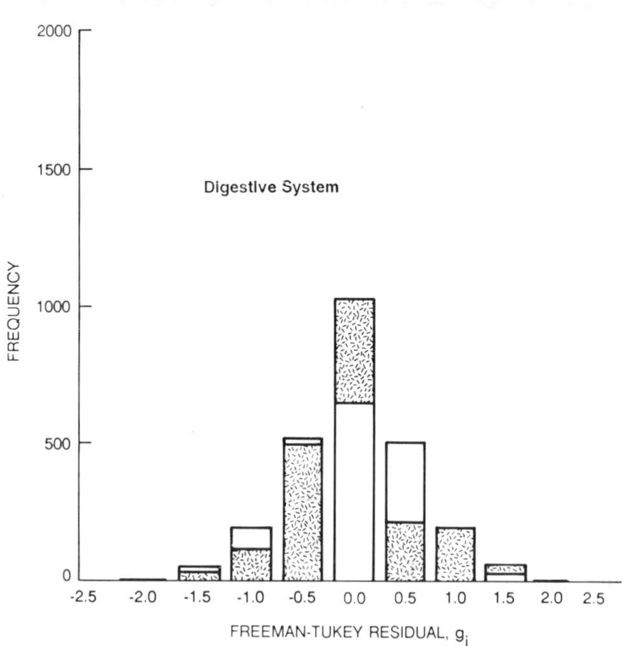

FIGURE 4F-1 Distribution of Freeman-Tukey residuals under the committee's models for leukemia, cancers of the respiratory and digestive systems, and the group "other cancers" (stippled) compared to a normal distribution with the same mean, variance, and sample size.

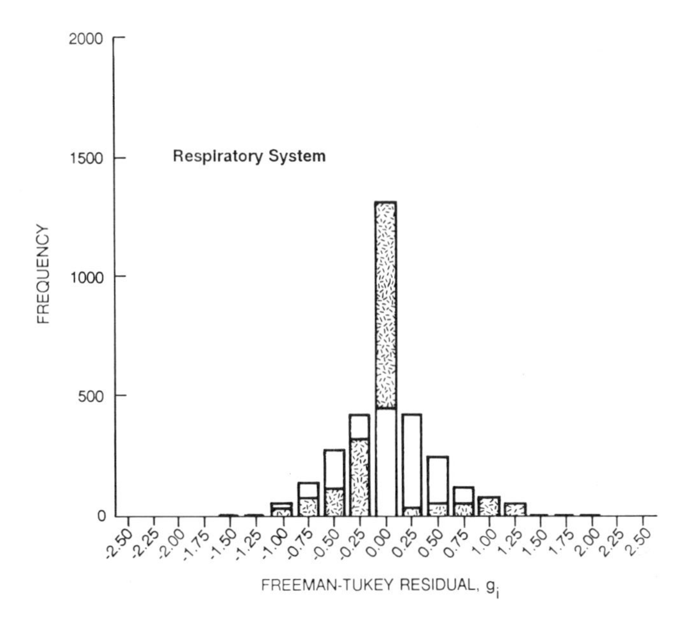
2000
1500
1000
500
0
Respiratory System
FREQUENCY
-2.50
-2.25
-2.00
-1.75
-1.50
-1.25
-1.00
-0.75
-0.50
-0.25
0.00
0.25
0.50
0.75
1.00
1.25
1.50
1.75
2.00
2.25
2.50
FREEMAN-TUKEY RESIDUAL, g_i

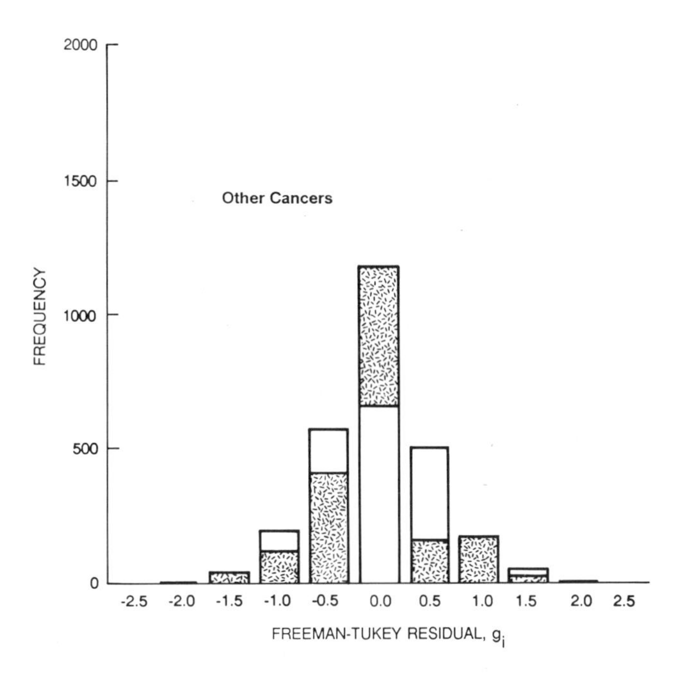
2000
1500
1000
500
0
Other Cancers
FREQUENCY
-2.5
-2.0
-1.5
-1.0
-0.5
0.0
0.5
1.0
1.5
2.0
2.5
FREEMAN-TUKEY RESIDUAL, g_i

TABLE 4F-2 Summary of Residual Analysis for BEIR V Models

Tumors	d(Min)	d(Max)	χ(Min)	χ(Max)	g(Min)	g(Max)
Leukemia	−1.375	2.389	−0.973	5.853	−1.187	1.812
Digestive	−2.739	3.309	−1.937	25.418	−3.001	2.393
Respiratory	−2.143	3.197	−1.515	10.862	−2.191	2.074
Other	−2.220	3.127	−1.685	12.091	−2.295	2.301
	No. of $d_i \leq -2$	No. of $d_i > 2$	No. of $\chi_i \leq -2$	No. of $\chi_i > 2$	No. of $g_i \leq -2$	No. of $g_i > 2$
Leukemia	0	8	0	56	0	0
Digestive	5	21	0	69	5	7
Respiratory	2	18	0	54	2	2
Other	2	29	0	86	2	5

	Sum of Squared Residuals			
	df	Σd_i^2	$\Sigma \chi_i^2$	Σg_i^2
Leukemia	2,266	498	811	244
Digestive	1,909	1,191	2,159	806
Respiratory	1,888	710	1,203	432
Other	1,904	1,124	1,774	712

NOTE: a) Deviance residual: $d_i = \text{sgn}(y_i - \hat{\mu}_i)\{2[y_i\log(y_i/\hat{\mu}_i) - (y_i - \hat{\mu}_i)]\}^{1/2}$

b) Pearson chi-squared residual: $\chi_i = (y_i - \hat{\mu}_i)/\sqrt{\hat{\mu}_i}$.

c) Freeman-Tukey residual: $g_i = \sqrt{y_i} + \sqrt{y_i + 1} - \sqrt{4\hat{\mu}_i + 1}$.

y_i = observed cases. $\hat{\mu}_i$ = fitted cases. $\hat{\mu}_i = \hat{\lambda} c_i$. c_i = person years at risk for i^{th} record.

Deviance $= \sum^{n''} d_i^2$.

Chi-squared $= \sum^{n''} \chi_i^2$.

n'' = number of records for which $\hat{\mu}_i > 0$. See Table 4F-3.

$\sum^{n} g_i^2$ = sum of squared Freeman-Tukey residuals.

n = total number of records. See Table 4F-3.

respective sample as listed in Table 4F-3. These are (1) the n Freeman-Tukey residuals, g_i, of the BEIR V models of the sample (stippled); and (2) the n random variates drawn from a Normal population with the mean and variance equal to those of the sample of Freeman-Tukey residuals.

Note in Figure 4F-1 there is an excess (with respect to the Normal) of g_i in the vicinity of $g_i = 0$. This is evidence of the extreme sparseness of these μ_i data, where there are many records for which $y_i = 0$. Since $\Sigma y_i = \Sigma \hat{\mu}_i$, it follows that there are, as well, many small residuals, $g_i = \sqrt{y_i} + \sqrt{y_{i+1}} - \sqrt{4\hat{\mu}_i + 1}$. The respective distributions of Freeman-Tukey residuals are described more precisely in Table 4F-4.

TABLE 4F-3 Records, Strata, Parameters, Degrees of Freedom, and Cases for BEIR V Models

Tumor	n (Records)	s (Strata)	p (Parameters)	df[a]	Total y_i (Cases)	Total n'[b]	n''[c]
Leukemia	2,575	302	7	2,266	173	158	1,010
Digestive	2,162	250	3	1,909	2,193	726	1,789
Respiratory	2,141	250	3	1,888	555	331	1,181
Other	2,156	250	2	1,904	1,162	558	1,741
				TOTAL	4,083		

[a] df = degrees of freedom = $n - s - p$.
[b] n' = number of records for which $y_i \geq 1$.
[c] n'' = number of records for which $\hat{\mu}_i > 0$.

NOTE: Deviance, $\sum^{n''} d_i^2$, and Pearson chi-squared, $\sum^{n''} \chi_i^2$, are calculated from those records, n'' in number, for which $\hat{\mu}_i > 0$, since d_i and χ_i are *not* defined at $\hat{\mu}_i = 0$. However, the Freeman-Tukey residuals are defined at $\hat{\mu}_i = 0$ and hence the sum Σg_i^2 is over all n records.

It should be noted in Table 4F-2 that for Leukemia, no value of g_i exceeds |2|. For Digestive tumors, only 12 g_i exceed |2|. For Respiratory and Other tumors, the respective numbers having $g_i > 2$ are also acceptably small, see Frome (Fr83a). Thus, on the evidence of the distributions of the Freeman-Tukey residuals (Mc83), the BEIR V models are not inconsistent with the LSS (DS86) data: the number of g_i exceeding |2.0| is very small compared to the number of records, *n*, and there is no strong pattern (suggestive of model mis-specification) in plots of g_i against either the response or predictor variables (Gi84).

The Bias and Variance of the Sample Estimate of the Cross-Over Dose Θ_1 and Dose-Rate Effectiveness Factor Θ_2 for Leukemia Dose-Response

There are two important classes of problems in the study of somatic responses to low doses of low-LET radiation for which the solutions devolve into inferences on a ratio, say Θ, of two regression parameters. These ratios are the cross-over dose, Θ_1, and the dose-rate effectiveness factor, Θ_2.

1. The dose at which the linear and quadratic terms in a linear quadratic (*LQ*) dose-response function are equal is called the *cross-over dose*. This dose is defined by the ratio, $\Theta = \beta_1/\beta_2$, where β_1 is the coefficient of the dose, *D*, and β_2 is the coefficient of D^2 in the *LQ* model. It should be noted that for the BEIR V *LQ* model of leukemia mortality the precision of the respective estimates, $\hat{\beta}_1$ and $\hat{\beta}_2$ is quite low:

$$\hat{\beta}_1/\sqrt{\mathrm{Var}(\hat{\beta}_1)} = 0.864 \text{ and } \hat{\beta}_2/\sqrt{\mathrm{Var}(\hat{\beta}_2)} = 0.865.$$

Note also that these are rather less than are the cognate precisions of the LQ-L model of leukemia incidence described in Table V-8 of the BEIR III Report (Na80): $\hat{\beta}_1/\sqrt{\mathrm{Var}(\hat{\beta}_1)} = 1.065; \hat{\beta}_2/\sqrt{\mathrm{Var}(\hat{\beta}_2)} = 1.518.$

2. The ratio $\Theta_2 = \beta_1(L)/\beta_1(LQ)$ where β_1 (*L*) is the coefficient of dose, *D*, in the linear model, and β_1(*LQ*) is the coefficient of dose in the linear-quadratic model (of the same set of observations) is taken to be a measure of the dose-rate effectiveness factor (DREF).

It should be noted that for the BEIR V models of leukemia mortality the precision of the respective estimates, $\hat{\beta}_1(L)$ and $\hat{\beta}_1(LQ)$ is quite low: $\hat{\beta}_1(L)/\sqrt{\mathrm{Var}(\hat{\beta}_1(L))} = 0.878$ and $\hat{\beta}_1(LQ)/\sqrt{\mathrm{Var}(\hat{\beta}_1(LQ))} = 0.834$. Note also that these are rather less than the cognate precisions of the BEIR III models of leukemia incidence: $\hat{\beta}_1(L)/\sqrt{\mathrm{Var}(\hat{\beta}_1(L))} = 3.647$ and $\hat{\beta}_1(LQ)/\sqrt{\mathrm{Var}(\hat{\beta}_1(LQ))} = 1.065$.

TABLE 4F-4 Distribution of Freeman-Tukey Residuals for BEIR V Models of LSS (DS86) Data

Tumor	n	Mean	Standard Deviation	Skewness	Kurtosis	Quantiles				
						0[a]	0.25	0.50	0.75	1.00[b]
Leukemia	2,575	−0.02	0.31	2.20	8.75	−1.18	−0.10	0	0	1.81
Digestive	2,162	−0.03	0.61	0.36	1.71	−3.00	−0.32	−0.07	0.05	2.39
Respiratory	2,141	−0.02	0.45	0.66	4.31	−2.19	−0.13	0	0	2.07
Other	2,156	−0.03	0.57	0.57	1.87	−2.29	−0.28	−0.07	0	2.30

[a] Minimum.
[b] Maximum.

Since these ratios, Θ, are non-linear functions of the regression parameters, say β_j and β_k, the maximum likelihood (ML) estimates, $\hat{\Theta} = \hat{\beta}_j/\hat{\beta}_k$, are *biased*: $E(\hat{\Theta}) - \Theta \neq 0$ (Co74, Ef82, Hi77, We83). If the respective precisions of the sample estimates, $\hat{\beta}_j$ and $\hat{\beta}_k$, are quite poor *and* the correlation, say ρ, between $\hat{\beta}_j$ and $\hat{\beta}_k$ is *negative* ($\rho < 0$), then the *bias*, as well as the *variance* of the sample estimate, $\hat{\Theta} = \hat{\beta}_j/\hat{\beta}_k$ of Θ, may be quite large. Estimates of the bias and variance of Θ can be obtained by several methods: the delta method (Co74, Hi77) and the weighted jackknife method (Hi77, We83) are two. Estimates of the bias can also be obtained by the MELO method (Ze78). All three methods yield comparable estimates of Θ for which the bias is less than for the ML estimate, $\hat{\Theta}$, when $\hat{\beta}_j/\sqrt{\mathrm{Var}(\hat{\beta}_j)} > 1.0$. However, only the weighted jackknife methods (Hi73, We83) provide useful estimates of Θ when $\hat{\beta}_j/\sqrt{\mathrm{Var}(\hat{\beta}_j)} < 1.0$ and/or $\hat{\beta}_k/\sqrt{\mathrm{Var}(\hat{\beta}_k)} < 1.0$.

Table 4F-5 presents estimates of the bias and variance of $\hat{\Theta}_1$ and $\hat{\Theta}_2$ for the preferred (non-linear) Poisson models of leukemia mortality. Cognate estimates for the Poisson (linear) models of leukemia incidence in the BEIR III report, (Table V-8; NRC80) are included for comparison (He86, He89).

The sample estimate of the parameter variance-covariance matrix, $\mathrm{Var}(\hat{\beta})$, for the BEIR V model is conservative and hence the confidence limits are wide. In this regard it should be noted that the dispersion factor (Mc83), $\sigma^2 = \chi^2/df = 0.358$, *is not* included in the estimates given in Table 4F-5. However, a dispersion factor is included in the estimates given by Table V-8 in the BEIR III report (NRC80, He86).

It is well-known that the statistical theory and measures for assessing the adequacy (e.g., goodness-of-fit) of a regression model and the precision of the parameter estimates that are adequate for models that are *linear* in the parameter vectors (e.g., the Poisson regression models of the BEIR III data) are only approximately valid for models that are non-linear in the parameters (e.g., the Poisson regression models of the BEIR V data). For instance, the exact likelihood ($1 - \alpha$) confidence regions on the parameters of non-linear models differ considerably in both size and symmetry from the familiar ellipsoids of linear models as $\alpha \to 0$. There has been some work in the development of indices of the degree of non-linearity that would identify those combinations of model and data in which the measures (e.g., confidence regions) for linear models provided adequate approximations for non-linear models (Ba80; Be60; Gu65). However, these measures have been developed only for non-linear models of observed responses in which

TABLE 4F-5 Maximum Likelihood and Reduced Bias Estimates of the Ratios Θ_1 and Θ_2 for Poisson Regression Models of Leukemia

	$\hat{\Theta}_j$ (ML Est.)	$\hat{\Theta}_j^*$ (Delta Est.)[a]	Standard Error $\sqrt{\text{Var}(\hat{\Theta}_j)}$ (Delta Est.)	Ratio $\hat{\Theta}_j/\sqrt{\text{Var}\,\hat{\Theta}_j}$
Θ_1, Cross-over dose (Gy)				
BEIR V ($\rho > 0$)	0.89	1.12	0.86	1.04
BEIR III ($\rho < 0$)	1.18	0.31	1.82	0.6
Θ_2, DREF				
BEIR V	1.99	1.92	2.33	0.85
BEIR III	2.24	1.51	1.92	1.17

[a] ML estimate with a first-order correction for bias.

NOTE: The estimates of Θ_2 were based on an assumed value of the correlation coefficient, ρ^*, for $\hat{\boldsymbol{\beta}}_1(L)$ and $\hat{\boldsymbol{\beta}}_1(LQ)$. This value is $\rho^* = 0.50$. This value of ρ^* was obtained from the observed correlation of $\hat{\boldsymbol{\beta}}_1(L)_{(i)}$ in the set of n row-deleted estimates $\hat{\boldsymbol{\beta}}_{(i)}$, $1 \leq i \leq n$ (Be80, Co82) of the respective parameter vectors, $\boldsymbol{\beta}$, of the BEIR III $L-L$ and $LQ-L$ models of the BEIR III leukemia incidence data. The estimate of Θ_2 is much more sensitive to the size and sign of ρ^* for the models of the BEIR V data than for those of the BEIR III data. The estimates of bias and variance obtained by the delta method are conservative. Cognate estimates obtained by the jackknife method will be larger.

the random part has a Normal distribution, and hence are not directly applicable to the non-linear Poisson models in the BEIR V report.

Nonetheless, the comparison of the estimated parameters of non-linear models with their respective standard errors provides a useful appreciation of the precision of the estimates. And indeed, for small values of α, the exact confidence regions on the parameters of a non-linear model are frequently well-approximated by those obtained from linear theory. For example, the exact 50% confidence regions ($\alpha = 0.50$) on the parameters of a non-linear (Normal theory) model often are nearly coincident with the cognate ellipsoids of linear theory (Be77).

Therefore, the comparison of the estimates of non-linear functions of parameters, such as DREF $= \hat{\Theta}_2 = \hat{\beta}_1(L)/\hat{\beta}_1(LQ)$, with their respective standard errors will provide a useful appreciation of the precision (or, perhaps more precisely, the lack thereof) with which estimates of these important ratios can be obtained from the L and LQ regression models of a given set of data.

Such comparisons disclose that the respective *standard errors* of the two ratios are about equal to the ML point estimates: $\hat{\Theta}_j/\sqrt{\text{Var}(\hat{\Theta}_j)} \simeq 1.0$

for $j = 1, 2$. This is consistent with the precision of the estimates of the numerator and denominator: $\hat{\beta}_j/\sqrt{\mathrm{Var}(\hat{\beta}_j)} \simeq 1.0$ for $j = 1, 2$.

The bias in the ML point estimates varies from about 5% to about 50% of the size of the respective standard errors in each case. The negative sign of the bias in the BEIR V models may be due to the presence of large random errors in the sample estimates of the respective covariances, Cov $(\hat{\beta}_j,\hat{\beta}_k)$ or to the presence of the covariates in time, age, etc. which, of course, also inflates $\mathrm{Var}(\hat{\beta})$.

References

Ba80 Bates, D. M., and D. G. Watts. 1980. Relative curvature measures of non-linearity. J. Royal Statist. Soc. B. 42(1):1-25.

Be60 Beale, E. M. L. 1960. Confidence regions in non-linear estimation. J. Royal Statist. Soc. 22(1):41-88.

Be77 Beck, J. V., and K. J. Arnold. 1977. Parameter Estimation in Engineering and Science. New York: John Wiley & Sons.

Be80 Belsley, D. A., E. Kuh, and R. E. Welsch. 1980. Regression diagnostics. New York: John Wiley & Sons.

Bi75 Bishop, Y. M. M., S. E. Fienberg, and P. W. Holland. 1974. Discrete Multivariate Analysis: Theory and Practice. Cambridge, Mass.: MIT Press.

Br65 Brownlee, H. 1965. Statistical Theory and Methodology in Science and Engineering. New York: John Wiley & Sons.

Br85 Breslow, N. E., and B. E. Storer. 1985. General relative risk functions for core control studies. Am. J. Epidemiol. 122(1):149-162.

Co74 Cox, D. R., and D. V. Hinkley. 1974. Theoretical Statistics. New York: Chapman and Hall.

Ef82 Efron, B. 1982. The Jackknife, The Bootstrap and Other Resampling Plans. Philadelphia, Pa.: SIAM.

Fr50 Freeman, M. F., and J. W. Tukey. 1950. Transformations related to the angular and the square root. Ann. Math. Statist. 21:607-611.

Fr83a Frome, E. L. 1983. The analysis of rates using Poisson regression models. Biometrics 39:665-674.

Fr83b Frome, E. L., and R. J. DuFrain. 1983. Maximum likelihood estimation for cytogenetic dose-response curves. ORNL/CSD-123. Office of Health and Environ. Research and U.S. Dept. of Energy and Oak Ridge Assoc. Univ.

Gi84 Gilchrist, W. 1984. Statistical Modelling. New York: John Wiley & Sons.

Gu65 Guttman, I., and D. A. Meeter. 1965. On Beale's measures of non-linearity. Technometrics 7:623-637.

He86 Herbert, D. E. 1986. Clinical Radiocarcinogenesis. Applications of regression diagnostics and Bayesian methods to Poisson regression models. Pp. 307-364 in Multiple Regression Analysis:2 Applications in the Health Sciences, D. E. Herbert and R. H. Myers, eds. AAPM Med. Phys. Monograph No. 13. New York: AIP.

He89 Herbert, D. E. 1989. Dose response models: construction, criticism, discrimination, validation and deployment. Pp. 534-630 in Prediction of Response in Radiation Therapy: Analytical Models and Modelling, B. R. Paliwal, J. F. Fowler, D. E. Herbert, T. J. Kinsella, and C. G. Orton, eds. AAPM Symposium Proceedings No.7, Part 2. New York: AIP. In press.

Hi77 Hinkley, D. V. 1977. Jackknifing in unbalanced situations. Technometrics. 19(3):285-292.

Ho85 Hoaglin, D. C., F. Mosteller, and J. W. Tukey. 1985. Exploring Data. Tables, Trends, and Shapes. New York: John Wiley & Sons.

IC86 International Commission on Radiation Units and Measurements. 1986. The Quality Factor in Radiation Protection. ICRU Report 40. Report to the ICRP and ICRU of a joint task group. Bethesda, Md.: International Commission on Radiation Units and Measurements.

Mc83 McCullagh, P., and J. A. Nelder. 1983. Generalized Linear Models. New York: Chapman and Hall.

NRC80 National Academy of Sciences/National Research Council. 1980. The Effects on Populations of Exposure to Low Levels of Ionizing Radiation: 1980 (BEIR III). Washington, D.C.: National Academy Press.

NIH85 National Institutes of Health. 1985. Report of the Ad Hoc Working Group to Develop Radioepidemiological Tables. NIH Publication 85-2748. Washington, D.C.: U.S. Government Printing Office.

Pi89 Pierce, D.A., D.O. Stram, and M. Vaeth. 1989. Allowing for random errors in radiation exposure estimates for the atomic bombs RERF TR 2-89. Hiroshima: RERF

Ro86 Robins, J. M., and S. Greenland. 1986. The role of model selection in causal inference from nonexperimental data. Am. J. Epidemiol. 123(3):392-402.

Ve81 Velleman, P. F., and D. C. Hoaglin. 1981. Applications, Basics, and Computing Exploratory Data Analysis. Boston: Duxbury Press.

We83 Weber, N.C., and A. H. Welsh. 1983. Jackknifing the general linear model. Austral. J. Statist. 24(3):425-436.

Ze78 Zellner, A. 1978. Estimations of functions of population means and regression coefficients including structural coefficients: a minimum expected loss (MELO) approach. J. Econometrics 8:127-158.

ANNEX 4G: THE BEIR IV COMMITTEE'S MODEL AND RISK ESTIMATES FOR LUNG CANCER DUE TO RADON PROGENY

The BEIR IV Committee's risk model is based on analyses of the lung cancer mortality experience of four cohorts of underground miners. These analyses indicated a decline in the excess relative risk with both attained age and time since exposure. The Committee modeled these temporal parameters as step functions as indicated in the equation below, where $r(a)$ is the age specific lung cancer mortality rate.

$$r(a) = r_o(a)[1 + 0.025\gamma(a)(W_1 + 0.5W_2)],$$

where $r_o(a)$ is the age specific ambient lung cancer rate for persons of a given sex and smoking status; $\gamma(a)$ is 1.2 when age a is less than 55 yr, 1.0 when a is 55-64 yr, and 0.4 when a is 65 yr or more. W_1 is the cumulative exposure in Working Level Month (WLM) incurred between 5 and 15 yr before this age and W_2 is the WLM incurred 15 or more years before this age.

TABLE 4G-1 Ratio of Lifetime Risks[a] (R_e/R_o), Lifetime Risk of Lung Cancer Mortality (R_e), and Years of Life Lost $(L_o - L_e)$[b] for Lifetime Exposure at Various Rates of Annual Exposure (NAS88)[c]

Exposure Rate	Males Nonsmokers			Smokers		
(WLM/yr)	R_e/R_o	R_e	$L_o - L_e$	R_e/R_o	R_e	$L_o - L_e$
0	1.0	0.0112	0	1.0	0.123	1.50
0.1	1.06	0.0118	0.00907	1.05	0.129	1.59
0.2	1.11	0.0124	0.0181	1.10	0.135	1.69
0.3	1.16	0.0131	0.0272	1.15	0.141	1.79
0.4	1.22	0.0137	0.0362	1.20	0.147	1.88
0.5	1.27	0.0143	0.0453	1.24	0.153	1.98
0.6	1.33	0.0149	0.0544	1.29	0.159	2.07
0.8	1.44	0.0161	0.0724	1.39	0.170	2.26
1.0	1.54	0.0173	0.0905	1.48	0.182	2.44
1.5	1.82	0.0204	0.136	1.70	0.209	2.89
2.0	2.08	0.0234	0.180	1.91	0.235	3.33
2.5	2.35	0.0264	0.225	2.12	0.260	3.75
3.0	2.62	0.0294	0.270	2.31	0.284	4.16
3.5	2.89	0.0324	0.314	2.49	0.306	4.56
4.0	3.15	0.0354	0.359	2.66	0.328	4.95
4.5	3.41	0.0383	0.403	2.83	0.348	5.32
5.0	3.68	0.0413	0.447	2.99	0.368	5.68
10.0	6.24	0.0700	0.883	4.24	0.521	8.77

This model is applied as follows. First, exposures are separated into two intervals as indicated above for each year in the period of interest, and then the total annual risk is calculated, using the appropriate age specific ambient rate. This age-specific mortality rate for lung cancer, $r(a)$, is multiplied by the chance of surviving all causes of death to that age, including the risk due to exposure, and these products are summed over successive ages of interest. Lifetime risks of lung cancer mortality due to radon exposure over a full lifetime are presented in Table 4G-1. Three measures of risk are listed: R_e/R_o, the ratio of lifetime risk relative to that of an unexposed person of the same sex and smoking status; R_e, the lifetime risk of lung cancer; and the average years of life lost compared to the longevity of a *nonsmoker* of the same sex.

The BEIR IV Committee pointed out a number of uncertainties in these risk estimates. These include the model for the effect of smoking used by the committee, the statistical uncertainty and possible biases in the

TABLE 4G-1 *Continued*

Exposure Rate (WLM/yr)	Females Nonsmokers R_e/R_o	R_e	$L_o - L_e$	Smokers R_e/R_o	R_e	$L_o - L_e$
0	1.0	0.00603	0	1.0	0.582	0.809
0.1	1.06	0.00637	0.00606	1.06	0.0614	0.867
0.2	1.11	0.00672	0.0121	1.11	0.0646	0.925
0.3	1.17	0.00706	0.0182	1.16	0.0678	0.983
0.4	1.23	0.00741	0.0242	1.22	0.0710	1.04
0.5	1.28	0.00775	0.0303	1.27	0.0742	1.10
0.6	1.34	0.00809	0.0363	1.33	0.0773	1.16
0.8	1.46	0.00878	0.0484	1.44	0.0836	1.27
1.0	1.57	0.00946	0.0605	1.54	0.0898	1.38
1.5	1.85	0.0112	0.0907	1.80	0.105	1.67
2.0	2.14	0.0129	0.121	2.06	0.120	1.95
2.5	2.42	0.0146	0.151	2.32	0.135	2.22
3.0	2.70	0.0163	0.181	2.56	0.149	2.49
3.5	2.98	0.0180	0.211	2.81	0.163	2.76
4.0	3.26	0.0197	0.241	3.04	0.177	3.03
4.5	3.55	0.0214	0.271	3.28	0.191	3.29
5.0	3.83	0.0231	0.301	3.51	0.204	3.55
10.0	6.59	0.0398	0.598	5.56	0.324	5.98

[a] Relative to persons of the same sex and smoking status.
[b] L_o is the average lifetime of nonsmokers of the same sex.
[c] Estimated with the committee's TSE model and a multiplicative interaction between smoking and exposure to radon progeny.

cohort data, the modeling uncertainty, and the uncertainty introduced by using data for occupationally exposed males to project the risks to persons in the general population having a wide range of ages and differing exposure situations. All of these factors are discussed at some length in the BEIR IV Committee's Report (NRC 88).

References

NRC88 National Research Council, Committee on the Biological Effects of Ionizing Radiations. Health Risks of Radon and Other Internally Deposited Alpha-Emitters (BEIR IV). Washington, D.C.: National Academy Press. 602 pp.

5

Radiogenic Cancer at Specific Sites

LEUKEMIA

The induction of leukemia by ionizing radiation has been well documented in humans and laboratory animals. The types of leukemia induced and their rates of induction vary markedly, depending on the species, strain, age at irradiation, sex, and physiological state of the exposed individuals. They also depend on the dose, dose rate, anatomical distribution, and LET of the radiation, among other variables. The early literature has been summarized elsewhere (NRC80, UN77, UN82, UN86, UN88).

Human Data

The most extensive human data on the dose-incidence relationship come from studies of the Japanese atomic-bomb survivors and patients treated with x rays for ankylosing spondylitis. In the atomic-bomb survivors of the Life Span Study Cohort, a total of 202 deaths from leukemia were recorded for the period from 1950 to 1985, during which there were an estimated 2,185,335 person-years of follow-up. Analyzed in terms of absorbed dose to the bone marrow as estimated with the new DS86 dosimetry, the dose response for Nagasaki rises less steeply than for Hiroshima, especially in the dose range below 0.5 Gy, but the difference between the two cities is smaller with the DS86 dosimetry than with the T65D dosimetry and is no longer significant (Sh87). For the combined data, the rate of mortality is significantly elevated at 0.4 Gy and above but not at lesser doses. At bone marrow doses of 3-4 Gy, the estimated dose-response curve peaks and

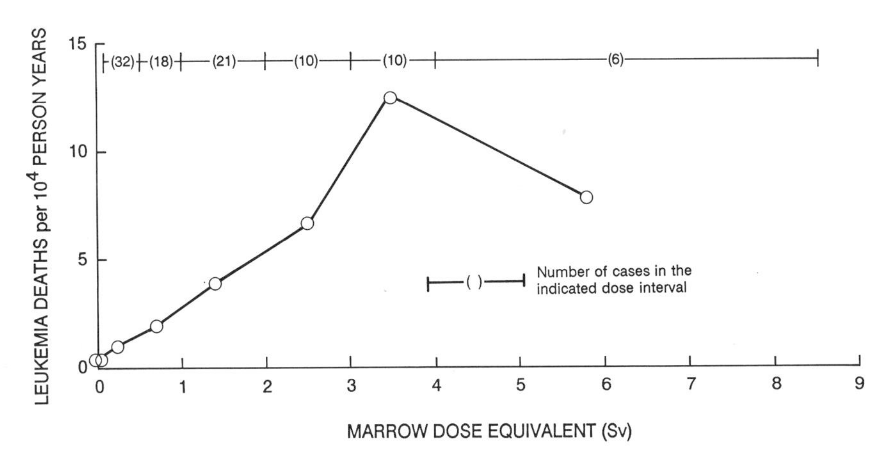

FIGURE 5-1 Cumulative leukemia mortality in Hiroshima and Nagasaki as a function of the estimated dose equivalent to the bone marrow under DS86. By 1985, there were 51 cases in the 0 Sv category and 31 cases in the 0.01-0.1 Sv stratum.

turns downward (Figure 5-1). As noted below, this pattern is characteristic of the leukemia response in other irradiated populations. The saturation of the leukemia response at high doses has been attributed to the reduced survival of potentially transformed myeloblasts in the range above 3-4 Gy (Un86).

Based on a simple linear dose-response model, which in the opinion of RERF analysts fit the LSS data for leukemia mortality as well as a linear-quadratic model and better than a simple quadratic model, the excess relative risk per Sievert was estimated to range from 4.24 to 5.21, and the number of excess deaths per 10^4 person-year-Sv was estimated to range from 2.40 for a neutron RBE of 20 to 2.95 for an RBE of 1 (Sh87). The excess mortality from leukemia reached a peak within 10 years after irradiation and has persisted at a diminished level (Figure 5-2). No excess cases of chronic lymphocytic leukemia have been observed (Pr87a).

Among 14,106 patients who were followed for up to 48 years after a single course of x-ray therapy for ankylosing spondylitis, 39 deaths from leukemia were recorded versus a total of 12.29 expected cases (ratio of observed to expected deaths, 3.17) (Da87). The excess deaths became detectable within two years after irradiation, reached a peak within the first 5 years, and declined thereafter; however, the excess death rate remained significantly elevated (relative risk, 1.87) for more than 15 years, after which it appeared to persist with little change (Da87). The relative risk did not vary significantly with age at the time of treatment, but it was higher in males (3.43) than in females (1.79). The relative risk also varied with the hematologic type of the disease, being higher for those with acute myeloid

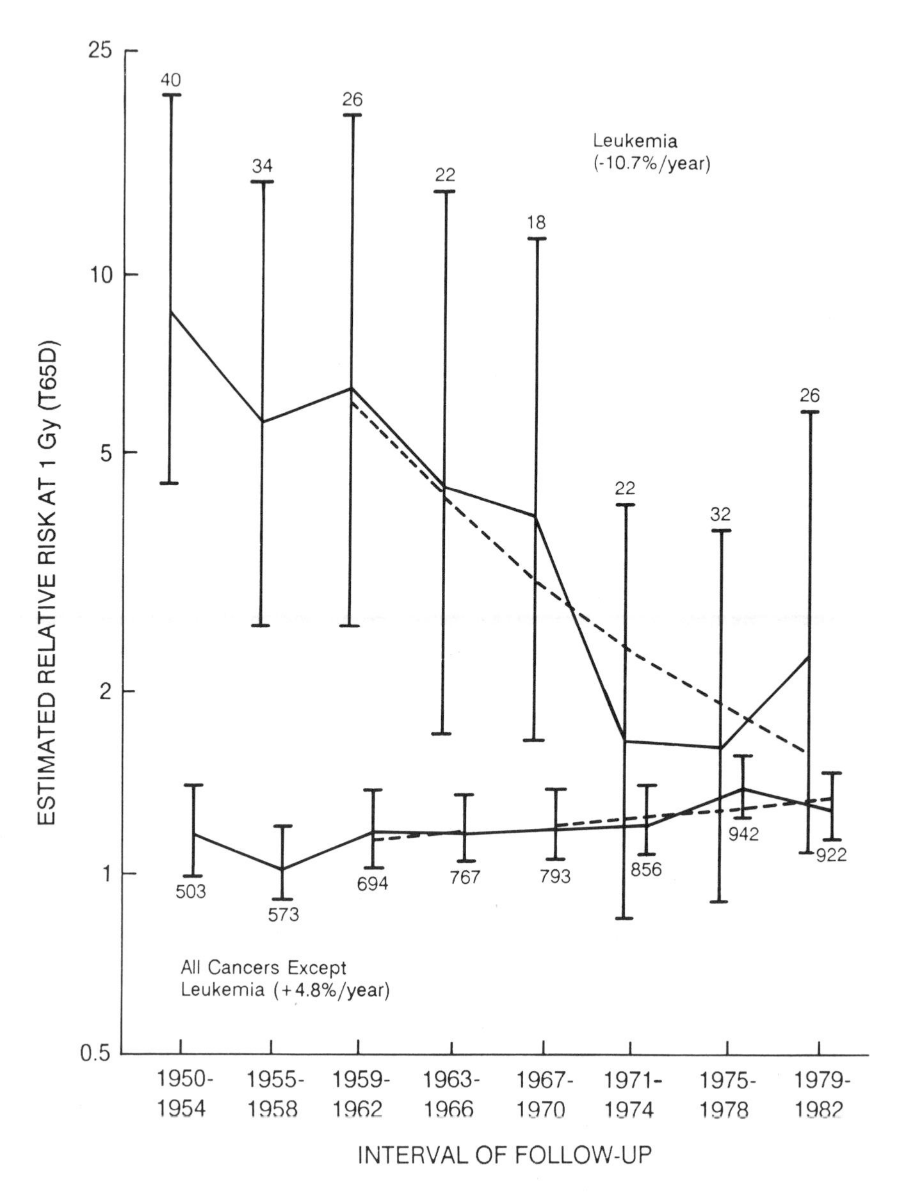

FIGURE 5-2 Relative risk of mortality from leukemia and all cancers other than leukemia in A-bomb survivors, 1950-1982, in relation to time after irradiation. The number of deaths in each interval of follow-up and 99% confidence intervals are indicated (Pr87).

TABLE 5-1 Observed, as Compared with Expected, Numbers of Deaths from Leukemia in Persons Treated with Spinal Irradiation for Ankylosing Spondylitis[a]

Type of Leukemia	Number of Deaths[b]		Ratio of Observed/Expected
	Observed	Expected	
Myeloid leukemia			
Acute	17	4.34	3.92
Chronic	3	2.05	1.46
Unspecified	4	0.71	5.63
All types	24	7.10	3.38
Lymphatic leukemia			
Acute	2	0.93	2.15
Chronic	2	2.38	0.84
Unspecified	3	0.38	7.89
All types	7	3.69	1.89
Unspecified leukemia	3	0.28	10.71
All types	36	11.29	3.19

[a] From Darby et al. (Da87).

[b] Observed and expected deaths from leukemia occurring more than one year after first treatment at ages less than 85 years by age at first treatment and by type of leukemia as recorded on the death certificate. Retreated patients were included for 12 months following treatment.

leukemia than for those with other types of leukemia. It was not elevated for those with chronic lymphatic leukemia (Table 5-1).

Analyzed in relation to the average dose to the bone marrow, which was estimated to be 3.21 Gy, the excess relative risk amounted to 0.98/Gy, or 0.45 additional cases of leukemia per 10^4 PYGy (Sm82). The smaller magnitude of the risk per Gy in patients with ankylosing spondylitis, compared with that in atomic-bomb survivors, may be ascribable to the younger average age of atomic-bomb survivors at the time of exposure and to the fact that they received instantaneous whole-body irradiation, whereas in the patients with ankylosing spondylitis only a portion of the active marrow was irradiated and the dose was received in fractionated exposures that usually totaled more than 5 Gy within a given treatment field (Le88). Muirhead and Darby have proposed different models of leukemia risk for the spondylitics and the A-bomb survivors. They proposed a relative risk model for the spondylitics and an absolute risk model for the atomic-bomb survivors (Mu87).

In an international case-control study of 30,000 women treated with fractionated doses of radiation for carcinoma of the uterine cervix, the risk was estimated to be increased by about 70%/Gy, corresponding to an excess of 0.48 cases of leukemia/10^4 PYGy (Bo87, Bo88). As in the

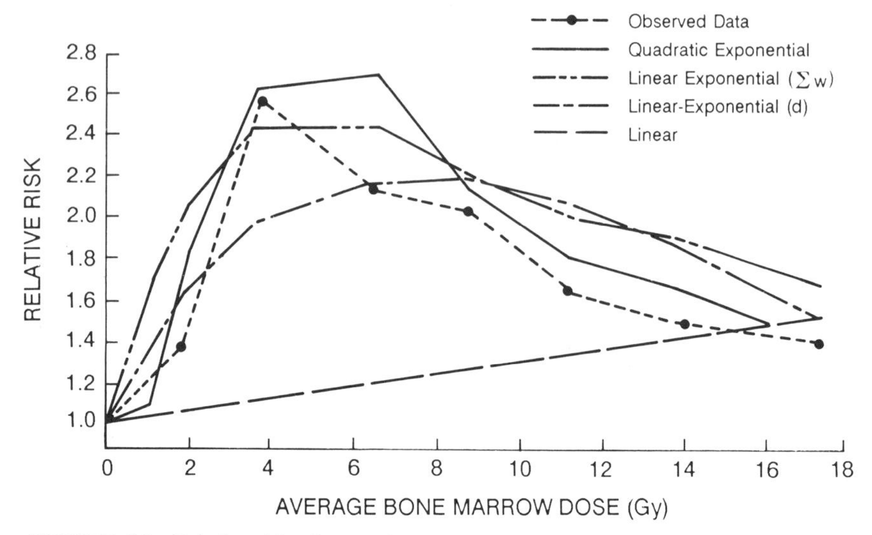

FIGURE 5-3 Relative risk of acute leukemia and chronic myeloid leukemia in women treated with radiation for carcinoma of the uterine cervix, as influenced by the average dose to the bone marrow. A better fit was obtained with a linear exponential model (ΣW) which considered the weighted dose to each marrow component as opposed to the average dose over all compartments (d) (Bo87).

A-bomb survivors mentioned previously, the excess cases were confined to leukemias other than those of the chronic lymphatic type. The relative risk was maximal within the first 5 years after irradiation, was larger in women who were irradiated when they were under age 45 than in those who were irradiated when they were over age 45, and reached a peak at an average bone marrow dose of 2.5-5.0 Gy, above which it decreased (Figure 5-3). The data conformed to a linear-exponential model in which the total risk equaled the sum of incremental risks to individually irradiated masses of marrow. The latter risks, in turn, were taken to increase linearly with the mass exposed and inversely with the total mass of marrow in the body; they were also taken to increase curvilinearly in a manner consistent with the dose-dependent killing of marrow cells (Bo87). In view of the decreased risk per Gy at high doses, it is not surprising that the average risk per Gy in the women of this series was appreciably lower than that which has been observed in women treated with smaller doses of x rays for benign gynecologic disorders (Bo86).

The incidence of leukemia has been observed to be elevated similarly in patients treated with radiation for cancers of other sites (Bo84, Cu84, Wa84). An association between previous diagnostic irradiation and adult onset myeloid or monocytic leukemia has been suggested by three case-control studies (St62, Gu64, Gi72); however, the data in the first and largest of the three studies (St62) have since been reinterpreted to argue against a causal relationship on the grounds that "the 'extra' examinations all happened within 5 years of the onset" of symptoms of leukemia (St73). No association between previous diagnostic irradiation and adult-onset myeloid or monocytic leukemia was observed in a fourth case-control study (Li80). On the basis of extrapolation from the leukemogenic effects of irradiation in atomic-bomb survivors and other relatively heavily irradiated groups, it has been estimated that about 1% of all leukemia cases in the general population may be attributable to diagnostic radiography (Ev86).

The risk has not been confined to acutely irradiated populations, such as those mentioned above. Early cohorts of radiologists in the United States (Le63, Ma84), the United Kingdom (Co58), and the People's Republic of China (Wa88), who were exposed to x rays occupationally in the days preceding modern safety standards, also have shown an increased incidence of acute leukemia and chronic granulocytic leukemia. These diseases have, likewise, been observed to occur with increased frequency in patients previously injected with radium-224 or Thorotrast (NRC80). Because of uncertainty about the doses to the bone marrow in the occupationally and internally irradiated populations, it is not clear how their risks per unit dose compare with those in the more acutely irradiated populations described above.

An excess number of cases of leukemia have been observed in children who were exposed to diagnostic x-irradiation in utero; the excess is larger per unit dose than that in children who were irradiated during postnatal life. The magnitude of the excess and the extent to which it may signify an unusually high susceptibility of the embryo and fetus are discussed in Chapter 6 of this report. Reports of an increased incidence of leukemia in children residing in the vicinity of nuclear installations in the United Kingdom are reviewed in Chapter 7.

Committee Analysis

For purposes of risk estimation, the Committee's analysis was restricted to the total mortality from leukemias of all hematologic types combined, excluding chronic lymphocytic leukemia. Modeling in terms of the various types of leukemia was not possible because of limitations in the available data. The different types vary markedly in the age distributions of their occurrence in the general population and in their relative frequencies with time after irradiation, depending on age at the time of exposure. To this extent, the Committee's risk model for leukemia is a gross simplification.

For both the Life Span Study (LSS) and the Ankylosing Spondylitis

(ASS) data, essentially comparable fits could be obtained using either additive or relative risk models, although somewhat different modifying effects were required in the two models and the relative risk model was consistently more parsimonious. It must be remembered that follow-up of the LSS cohort did not begin until five years after exposure, by which time the peak in the excess rate had already occurred in the ASS data. Despite this and other differences between the two studies, the modifying effects are reasonably consistent. The preferred model from the ASS data is a relative risk model with a decreasing effect in time after exposure. However, the addition of an effect of age at exposure significantly improves the fit of the LSS data. The magnitude of this effect and also the effect of time after exposure depends on whether exposure occurred before or after age 20. The ASS cohort did not include individuals younger than 20 years of age at exposure, so the age factor could not be tested in that data set.

Dose-response in the LSS data was significantly improved by the addition of a quadratic term in dose. (Here, the linear term includes both the gamma and neutron components, the latter weighted by the assumed RBE of 20; the quadratic component includes only the gamma component.) The "cross-over dose" (the dose at which the linear and quadratic contributions are equal) was estimated to be about 0.9 Gy. However, ratios of log likelihood estimates are biased and for these data the uncertainty is very large (see Annex 4F). Similarly, the "dose rate effectiveness factor" (DREF, the ratio of the fitted slopes of the pure linear and the linear-quadratic models) is estimated as 2 but again with a very large uncertainty.

The final preferred model for leukemia mortality used in the risk projections is given by equation 4-3 reproduced below.

$$f(d) = \alpha_2 D + \alpha_3 d^2$$

$$g(\beta) = \begin{cases} \exp[\beta_1 I(T \leq 15) + \beta_2 I(15 < T \leq 25)] \text{ if } E \leq 20 \\ \exp[\beta_3 I(T \leq 25) + \beta_4 I(25 < T \leq 30)] \text{ if } E > 20 \end{cases}$$

This model is plotted as a function of attained age in Figure 5-4 and excess risk as a function of time after exposure for males is shown in Figure 5-5.

The abrupt changes in risk with age at the time of irradiation that are specified in the model reflect simplifying compromises in model fitting and are not based on hypotheses concerning the biological mechanism of age-dependent changes in susceptibility. Insofar as different types of leukemia vary in age distribution in the general population, their causative mechanisms and temporal distributions in irradiated populations might be expected to vary as well.

This leukemia model is based on LSS data, which do not include information prior to five years post exposure. A number of fitted models

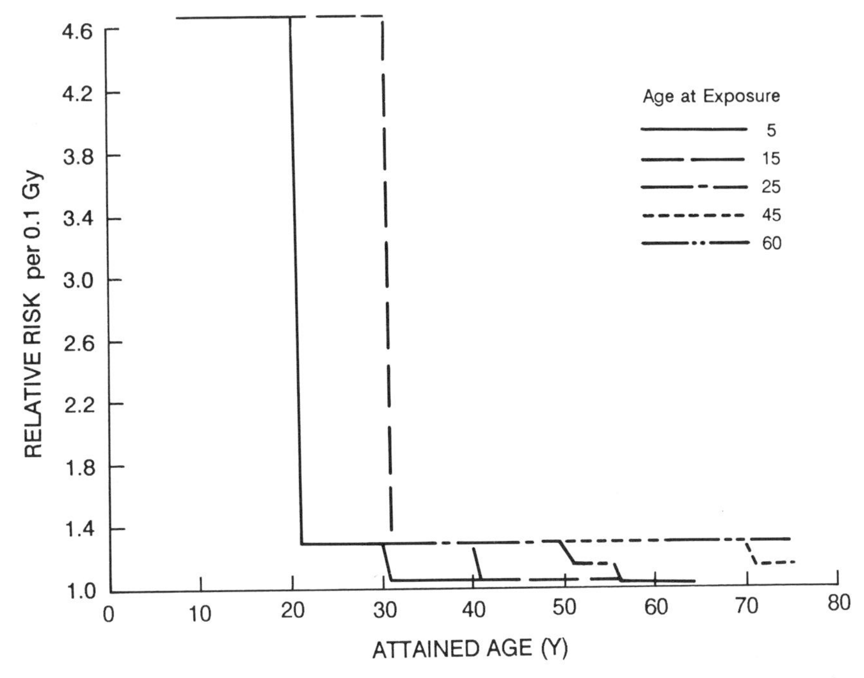

FIGURE 5-4 The relative risk of leukemia due to low LET radiation for both sexes by attained age from age 7 to age 75 for exposure at various ages.

were tested but these produced rather varied and unreliable risk estimates in extrapolations to this early, first 5-year period. Sources of data, other than that from A-bomb survivors, provide some guidance on this point. The cervical cancer study by Boice et al. (Bo87) indicates that excess leukemia cases were observed only within the first five years post exposure. On the other hand, the spondylitic cohort shows a mixture of excess cases before and after five years post exposure (Da87). In that study, 14 cases with 1.6 expected were observed in the first five years, and 25 cases with 10.7 expected after five years post exposure. One could then reasonably argue that nearly one-half of the excess leukemias would be observed within the first five years after exposure. The Committee chose to model the 2- to 5-year post-exposure period by extrapolating to two years the excess relative risk observed for the 5- to 10-year post-exposure period. This method resulted in an approximately 15% increase in the lifetime risks. The Committee's extrapolation procedure for the 2- to 5-year post-exposure period may lead to an underestimate of the actual risk, and this should be kept in mind when interpreting the Committee's risk estimates for leukemia.

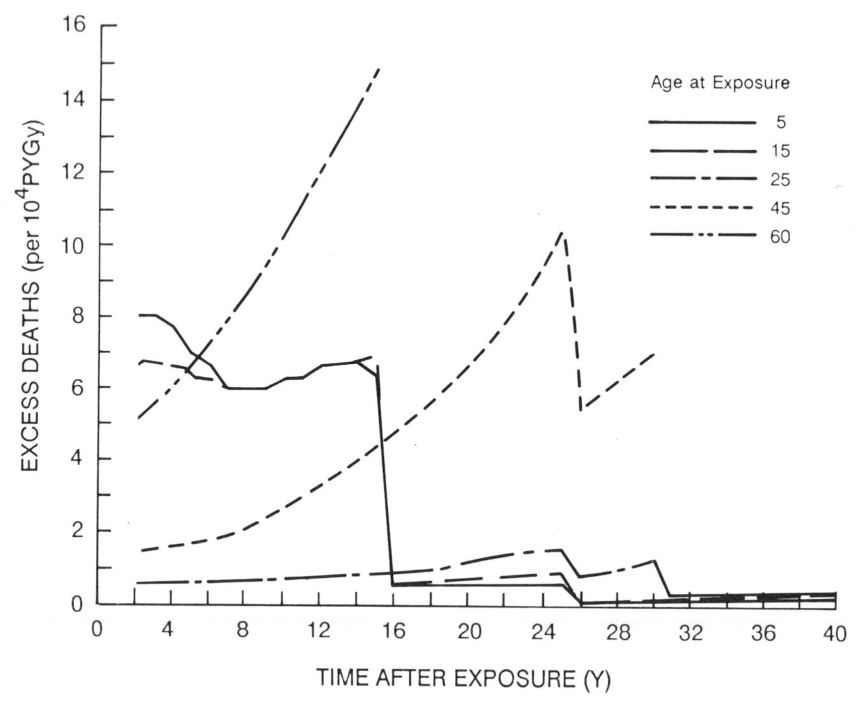

FIGURE 5-5 Excess leukemia deaths by time after exposure to low-LET radiation for U.S. males at various ages of exposure.

Leukemia Studies in Animals

In mice, rats, dogs, swine, and other laboratory animals, a variety of lymphoid and myeloid leukemias have been induced by irradiation (UN77, UN86, NRC80). In such animals, the dose-incidence relationship has been observed to vary from one type of leukemia to another, but in no instance does it conform to a simple, linear nonthreshold function. The most extensively studied of the experimental leukemias are T-cell neoplasms that arise in the mouse thymus. The induction of these growths is inhibited drastically by shielding a portion of the hemopoietic marrow (UN77) and may involve thc activation of a latent leukemia virus (Rad LV) (Yo86). The dose-incidence curve for the disease is of the threshold type in mice of certain strains (UN86). In the range of 0.5-1.0 Gy, the RBE of fast neutrons for induction of these neoplasms has been observed to range from a value of 1.0-2.0 with single or fractionated exposures to a value exceeding 10 with continuous, duration-of-life irradiation (UN77, UN86 Fe87).

Less thoroughly investigated are experimentally induced myeloid leukemias, which have been observed in mice (Up70, Ma78, Hu87), dogs (Fr73), and swine (Ho70) that were subjected to various regimens of external or internal irradiation. The dose-incidence curve for myeloid leukemia in mice rises with increasing dose of acute whole-body x or gamma radiation, passes through a maximum at 2-3 Gy of x or gamma rays (lower dose of neutrons), and decreases at higher doses (Figure 5-6); in the dose range below 1 Gy, the shape of the curve appears to vary among strains (UN86, Ul87). The downturn in the dose-incidence curve at doses above 2-3 Gy is consistent with the reduction in numbers of potentially transformed myelopoietic cells surviving such doses (Gr65, Ba78, Ro78, Ma78, UN86). In the low to intermediate dose range, the curve rises more steeply with fast neutrons than with x rays or gamma rays (Up70, Mo82, Ul87, Pr87a), and on fractionation or protraction, the incidence per Gy decreases markedly with x or gamma irradiation but decreases less markedly, if at all, with fast neutron irradiation (Figure 5-6). As a result, the neutron RBE increases with decreasing dose rate, from a value of 2-3 at dose rates exceeding 0.1 Gy/minute to a value as high as 16 at dose rates of less than 0.01 Gy/minute (Up70). Various models have been fitted to the observed dose-incidence data, all of which have included cell-killing terms to account for the diminution of the response at intermediate to high dose levels (UN86). Although the data do not exclude a linear dose term in the low to intermediate dose range, all models also include higher power dose terms to account for the fact that the incidence per Gy of low-LET radiation increases with increasing dose at high dose rates in the intermediate dose range but is substantially reduced at low dose rates (UN86). The induction of myeloid leukemia, in contrast to induction of thymic lymphoma, is not inhibited disproportionately by shielding part of the hemopoietic system (Up64).

The incidence of myeloid leukemia per Gy has been observed to be increased in mice in which granulocyte turnover is accelerated by injection of turpentine and decreased in mice in which granulocyte turnover is reduced by the elimination of microflora, implying that induction of the disease is promoted by proliferation of granulocyte precursors (Up64). Susceptibility to the induction of lymphoid and myeloid leukemias also varies among mice of different strains and in relation to age at the time of irradiation (UN77). There is no evidence, however, that susceptibility in mice is unusually high during prenatal life; on the contrary, the data imply that it may be substantially reduced at that time of life (Up66, Si81, UN86). Whereas the incidence of lymphoid and myeloid leukemias is typically increased by whole-body irradiation in most strains of mice, depending on the conditions of irradiation, the incidence of reticulum cell

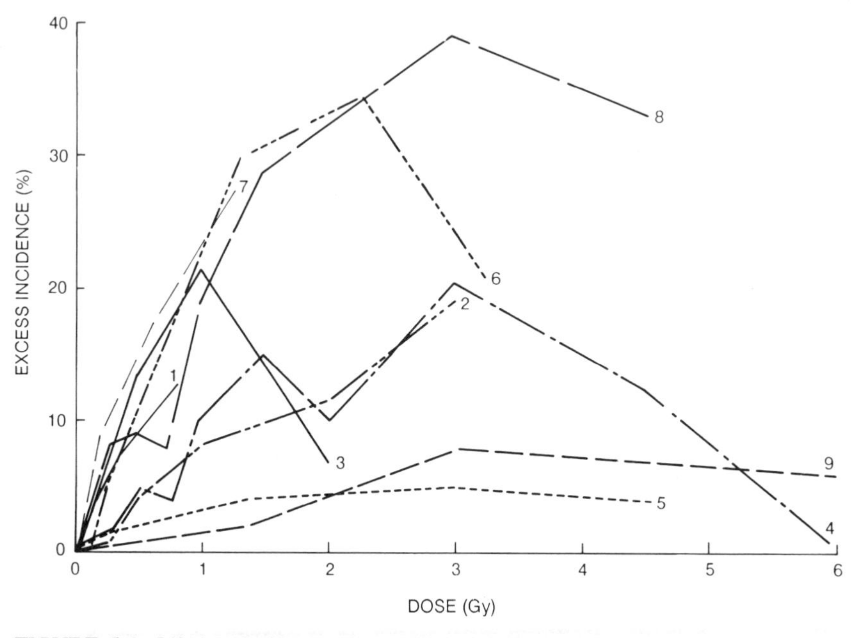

FIGURE 5-6 Lifetime incidence of myeloid leukemia (in excess of control incidence) in male mice of different strains, in relation to dose and dose rate of whole body neutron-, x-, or γ-irradiation. RFM mice (Ul87): acute neutron irradiation (curve 1); acute γ-irradiation (curve 2); CBA mice (Mo82, Mo83a, Mo83b). acute neutron irradiation (curve 3); acute x-irradiation (curve 4); protracted γ-irradiation (curve 5). RF/Up mice (Up70): acute neutron irradiation (curve 6); protracted neutron irradiation (curve 7); acute x-irradiation (curve 8); protracted γ-irradiation (curve 9).

neoplasms in such animals has usually been observed to decrease with increasing dose (UN77, UN86).

Summary

The risks of acute leukemia and of chronic myeloid leukemia are increased by irradiation of hemopoietic cells, the magnitude of the increase depending on the dose of radiation, its distribution in time and space, and the age and sex of the exposed individuals, among other variables. The mean latent period preceding the clinical onset of the leukemia also varies, depending on the hematologic type of the disease as well as age at the time of irradiation. The data do not suffice to define the dose-incidence relationship precisely, but the dose-response curve for the total excess cases of leukemia appears to increase in slope with increasing mean dose to the marrow, to pass through a maximum in the dose range of 3-4 Gy, and to decrease with a further increase in the dose.

Age at exposure is an important modifier of risk. From the LSS data it is clear that risks are initially higher for those exposed at under 20 years of age but decrease somewhat more rapidly with time after exposure than for those exposed at older ages. There was no clear indication that the risks for those under 10 years of age were significantly greater than for persons 10-20 years old at the time of exposure. When data become available that will allow the analysis of human leukemia in terms of specific hematologic types, it may be possible to develop more precise risk models that capture the age and time modifying factors in more detail.

BREAST

Introduction

The sensitivity of the mammary gland to the carcinogenic effects of ionizing radiation was first demonstrated in x-irradiated mice in 1936 (Fu36a, Fu36b). and has since been described in other species of laboratory animal, including guinea pigs, dogs, and rats (Sh86a). An increase in the incidence of breast cancer in irradiated humans was first recognized in 1965 in women who had received repeated fluoroscopic examinations (Ma65), and subsequently in Japanese atomic-bomb survivors in 1968 (Wa68). During recent decades, mammary cancer has been studied extensively in irradiated animals and in several large series of irradiated women.

Although a number of questions about radiation-induced breast cancer still remain, the data are consistent with the following generalizations:

1. The development of overt cancer from the radiogenically damaged mammary target cells is critically dependent upon the hormonal status of the cells over time.
2. Radiation-related breast cancers are similar in age distribution and histopathological types to breast cancers resulting from other or unknown causes.
3. Women who are irradiated at less than 20 years of age are at a higher relative risk for breast cancer than those who are irradiated later in life.
4. The epidemiological data reveal little or no decrease in the yield of tumors when the total radiation dose is received in multiple exposures rather than in a single, brief exposure.

Parallel Analyses of Breast Cancer Incidence and Mortality

The Committee had available for analysis the original data from two mortality series and three incidence series. The mortality series were the

Canadian Tuberculosis Fluoroscopy (CAN-TB) Study (Mi89) (473 deaths) and the subcohort of the Radiation Effects Research Foundation (RERF) Life Span Study (LSS) of atomic bomb survivors for which DS86 doses were available with follow-up through 1985 (151 deaths) (Sh87, Sh88). The incidence series included data on women in the LSS for whom DS86 doses were available with follow-up through 1980 (367 cases), data on women in the New York Acute Postpartum Mastitis Study (NY-APM) (118 cases) (Sh86b), and data on women in the Massachusetts Tuberculosis Fluoroscopy (MASS-TB) cohort (65 cases) (Hr89).

In the Committee's analyses of breast cancer, the data from the first 5 years of follow-up have been omitted. As there were no cases of breast cancer in women less than 25 years of age, expression of risk in the 0-24 age group was excluded in the analysis. This made virtually no difference in the risk modeling.

In the LSS data, breast dose equivalents were computed by using an assumed relative biological effectiveness (RBE) of 20 for neutrons. As discussed in Annex 4E, women who received doses in excess of 4 Gray (Gy) were excluded from both the incidence and mortality analyses of the LSS data. For the NY-APM and MASS-TB cohorts, women with doses in excess of 6.5 and 4 Gy, respectively, were excluded from the analyses.

Second breast primaries were not included in the analyses. Since for the NY-APM series, the dose received by each breast could differ, the follow-up time was computed in terms of breast-years using the procedures described in Shore et al. (Sh86b). All results are presented in terms of person-years. Breast-years in the NY-APM series were converted to person-years by dividing by 2. For the LSS incidence data, the person-years were adjusted for the effects of migration by using the factors given by Tokunaga et al. (To87).

The AMFIT computer program, described in Annex 4C, was used to fit various models of the radiation effects for each of the individual series, and separately for the combined mortality and combined incidence data sets. The patterns seen in the combined analyses were generally present in the individual series, and results are presented only for the combined analyses. Those studies which depart significantly from the results of the combined analyses are described in Annex 4E.

The patterns of breast cancer mortality or incidence, in the absence of radiation exposure, were first modeled for each of the populations from which the cohorts were drawn. These background rates were then either multiplied by a function of dose, age at exposure, and time since exposure (relative risk model) or added to an appropriate function of these covariates (additive excess risk model). Details of the procedures used in modeling the background rate for the various cohorts are described in Annex 4E.

The Committee's Preferred Model

The Committee has investigated a number of models for lifetime excess risk of breast cancer incidence and mortality, and its preferred models are described here in general terms. More detailed information on the parameter estimates for the preferred models and on issues which arose as these models were developed is presented in Annex 4E.

The Committee's preferred models for both incidence and mortality are relative risk models in which the excess relative risk is linear in dose and varies with both age-at-exposure and time-since exposure. A relative risk model was chosen for the incidence data because it was found that the A-bomb survivors and the U.S. relative risks for breast cancer did not differ significantly ($p = 0.3$) while the additive excess risks among the A-bomb survivors were significantly lower than those in the NY-APM and MASS-TB cohorts ($p = 0.001$).

The choice between relative and absolute risk models for the mortality data was less clear-cut. Within the CAN-TB cohort the estimated risk per Gy for women treated in Nova Scotia was about six times that for women treated in other provinces. This difference is highly significant ($p < 0.001$). Women treated in Nova Scotia faced the x-ray beam and thus received higher doses than other women in the CAN-TB cohort. However, the analyses described in Annex 4E indicate the higher risk observed among Nova Scotia women is not attributable to non-linearities in the dose response. Since there is currently no explanation for the differences within the CAN-TB cohort and since the Committee was generally interested in low dose effects it was decided to use the data on the CAN-TB cohort without the Nova Scotia women as the basis for risk estimates in the parallel analysis.

Although the relative risk estimate for mortality in the LSS was about three times the estimate in the non-Nova Scotia CAN-TB cohort the difference was not statistically significant ($p = 0.1$). The estimated absolute excess risks were about equal in the two cohorts. Since the relative risks for the two cohorts were not significantly different and because of the evidence against equal excess absolute risks in the incidence data, the Committee's preferred model for breast cancer mortality is a relative risk model in which the level of risk was determined from the combined LSS/non-Nova Scotia CAN-TB data.

It should be noted that women in the LSS and NY-APM study received acute exposures whereas the women in both TB series received highly fractionated exposures, usually over several years. Despite this difference in the pattern of exposure, on a relative risk scale there are no significant differences among cohorts in risk of breast cancer incidence or mortality.

For the incidence data, the variation in relative risk with age-at-exposure has been modeled as a step function with separate values for different age-at-exposure groups under 20, 20-40, and over 40 years of age. The relative risk for the under 20 age group was estimated as four times that in the 20-40 age group while the risk for those over 40 was only about 40% of that in the 20-40 age group. As discussed in Annex 4E, 15- to 19-year-old women in the NY-APM cohort had a significantly lower risk of radiation-induced breast cancer than young women in the other two cohorts ($p = 0.01$). Therefore the model for those less than 20 years of age included a separate parameter for the New York cohort.

In the mortality analyses, the excess relative risk for the 10-14 year olds was found to be significantly greater than that for older women. For those under 10 years of age, the relative risk was somewhat smaller but not significantly so ($p = 0.2$) and the groups were combined. For the older women, the relative risks appeared to decrease with increasing age-at-exposure ($p = 0.05$). The Committee's final model allows for the increased risks seen among the young women, for the decreasing trend in risk with age-at-exposure in older women, and for a discontinuous drop in relative risk estimate at 15 years of age. The use of additional steps for other age groups did not significantly improve the fit of the model for mortality due to breast cancer.

In the incidence data there is evidence that the excess relative risk varies with time ($p = 0.01$). In particular it was found that, for women over the age of 25, it increases to a maximum value at about 15 to 20 years after exposure and decreases slowly thereafter. In the Committee's preferred model the change in the log relative risk with time-since-exposure was modeled as a linear spline in log time with a single knot at 15 years after exposure. There is some evidence ($p = 0.12$) of a similar pattern in the mortality data with the maximum relative risk occurring between 20 and 25 years after exposure. In the Committee's preferred model for the mortality data the log relative risk is modeled as a quadratic in log time. This model fit somewhat better than a linear spline. The Committee found no evidence in either the mortality or incidence data that the temporal pattern was affected by dose or age-at-exposure (though it should be borne in mind that tests for such effects lack power). For both incidence and mortality the models assume that there is no excess risk during the first five years after exposure and no excess risk occurs among women under the age of 25. In fact there is no evidence of a significant excess risk for at least 10 years post exposure. The Committee's preferred risk model for breast cancer mortality is given in equation 4-5, reproduced below.

$$f(d) = \alpha_1 d$$

$$g(\beta) = \begin{cases} \exp[\beta_1 + \beta_2 \ln(T/20) + \beta_3 \ln^2(T/20)] \text{ if } E \leq 15 \\ \exp[\beta_2 \ln(T/20) + \beta_3 \ln^2(T/20) + \beta_4(E - 15)] \text{ if } E > 15, \end{cases}$$

To illustrate the variation in excess risk with age-at-exposure and time in the preferred models, Figures 5-7 through 5-10 present risk estimates for specific ages-at-exposure by attained age and by time-since-exposure for breast cancer mortality. Figures 5-11 and 5-12 illustrate the Committee's model for breast cancer incidence by attained age in terms of relative risk and the estimated number of excess cases per 10,000 person year Gy.

The excess absolute risks for incidence are based upon fitted background rates derived from the Connecticut Tumor registry, while the excess absolute mortality estimates are based upon fitted Canadian mortality rates. It was assumed that the exposure took place in 1980 and that temporal trends in the age-specific baseline rates do not occur after 1985.With the fitted models, excess absolute incidence rates increase until the women reach the age of 50, after which they decrease. The youngest women generally have the highest absolute risks. The general pattern of risk predicted by the mortality models is similar, with absolute risks increasing with time for women under age 50 and decreasing after age 50.

In the Committee's final model, the relative risks at a given time after exposure are the same for all women under 20 years old at exposure. In fact, the data indicate that women who were 10-14 years old at exposure have higher relative risks than women who were older. However, risks of women who were less than 10 years old at exposure are poorly estimated. The data from this age group of atomic bomb survivors are as yet not adequate for precise characterization of risk.

The contrast between the incidence and mortality predictions with regard to the excess risks for women under 10 years of age at exposure is not surprising in view of the limited data currently available for this group. Among the cohorts available, only the RERF cohort had an appreciable amount of information on risks in women who were very young when exposed. The youngest women in the RERF cohort are just now reaching the age at which one would expect an appreciable incidence of breast cancer. Projections based upon the Committee's models for the youngest age group should be interpreted with caution. Additional follow-up is clearly important in order to clarify our understanding of excess breast cancer risks in women exposed under the age of 10.

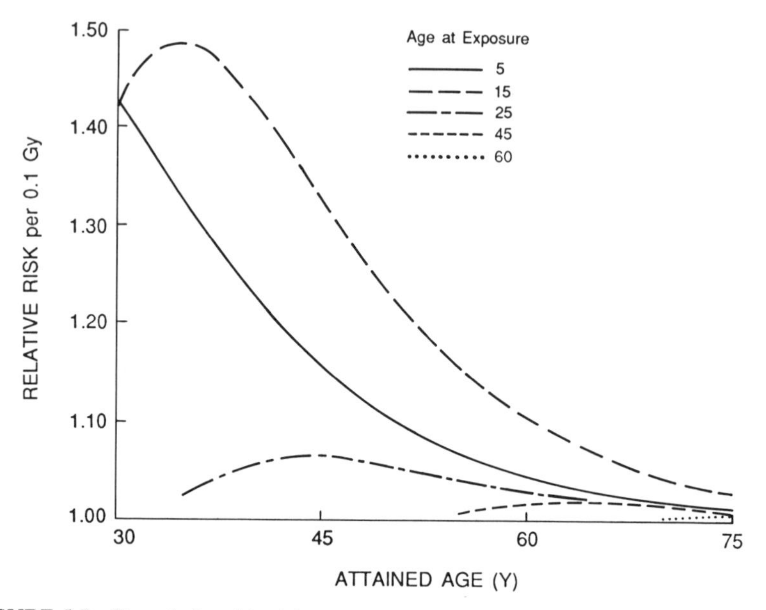

FIGURE 5-7 The relative risk of female breast cancer mortality due to low-LET radiation by attained age for exposure at various ages.

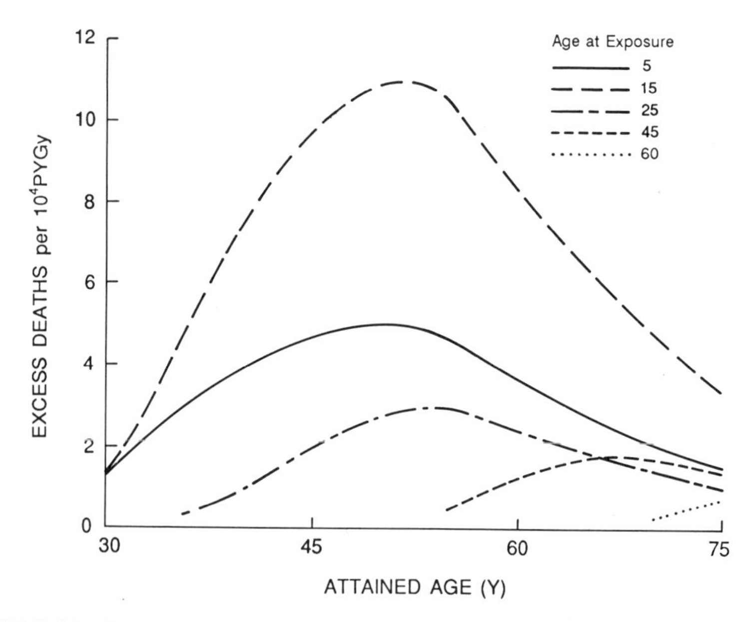

FIGURE 5-8 Excess breast cancer deaths due to low-LET radiation by attained age for U.S. females for exposure at various ages.

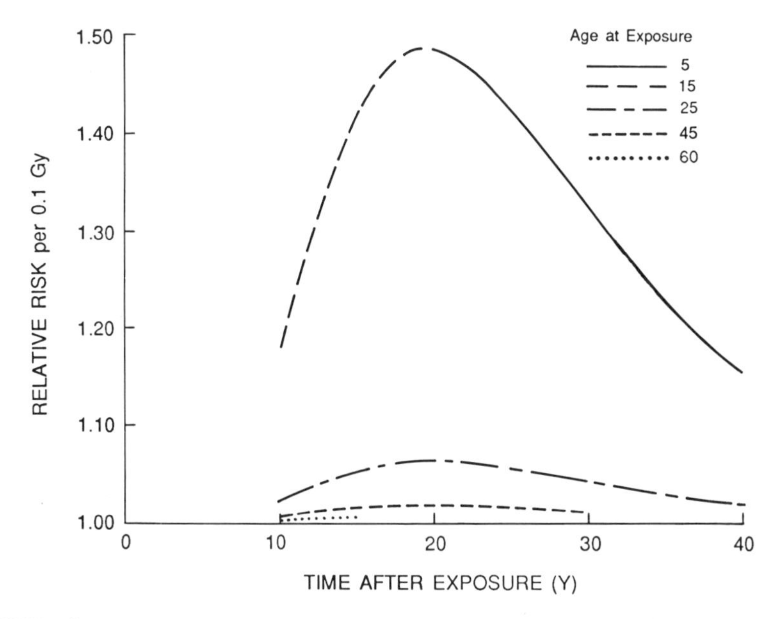

FIGURE 5-9 The relative risk due to low-LET radiation of female breast mortality as a function of time after exposure for exposure at various ages. Risk is not projected for an attained age greater than 75 years.

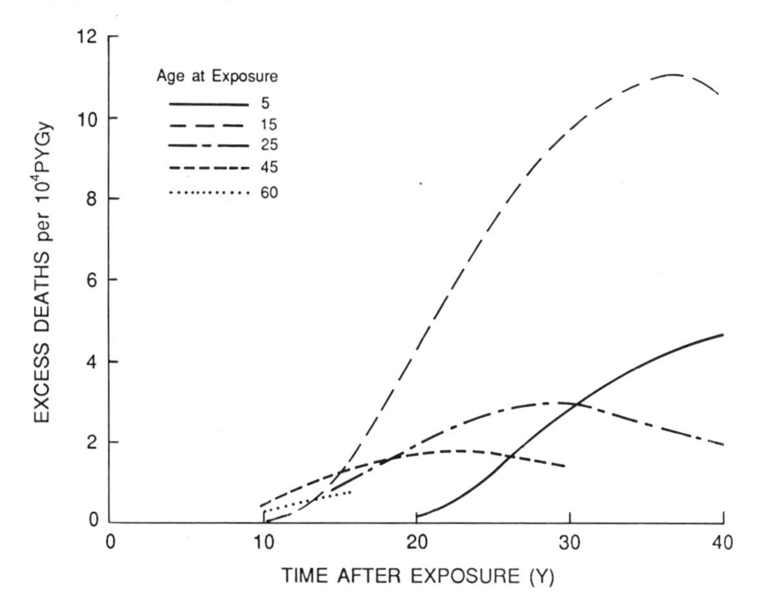

FIGURE 5-10 Excess breast cancer deaths as a function of time after exposure to low-LET radiation for U.S. females of various ages. No data are presented for an attained age greater than 75 years.

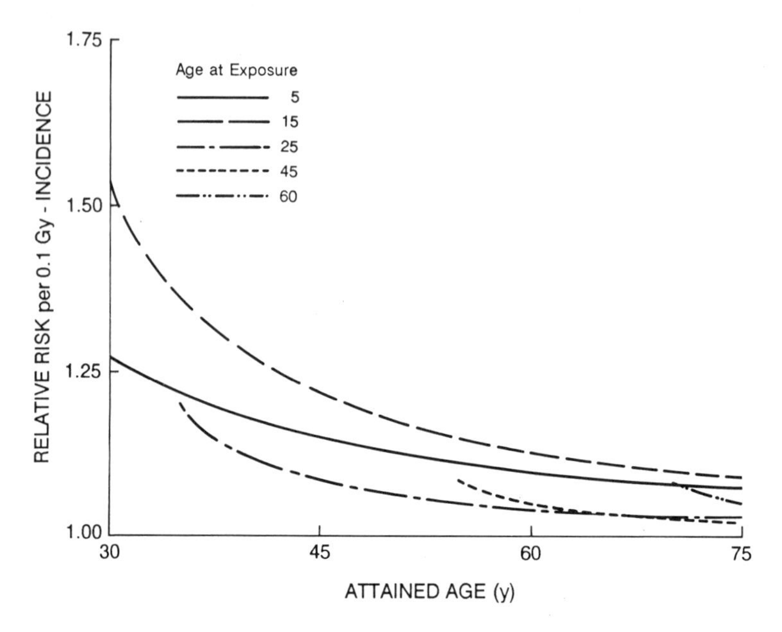

FIGURE 5-11 Excess relative risk of breast cancer incidence due to low-LET radiation by attained age for U.S. females exposed at various ages.

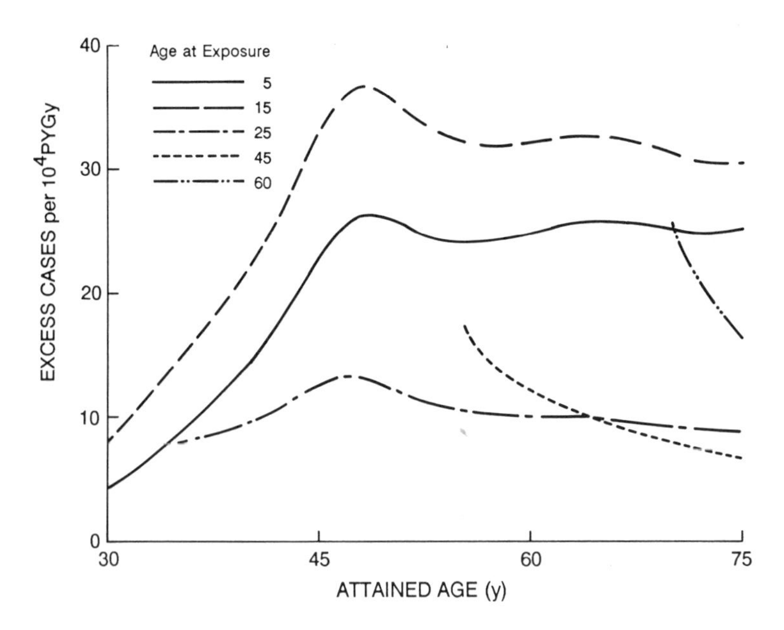

FIGURE 5-12 Excess cases of breast cancer by attained age for U.S. females exposed to low-LET radiation at various ages.

Experimental Data on Cancer Latency and Dose Response

Latency

The prolonged persistence of radiogenically initiated cells, suggested by the long latency of cancer in women who were exposed when they were children, has been confirmed experimentally. When rats are subjected to small doses of x rays or fission-spectrum neutrons and gamma rays, which alone produce few or no mammary neoplasms, and are then grafted with pituitary tumors that secrete high levels of prolactin, mammary neoplasms shortly appear in a high incidence (Yo77, Yo78). The time between irradiation and elevation of prolactin levels can be extended from a few days to as long as 12 months with little change in lag period from increased prolactin to appearance of tumors or in final tumor incidence.

Life-span studies of irradiated rats have illustrated that there is a marked inverse relationship between radiation dose and the latency of mammary neoplasms (Sh80, Sh82, Sh86a). This relationship may be related to the number of radiogenically initiated cells. The development of quantitative normal cell transplantation techniques has allowed the identification of a subpopulation of rat mammary cells that, when stimulated with appropriate hormones, can give rise to clonal multicellular glandular units (Cl85a). Following acute exposure, the acute post-irradiation survival and repair capacities of these mammary clonogens have been defined by transplantation assays (Cl85a). When grafts containing about 120 clonogens which had survived 7 Gy of ^{137}Cs gamma irradiation were transplanted to rats with marked prolactin and glucocorticoid deficiencies, mammary cancers arose in approximately 50% of the graft sites, indicating that there was one neoplastically initiated cell per 240-300 grafted clonogens (Cl86a). Consideration of these and other experimental data in rats suggests that there is a shortening of cancer latency as well as an increase in cancer incidence as irradiated mammary clonogen numbers are increased. The data also are consistent with the conclusion that a considerable period of hormonal promotion/progression is necessary for the development of overt cancer from radiation-initiated mammary target cells. In the women in the studies analyzed by the Committee, the normally functioning endocrine system supplied sufficient hormone, and a radiation dose-related shortening of latency was not observed. In the relatively small groups of experimental rats, more intense normal stimulation was often necessary to reveal radiogenic initiation, and an abbreviation of latency with increased dose was found.

Dose Response

A number of investigators have suggested that the low-LET radiation dose-response relationship for mammary tumors in rats is linear and that there is little effect of dose fractionation or protraction (Sh66, Sh86a).This appears to be true in Sprague-Dawley rats over the dose range of 0.28 to 4.0 Gy when the experiments are terminated at 10-12 months (Bo60, Sh57, Sh86a). In life-span studies of Sprague-Dawley rats (Sh80) and rats of the ACI strain (Sh82), inspection of the final numbers of animals with neoplasms has suggested a linear dose response over the ranges of 0.28-0.85 Gy and 0.37-3.0 Gy of x rays, respectively. Because of the effect of dose on the time of appearance of mammary neoplasms, more complex analyses that included both tumor incidence and latency have been employed (Sh80, Sh82). The results obtained are somewhat difficult to relate to human data. The designs of rat experiments have differed from laboratory to laboratory, hormonal manipulations were often used, experimental groups were often small, and benign fibroadenomas were often grouped with adenocarcinomas. The promotional effects of hormones on the induction of fibroadenomas differ from those on the induction of carcinomas (Cl78, Sh66, Sh82).

Ullrich's study of mammary carcinogenesis in otherwise untreated BALB/c mice exposed to ^{137}Cs gamma rays (low LET) at different dose rates and different dose fractions revealed a linear-quadratic dose-response relationship over the dose range 0-0.25 Gy administered at 0.35 Gy/minute (Ul87b). The ratio of the linear to the quadratic dose term is very low, 2.3; that is, the dose at which the effects governed by the linear dose component equals those governed by the quadratic function is 0.023 Gy (Ul87b). When the exposures were delivered at a low dose rate of 0.083 Gy/day, the dose response followed the linear term. When a total dose of 0.25 Gy was delivered in daily fractions of 0.01 Gy at a high dose rate, the carcinoma response fell on the linear curve, but when the same total dose was delivered in 0.05-Gy fractions at a high dose rate, the carcinoma incidence fell near the linear-quadratic curve (Ul87b).

Although these experimental results with mice are of considerable theoretical interest, their quantitative application to humans is problematic. As noted above, of the recent analyses of breast cancer mortality among irradiated women, only the Canadian fluoroscopy series (Mi89), with the Nova Scotia series included, presented any evidence of a positive quadratic component in the dose-response relationship. In that series, the ratio of the linear to the quadratic coefficient derived from all of the combined data is 205; and for those exposed to less than 6 Gy this ratio is 613 (Mi89). These ratios are 89- and 266-times larger respectively than the ratios from Ullrich's mouse data. Furthermore, the role of the mouse mammary tumor virus or its genomic equivalent in radiogenic mammary neoplasia is not clear (Sh86a). Thus, the nature of the dose response of mammary cancer to low-LET radiation deserves continuing investigation with respect to its underlying mechanisms, particularly at low doses, low dose rates, and small fractions.

Neutrons and Mammary Cancer

The results of the reevaluation of the atomic-bomb and radiation doses received by individuals in Hiroshima and Nagasaki (Ro87) have precluded the likelihood that the LSS data will yield useful information on the relative carcinogenic effect of neutrons (Sh87, Sh88). Experimental studies have, however, shown that per unit dose, neutrons have a significantly higher mammary neoplasm-inducing potential than low-LET radiations, and that this greater relative neoplastic potential is increased at small doses (Sh86a). An RBE of 20-60 was calculated for fission-spectrum neutrons from lifetime mammary neoplasm incidences in Sprague-Dawley rats irradiated with 0.05-2.5 Gy (Vo72). In this study, 0.05 Gy of fission-spectrum neutrons (average energy about 1 million electron volts [MeV]) was as effective in terms of the final yield of mammary tumors as were higher doses; that is, the tumor yield appeared to reach a plateau at 0.05 Gy. In contrast, in an experiment with the same rat line exposed to 14-MeV neutrons, mammary tumor incidence and the number of tumors per rat at 11 months after exposure was a near linear function of dose over the range 0.025-0.4 Gy (Mo77). The 14-MeV neutrons were about half as effective as reported for 0.43-MeV neutrons (Mo77). The RBE for mammary tumor induction by neutrons with different energies rank as follows: 2.0 MeV (fission spectrum) $>$ 14 MeV $>$ 0.025 MeV (thermal) (Ka85). Compared with x rays, the RBE of fission neutrons was $\sim$18.

In two of the most complete experimental life-span studies involving Sprague-Dawley and ACI rats, the latter with and without estrogen (diethylstilbestrol) supplementation (Sh80, Sh82), the analyses involved effects on both latency and incidence. In the estrogen-supplemented ACI rats, a significant increase in early tumor incidence was seen after they received 0.01 Gy of 0.43-MeV neutrons (Sh82). At low doses, the RBE increased in inverse proportion to the square root of the neutron dose and exceeded 100 at 0.01 Gy.

Unfortunately, the proper interpretation of many of the rat experiments is difficult because benign fibroadenomas were combined with carcinomas. These neoplasms differ fundamentally (Br85). For example, irradiated and unirradiated, but otherwise untreated ACI rats develop both mammary fibroadenomas and carcinomas. When either irradiated or unirradiated rats are given estrogen, the mammary tumors are virtually all adenocarcinomas (Sh82).

The effect of total dose and dose rate of fission-spectrum neutrons on mammary cancer incidence has been investigated in BALB/c mice (Ul84).

Lifetime mammary cancer incidence after exposure at a rate of 0.05-0.25 Gy/minute increased linearly with dose to a total of 0.1 Gy. At greater doses, tumor incidence increased less markedly over the range of 0-0.5 Gy, the dose-cancer response relationship was best fit by a model in which effect increased as the square root of dose (Ul84). When the neutron dose rate was reduced to 0.1 Gy/day or less, the mammary cancer incidence was approximately twice that at higher dose rates at total doses of up to 0.05 Gy. At higher total doses, cancer incidence plateaued, being very similar at 0.4 and 0.1 Gy (Ul84). When mammary cells from neutron-irradiated mouse glands were transplanted into gland-free fat pads, they gave rise to ductal dysplasias, that is, precancerous lesions (Ul86). Most such lesions derived from mice exposed to 0.025- or 0.2-Gy neutrons at 0.01 Gy/minute regressed by 16 weeks after grafting. In contrast, most such lesions derived from mice exposed to the same total doses at 0.01 Gy/day persisted, suggesting the promotion of initiated cells during the long exposure (Ul86). Finally, the decreasing effectiveness per unit dose of higher neutron doses given either at a high or a low dose rate may be related to the high sensitivity of the mouse ovaries to radiation damage.

In summary, the experimental data show that the neoplastic effect of neutrons on mammary tissue is higher than previously considered, particularly at low radiation doses ($\leq$ 0.01 Gy). Furthermore, in contrast to both human and experimental data with low-LET radiations, exposure to neutrons at low dose rates may be more damaging than exposures at high dose rates. The fine structure of the dose response for the production of mammary tumors at low dose and dose rate requires further mechanistic analysis.

Hormones and Breast Cancer

The growth, development, and function of the normal mammary gland is dependent upon hormonal regulation (Cl79, He88, Ro79, Ru82, Sh86a). The spectrum of hormones involved includes the steroids estrogen, progesterone, and glucocorticoid; the peptides prolactin, growth hormone, and placental lactogen; and perhaps other hypophyseal factors (Cl78, Kl87, Ru82). The actions of these hormones at a given time depend on the past hormonal exposure of the mammary tissue as well as the concurrent titers of other hormones. Hence, a given hormone may potentiate breast neoplasia under one set of circumstances and suppress it under others. Many breast carcinomas in women retain responsiveness to hormonal therapy (Cl77, Sh71). Although hormones are important to the promotion and progression of initiated mammary cells, experimental studies have unequivocally shown that radiogenic initiation is a scopal effect directly on mammary target cells (Cl86a, Sh71).

Gonadal estrogens play a dual role in mammary growth. They act directly as mitogenic agents on cells in the mammary gland and indirectly to induce the secretion of prolactin by the anterior pituitary gland. In women, estrogens are the primary mitogenic hormones; prolactin may facilitate the mitogenic action of estrogens and promotes differentiation and function (He88). In rats, prolactin is the primary mammary mitogen and agent for the promotion and progression of cancers (Cl79). Estrogens also induce progesterone receptors in mammary cells; progesterone, in turn, acts in synergy with estrogen and prolactin to control mammary growth and differentiation (Ro79, Ru82).

Finally, glucocorticoids are essential for milk secretion. The most efficient hormonal combination for promotion/progression of radiation-initiated rat mammary cells is a combination of elevated prolactin coupled with a glucocorticoid deficiency (Cl85c).

Given the profound role that is played by hormones in mammary cells, it is likely that most of the conditions that have been shown to enhance or suppress human breast cancer risk are mediated through effects on hormone levels and, in turn, on the number and condition of the mammary target cells. Most significant among these conditions are the ages at menarche and menopause. In addition, at first full-term pregnancy, the lactational history, and body weight are significant.

Circulating estrogens, progestins, and prolactin increase in women at menarche and decrease in the perimenopausal period (Pi83). Both an early age at menarche and a late age at menopause predispose an individual to breast cancer (Ho83, Ma73). Both of these conditions lengthen the total period of time during which the breast is subjected to the mitogenic stimuli of gonadal steroids. For example, the relative risk of breast cancer increased nearly linearly ($p < 0.004$) to 2.2 in women in Shanghai who entered menarche at ages ≤ 12 years compared with women who reached menarche at ages ≥ 18 years (Yu88). Surgically induced menopause (ovariectomy) before age 35 is strongly protective, but breast cancer risk is detectably reduced by ovariectomy at as late as 45-50 years of age (Ma73). In irradiated rats, even in the presence of high levels of prolactin plus glucocorticoid deficiency, ovariectomy reduces mammary cancer risk (Cl85c).

Breast cancer is most markedly reduced by the occurrence of a full-term pregnancy at an early age. Women of 15-16 years of age who carry a child to full term have 35-40% the risk of breast cancer of nulliparous women and ~30% the risk of women who first give birth when they are over 30 years of age (Ma73). When corrected for age of first full-term pregnancy and lactation duration, multiparity was found to decrease risks further; for example, the relative breast cancer risk of women in Shanghai with five children was 0.39 compared with those with just one child (Yu88).

Previous lactation decreases the breast cancer risk, particularly premenopausal disease (By85). In a study of Caucasian women in the United States, the risk of premenopausal breast cancer among those who had ever nursed a child compared with that among those who had not nursed a child was 0.49 (Mc86a). Among women in Shanghai of average age (~50.5 years), those who had lactated for a total duration of ≥9 years had a breast cancer risk of 0.37 compared with those who had lactated for ≤3 years (Yu88).

The effects of diet, and hence of body weight and body fat, are presumably likely hormonally mediated and appear to account to a large extent for differences in the geographic distribution of breast cancer rates. The incidence of breast cancer is five- to sixfold greater in North America and northern Europe than it is in Asia and Africa (Ma73). This effect is most likely related to life-style, and especially to the effect of diet on body weight and body fat. Recent experimental evidence suggests that total caloric intake is a more important risk factor for breast cancer than is the fat concentrations of the diet (He88). Among women in Shanghai, the breast cancer risk of those who weighed ≥60 kg was 2.4 times that of women who weighed ≤45 kg (Yu88). Second- and third-generation American offspring of Oriental immigrants have increased breast cancer risks (Ho83); Hawaiians of Chinese ancestry have a risk pattern similar to that of Hawaiian Caucasians (Ma73). Age at menarche in Japanese women is inversely related to body weight, and age at menopause is directly related to body weight (Ho83). In Japan, age at menarche decreased 6 months per decade during the period 1900 to 1945; age at menopause increased 1 year during the same period. There was a marked upswing in the age at menarche among Japanese women born between 1930 and 1940; this birth cohort reached the age of puberty during the severe food shortage during and after World War II (Ho83). Thus, the incidence of radiogenic breast cancer in individuals in the various age cohorts of the Hiroshima-Nagasaki Life Span Study are being measured against a shifting background of breast cancer risk from other causes.

Fatty tissues contain aromatizing enzymes that convert adrenal androgens into estrogens. This leads to continued hormonal stimulation of the mammary glands after menopause, and likely accounts in part for the greater risk among the generally heavier postmenopausal women of North America and northern Europe than in the lighter Oriental population (Ho83, Ma73).

These findings are consistent with the conclusion that radiogenic initiation and the expression of such radiogenic damage in the formation of overt breast cancer is dependent on the number of mammary target cells and their degree of differentiation at the time of exposure and on subsequent promotion and progression of the initiated cells under hormonal control. Those conditions that induce functional mammary differentiation, and hence, that reduce target cell numbers (Ru82, Cl86a), for example, early and multiple pregnancies and lactation, reduce the risk of breast cancer. Those conditions that reduce or block full functional differentiation and increase mitogenesis, for example, nulliparity and glucocorticoid deficiency, increase the risk.

Summary

1. Animal experiments and human studies indicate that the induction of breast cancer is hormonally mediated and that hormonal status plays a critical role in radiation carcinogenesis of the breast.
2. A strong linear component is seen in the dose-response relationship for radiation-induced cancer.
3. There is little evidence of reduction in risk associated with dose fractionation in the human cohorts considered, even though these cohorts included both fractionated and acute exposures.
4. There is no evidence in humans of the occurrence of radiation-induced cancers until after the age of 25, which is about the youngest age at which breast cancer is seen in the general population.
5. Age at exposure strongly influences susceptibility, with risk being highest among women under 20 years of age at the time of exposure, suggesting that the time of puberty corresponds to a period of elevated risk. The level of risk among those under age 10 at the time of exposure is uncertain. There is little evidence of any excess risk in women over age 40 at the time of exposure.
6. The low relative risk for women in the NY-APM study exposed between the ages of 15 and 17 suggests that the occurrence of a full-term pregnancy before age 20 may reduce susceptibility to radiation-induced breast cancer.
7. There is no evidence that radiogenic breast cancers appear during the first 10 years following exposure, but after this time the number of such cancers appears to increase rapidly. On the relative-risk scale, the data suggest that the incidence peaks at 15 to 20 years after exposure and the mortality about 5 years later. Observations to date indicate that the absolute risk continues to increase until 50 but may decrease at older ages.
8. Although animal data suggest that there is a relationship between dose and latency, the human data show no such relationship.

LUNG

There are three main sources of epidemiologic data on the induction of human lung cancer by radiation: (1) the survivors of atomic-bomb explosions in Hiroshima and Nagasaki, (2) patients who were treated with x rays for ankylosing spondylitis, and (3) uranium miners and other underground miners exposed chronically to high-LET (alpha) radiation from inhaled ^{222}Rn and its progeny. Each of these populations has received detailed long-term follow-up to ascertain the health risks associated with these diverse types of exposure.

Japanese Atomic-Bomb Survivors

Since the publication of the BEIR III report (NRC80), there have been three follow-up reports on the Life Span Study (LSS) of survivors of the atomic bomb explosions in Hiroshima and Nagasaki (Ka82, Pr87a, Sh88) in whom lung cancer has been a prominent late effect. These reports provide analyses of the excess cases of lung cancer in the period 1950-1985, or a maximum of 40 years after the detonations. In the 1950-1978 LSS report (Ka82), the absolute risk of lung cancer showed a significant increase during the period from 1950-1974, and this increase continued during the interval from 1974 to 1978. Overall, the absolute risk of lung cancer occurring between 1950 and 1978, based on T65D dosimetry, was 0.61 lung cancer deaths/10^4 person-year Gy (PYGy) (90% confidence interval, 0.37-0.86). When the relative risk and 90% confidence interval for lung cancer were compared with similar computations for other radiation-related cancers (viz., breast, stomach, and colon), the relative risks for these four sites fell within the same confidence intervals, suggesting that relative risks may not differ by target organ.

In the 1950-1982 LSS report (Pr87a), which used the T65 dosimetry, mortality from cancers of the trachea, bronchus and lung was significantly associated with the radiation dose ($p < 0.001$). The estimated relative risk at 100 rad (1 Gy) was 1.33 (90% confidence interval, 1.19-1.50) and the average excess risk was 0.82 cancers per 10^4 PYGy (90% confidence interval, 0.48-1.19). When the relative risk at 100 rad (1 Gy) for cancer of the trachea, bronchus, and lung was examined in 4-year intervals to 1982, the values remained approximately constant from 1955 to 1982.

Darby et al. (Da85) conducted a parallel analysis of the cancer mortality seen in patients with ankylosing spondylitis and Japanese atomic-bomb survivors who received a T65 dose of at least 100 rad (1 Gy); statistically significant excess deaths from lung cancer were observed when the results of these two studies were compared. In the ankylosing spondylitis series, using lung cancer deaths that occurred more than 3 years after treatment, the relative risk was 1.41 (90% confidence interval, 1.20-1.65); for the Japanese atomic-bomb survivors who received more than 100 rad (1 Gy) compared with the group who received <9 rad (0.09 Gy), the relative risk was 1.94 (90% confidence interval, 1.53-2.45).

In the most recent LSS report (Sh88), cancer mortality was analyzed for the period 1950-1985 as a function of the revised DS86 doses in a DS86 subcohort of approximately 76,000 individuals. In general, the number of excess deaths from all cancers other than leukemia was observed to increase proportional to the background cancer rate for attained age, with the result that the relative risks tended to remain constant for specific age cohorts, except for those who were 0-9 years old at the time of bombing. Lung cancer deaths deviated somewhat from this pattern because of a slightly decreasing trend with time. Using the DS86 organ dose values, the relative risk of lung cancer mortality at 100 rad (1 Gy) was 1.63 (90% confidence interval, 1.35-1.97); the absolute risk was 1.68 excess lung cancer deaths/10^6 PYR (90% confidence interval, 0.97-2.49).

Patients with Ankylosing Spondylitis in England and Wales

Results of earlier studies on over 14,000 patients treated with x rays for ankylosing spondylitis in England and Wales from 1935 to 1955 were summarized in the BEIR III report (NRC80). The BEIR III Committee estimated the average dose to the bronchus in such patients to be approximately 197 rad (1.97 Gy) on the assumption that 80% of the bronchial epithelium was irradiated. Because no dose values for individuals were available, no dose-response models were tested at that time.

Smith and Doll (Sm82) extended the follow-up to January 1, 1970. To avoid uncertainties caused by multiple radiation exposures, their analysis was restricted to patients receiving a single course of treatment. There were 133,874 person-years at risk available for analysis, 83% of which were from male subjects. Cancers arising in the heavily irradiated sites became prominent beginning about 9 years after the first treatment and continued at an elevated level up to 20 years or more after treatment. Lung cancer was the most frequent type of cancer observed in the heavily irradiated sites, accounting for 37 of the 92 excess cancers (ratio of observed to expected cancer, 1.42; $p < 0.001$).

The follow-up was subsequently extended to January 1, 1983, by Darby and colleagues (Da87) who observed 224 lung cancer deaths occurring at times >5 years after the first treatment, while 184.5 lung cancer deaths were expected (ratio of observed to expected deaths, 1.21; $p < 0.01$) during the interval from 5 years to 24.9 years after the first exposure. The ratio of observed to expected deaths was 0.97 for lung cancers occurring after 25 years. No indication was given of the contribution of smoking to the risk of lung cancer in this population.

Dosimetric estimates have since been published by Lewis et al. (Le88), who used Monte Carlo techniques to calculate the average doses received by various organs and tissues, but individual dose estimates are still unavailable. For the lungs, the average dose was reported to be 1.79 Gy, while for the main bronchi, the average dose was 6.77 Gy. Because dose estimates for individuals are not available for this cohort, it was not possible to analyze the follow-up results in terms of various dose-response models.

Cervical Cancer Patients

In a multicenter international study of 182,040 women treated for cancer of the uterine cervix, the relative risk of lung cancer was observed to be increased after radiation therapy (Bo85). On the basis of preliminary dosimetry, the lung was estimated to have received an average dose of 35 rad (range, 10 to 60 rad) and the observed relative risk was 3.7 ($0.001 < p < 0.01$). When the relative risk was examined as a function of time after irradiation, the pattern suggested an influence of misclassified metastases and confounding by cigarette smoking.

Underground Miners

Detailed epidemiologic studies on radiation-induced lung cancer are being conducted on underground miners of uranium and other minerals who are exposed chronically to alpha radiation from inhaled ^{222}Rn, ^{220}Rn, and their radioactive progeny. The exposure of the miner cohorts differs from the exposures of the Japanese atomic-bomb survivors and patients with ankylosing spondylitis in being chronic, low-dose-rate irradiation from internally deposited alpha emitters, as opposed to single, acute, or short-term fractionated, high-dose-rate, low-LET external irradiation. Because of the complexity of the dosimetry of radon in the respiratory tract, it has been customary to measure and record miner exposures in terms of working levels (WL), and cumulative exposures in working level months (WLM). It is difficult, therefore, to compare the results of the studies directly unless an appropriate value for the absorbed dose, per WLM is estimated. This involves a number of dosimetric uncertainties, and reported values of dose (rad) to the bronchial epithelium per WLM have differed by a factor of 10 or more (NCRP84).

A number of analyses of cohorts of underground miners concerning radiation and lung cancer have been conducted to examine dose-response relationships and various dose-effect modifying factors (UN77, ICRP87, NCRP84, NRC88). The BEIR IV Committee (NRC88) examined lung cancer risks associated with four principal underground mining populations: two Canadian uranium miner cohorts at Eldorado, Beaverlodge, Saskatchewan, and Ontario; Swedish iron miners at Malmberget; and Colorado Plateau uranium miners. The data base analyzed by the BEIR

TABLE 5-2 Comparison of Estimates of Lifetime Risk of Lung Cancer Mortality due to a Lifetime Exposure to Radon Progeny[a]

Study		Excess Lifetime Lung Cancer Mortality (deaths/10^6 person WLM)
BEIR IV	1988	350
ICRP	1987	170–230[b]
		360[c]
NCRP	1984	130
BEIR III	1980	730
UNSCEAR	1977	200–450

[a] Adapted from NAS88 and ICRP87.
[b] Relative risk with ICRP population.
[c] Relative risk with 1980 U.S. population as in BEIR IV.

IV Committee contained a total of 360 lung cancer deaths and 425,614 person-years at risk.

The analyses were based on a descriptive analytical model of the pooled data in which the excess relative risk eventually decreases with time after exposure and also depends on age at risk. This is in contrast to the analyses of the Japanese atomic-bomb survivors, in which a definite decline in relative risk has not yet been observed. When the BEIR IV model was used to project lifetime risks of lung cancer from lifetime exposures to radon progeny, an overall value of 350 excess lung cancer deaths/10^6 person-WLM was obtained. Annex 4G provides an analytical description of the risk model developed by the BEIR IV Committee and summary tables of their risk estimates for lifetime exposure to radon progeny. Additional information regarding the data and methods of these analyses are described in detail in the BEIR IV report (NRC88).

Table 5-2 compares the risk values derived by the BEIR IV Committee with values derived by United Nations Scientific Committee on the Effects of Atomic Radiation (UNSCEAR) (UN77), the BEIR III Committee (NRC80), the National Council on Radiation Protection and Measurements (NCRP) (NCRP84), and the International Commission on Radiological Protection (ICRP87).

Considering that the risk estimates in Table 5-2 are based, for the most part, on essentially the same epidemiological studies of underground miners, the range of estimates is fairly broad. This is largely due to the difference in the models used to project lifetime risks. For example, the BEIR IV Committee used the time and age dependent relative risk model outlined in Annex 4G, while the ICRP used a simple relative risk projection. In contrast, the NCRP used an absolute risk model with exponentially decreasing risks with time after exposure. The BEIR III Committee used an absolute risk model that is constant over time. One reason the BEIR IV Committee's model is more elaborate then the others is that original data from miners studies were made available to this Committee while the other studies had to rely on published summaries which contain little information on how risks change with increasing follow-up time as the population at risk ages.

Effect of Smoking

Several attempts have been made to study the influence of cigarette smoking on the carcinogenic effects of irradiation in the Japanese atomic-bomb survivors. Prentice et al. (Pr83) assembled several subsets of known smokers into a study population of 40,498 subjects. The T65DR doses were used along with questionnaire results on smoking habits to analyze 281 lung cancer deaths in this cohort. Using a Cox proportional hazards model and stratifying on city, sex, age at the time of the bombing, and survey date, the relative risk of lung cancer in nonsmokers rose from 1.0 to 1.1 to 2.3 for exposure doses of <10, 10-100, and >100 rad (<0.1, 0.1-1.0, and >1.0 Gy), respectively, whereas the corresponding relative risks for cigarette smokers were 2.4, 2.4, and 3.6. Neither a multiplicative nor an additive interaction for lung cancer mortality could be distinguished clearly.

Kopecky et al. (Ko86) examined the combined effects of irradiation and cigarette smoking in a cohort of 29,332 Japanese atomic-bomb survivors among whom there were 351 lung cancer deaths. An additive excess-risk model fit the data without either superadditivity or subadditivity; no corresponding test of a multiplicative model was presented.

The BEIR IV Committee (NRC88) also examined the question of the possible combined effects of cigarette smoking and exposure to radiation, reviewing the available information for underground miners as well as the A-bomb survivors and analyzing three populations in detail. The data sets used were from case-control studies of New Mexico uranium miners, a cohort study of Colorado uranium miners with follow-up through 1982, and Japanese atomic-bomb survivors. In discussing the results of these analyses, the BEIR IV Committee noted that analyses of this type normally included only some measure of radiation exposure and duration or intensity of cigarette use. Other factors that need to be considered in later analyses include age at first exposure, dose rate, sex, diet, and genetic predisposition.

The BEIR IV Committee noted that although a multiplicative model for the interaction between exposure to radon progeny and cigarette smoking appears to have received the greater support in the literature, a submultiplicative model may provide a more accurate description of the underlying relationship. The Committee's analysis of the Japanese atomic-bomb survivor data indicated that neither an additive nor a multiplicative model could be rejected on statistical grounds (NRC88), a finding consistent with the earlier observations of Prentice et al. (Pr83) and Blot et al. (Bl84).

The present Committee's analysis of lung cancer mortality relies heavily on death certificate information from the Life Span Study (LSS) of Japanese A-bomb survivors whose deaths were classified as due to cancers of the respiratory system, International Classification of Disease (ICD) codes 160-163. Use of this broad classification minimizes the loss of case information due to any lack of precision in the death certificates. Deaths due to lung cancer occur relatively late in life and thus in a rather narrow age range. The LSS data are limited in this regard in that the reliability of Japanese death certificates apparently diminishes rapidly for persons over 75 years of age. Therefore, deaths occurring after that age were not used in the Committee's analyses. Moveover, those exposed as children are still too young to provide reliabile information on lung cancer mortality.

The Committee's preferred risk model for respiratory cancer mortality is given in Equation 4-4 which is reproduced below.

$$f(d) = \alpha_1 d$$
$$g(\beta) = \exp[\beta_1 \ln(T/20) + \beta_2 I(S)].$$

The Committee's analysis of respiratory cancer in A-bomb survivors showed little effect of age at exposure but did show a decrease with time after exposure (Figure 5-13). Therefore the relative risk also decreases with attained age as shown in Figure 5-14. The exponential coefficient for decreasing risk with time after exposure in the Committee's model is relatively large, -1.44, but has a rather broad standard error, ± 0.91. The change in deviance when time after exposure is included in the preferred model is modest, 1.75. Even so, the similarity of diminishing risk with time after exposure observed in the ankylosing spondylitic study reinforces the Committee's view that the modifying effect of time should be included in the preferred model.

Because respiratory cancer is mainly a disease of old age, most of the excess mortality projected by the Committee's relative risk model occurs among those exposed rather late in life (Figure 5-15).

The Committee also modeled additive excess risks for respiratory cancer. Under this model, mortality increased more strongly with attained age, age at exposure, and time after exposure for lung cancer than for other cancers, but was virtually identical for males and females. The additive model involved risks that increased as the square of the time after exposure (starting 10 years after) and as the 2.7 power of age at exposure. The additive and relative-risk models gave virtually indistinguishable fits to

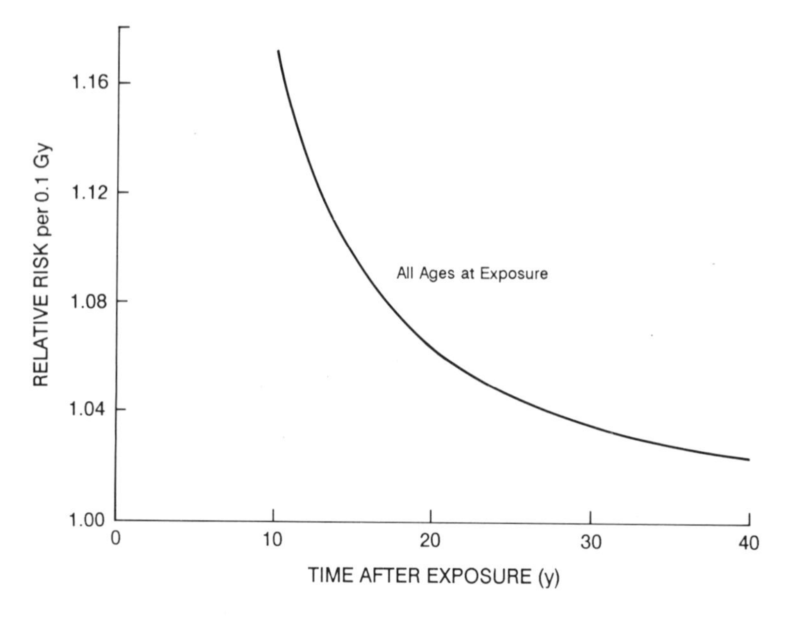

FIGURE 5-13 Relative risk of lung cancer mortality in males as a function of time after exposure to low-LET radiation.

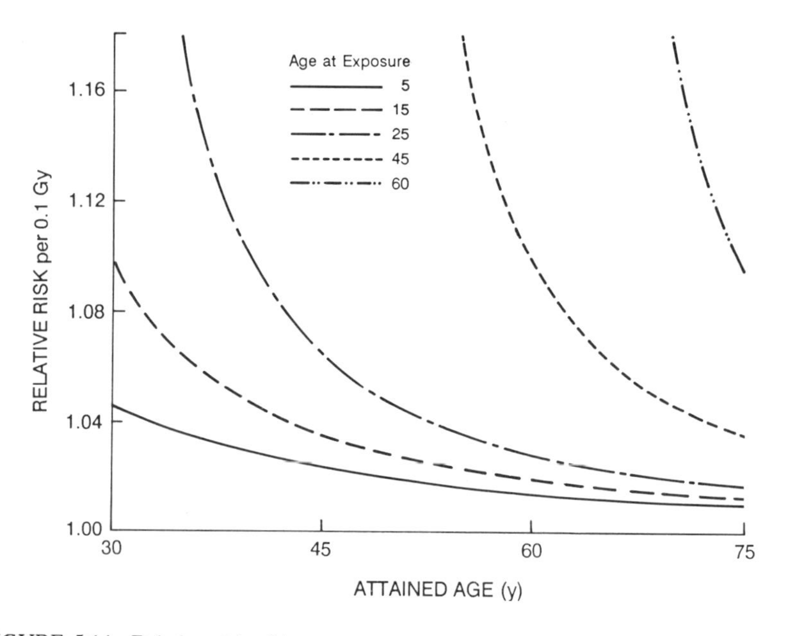

FIGURE 5-14 Relative risk of lung cancer in U.S. males by attained age at various ages due to low-LET radiation exposure.

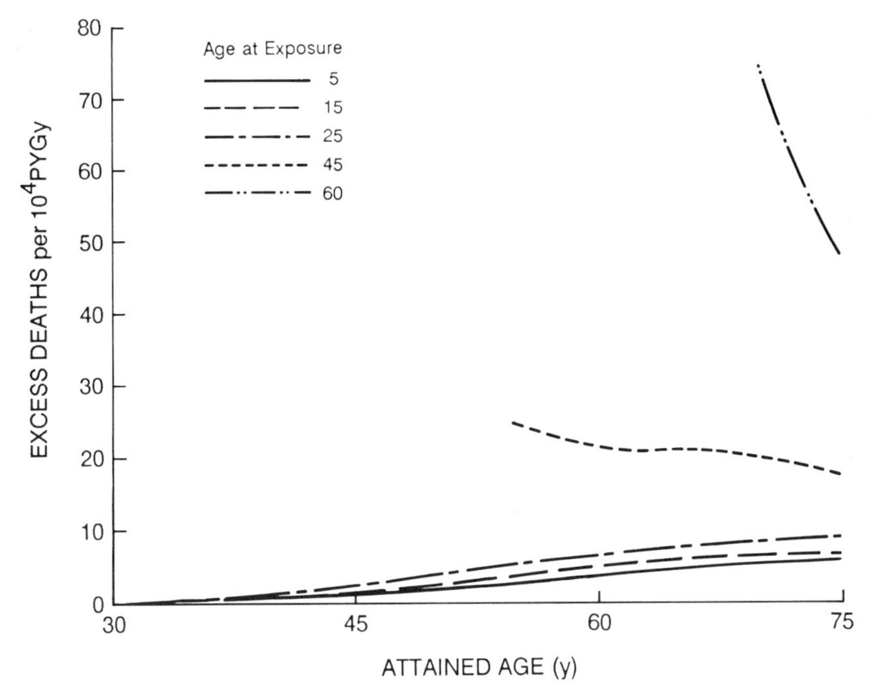

FIGURE 5-15 Excess lung cancer mortality due to low-LET radiation for U.S. males by attained age for various ages of exposure.

the observed mortality data, but the relative-risk model was preferred on the grounds of its simplicity and consistency with the Committee's treatment of cancer risks at other sites.

Studies in Laboratory Animals

The studies of populations described above are invaluable in providing information on the response of the lungs to certain kinds of exposure to ionizing radiation. However, except for radon progeny, there are no human data available on lung cancer due to internally deposited radionuclides. Therefore a number of life-span studies have been, and are being, conducted in various species of laboratory animals to supplement and extend the human data. Recent major summaries of research on the carcinogenic response of the respiratory tracts of laboratory animals to ionizing radiation include those by Kennedy and Little (Ke78), ICRP (ICRP80), Bair (Ba86), Thompson and Mahaffey (Th86), and the BEIR IV Committee (NRC88). These summaries provide a wealth of information on the influence of various factors on the temporal and spatial characteristics of the dose received by tissues of the respiratory tract from inhaled radionuclides and the resulting biological effects, particularly at long times after inhalation exposure. A few of these results are described below.

Various factors that influence the lifetime risks of inhaled beta-emitting radionuclides are being examined in a series of life-span studies in beagle dogs exposed once, briefly, by inhalation, to different beta-emitters in relatively soluble or insoluble forms (Mc86). The effect of dose protraction is being studied by using radionuclides with radioactive half-lives ranging from about 3 days to 29 years, encapsulated in fused aluminosilicate particles. Recent results from these studies have demonstrated that when the delivery of a dose of beta radiation to the lung is protracted from days to years, the pulmonary carcinogenic response is reduced by a factor of about 3 (Ha83a, Gr87).

The induction of lung tumors in mice from external x-irradiation has been compared with that from neutron irradiation by Ullrich and Storer (Ul79). The mice were sacrificed 9 months after they received 100-900 rad of x-irradiation or 5 to 150 rad of neutron irradiation localized to the thorax. The relationship between the number of lung tumors (adenomas) per mouse and the x-ray dose could be described adequately by a linear-quadratic model with a shallow initial slope or by a threshold model with a dose-squared response above the threshold. In contrast, the tumorigenic response of the lung to neutron irradiation could be described by a linear or threshold model, with the linear response being above the threshold. The relative biological effectiveness of the neutron irradiation increased with decreasing neutron dose: from 25 at 25 rad to 40 at 10 rad.

Lafuma et al. (1989) reported on the effectiveness of various radiation exposure modalities (radon-daughter inhalation, fission-neutron irradiation, or gamma irradiation) for inducing lung carcinomas in Sprague-Dawley rats. The observed equivalence ratio for radon daughter to neutrons was approximately 15 WLM to 10 mGy neutrons, and the ratio of neutron effectiveness to gamma rays from ^{60}Co was 50 or more at a gamma dose of 1 Gy (La89).

Coggle et al. (Co85) examined the tumorigenic effectiveness of uniform versus nonuniform external x-irradiation of the mouse thorax. The nonuniform irradiation was produced by 72 1-mm microbeams that irradiated about 20% of the total lung volume. Although a smaller study by these investigators had previously suggested that nonuniform x-irradiation might be more tumorigenic than uniform x-irradiation, the larger study demonstrated a nearly equal tumorigenic response of the lung to uniform and nonuniform x-irradiation.

A study of the lifetime relative biological effectiveness of chronic beta irradiation of the lung versus chronic alpha irradiation of the lung for the production of pulmonary carcinomas, primarily bronchioloalveolar carcinomas, was reported recently by Boecker et al. (Bo88b). Beagle dogs were exposed once, briefly, by inhalation, to the beta emitter ^{91}Y or to three different aerodynamic particle sizes of the alpha-emitter $^{239}PuO_2$. A proportional hazards model was used to estimate the relative risk coefficients for these various radiation exposure modalities. Comparison of the linear risk coefficients among the four studies indicated that all three exposure regimens with $^{239}PuO_2$ were more effective in producing lung cancer than was ^{91}Y. The ratios of the relative risk coefficients for $^{239}Pu/^{91}Y$ ranged from 10 to 18 for different sizes of $^{239}PuO_2$, with the more uniform irradiation being more effective in producing cancer.

Other studies on the lifetime biological effects of inhaled radon progeny, plutonium, and other actinide radionuclides were discussed in detail in the BEIR IV report (NRC88). Studies on the effects of inhaled radon progeny inhalation in laboratory animals, primarily rats and dogs, are being conducted in the United States and France. Although earlier studies focused primarily on acute effects, more recent studies have been directed to lung cancers resulting from chronic inhalation exposures. The use of laboratory animals has made it possible to study the impact of various modifying factors on the resulting lung cancer incidence, thereby broadening the knowledge obtained from human epidemiology studies. Factors discussed in this regard in the BEIR IV report include the effects of cumulative exposure, exposure rate, unattached fraction of radon progeny, disequilibrium of radon progeny, and concomitant exposure to cigarette smoke.

Major studies on the long-term effects of inhaled $^{238}PuO_2$ or $^{239}PuO_2$ in beagle dogs are continuing at two laboratories, the Battelle Pacific Northwest Laboratories, Richland, WA (Pa86), and the Lovelace Inhalation Toxicology Research Institute, Albuquerque, NM (Mc86). The lack of any human dose-response data for internally deposited plutonium radionuclides makes these studies particularly valuable for estimating the potential human health risks associated with such internal depositions. For instance, these studies provide firm evidence for the risk to the lung, skeleton, and liver following inhalation of $^{238}PuO_2$ and allows comparison with the risk to the lung following inhalation of $^{239}PuO_2$. Differences in risk are caused by differences in tissue distribution and retention of these two plutonium radionuclides.

Studies of the pulmonary tumorigenic responses of rats to inhaled low levels of $^{239}PuO_2$, by Sanders et al. (Sa88) indicate that the observed dose-response function is best fitted by a quadratic function, with a "practical" threshold of about 100 rad. These studies and those in progress by Lundgren et al. (Lu87a, Lu87b) will provide an important basis for comparing the relative effectiveness of low doses of brief external x-irradiation with those of chronic beta or chronic alpha irradiation. The Bayesian analysis performed in the BEIR IV report (NRC88) to estimate the human bone cancer risk from internally deposited ^{239}Pu, comparing available data for alpha-emitting radionuclides in humans and in laboratory animals, illustrates one way in which such data are useful.

Summary

Absolute radiogenic risks of radiation-induced lung cancer are similar for both sexes although baseline lung cancer risks are much higher for males than they are for females. Consequently, the excess relative risk for irradiated females is higher than for males. Because the relative risk attributable to radiation appears to be essentially constant with respect to age at the time of exposure the relative risk model is preferred for risk projection purposes. However, there is no clear evidence as to whether relative risks or additive risks are more consistent between populations. The constancy of additive risks between the sexes would support the use of an additive model for the comparing of populations.

Data suggesting greater than additive interactions between radiation and smoking are equivocal and are largely restricted to the effects of chronic exposure to radon progeny, which are of uncertain relevance to the effects of brief exposure to low-LET irradiation.

Studies in laboratory animals have provided much of what is known about the influence of various modifying factors on the long-term biological effects of chronic beta or chronic alpha irradiation on the respiratory tract. They have shown that prolongation of beta irradiation of the lung from a period of days to years reduces its tumorigenic effectiveness by a factor of about 3, and that chronic alpha irradiation of the lung from inhaled $^{239}PuO_2$ is 10 to 20 times more carcinogenic than is chronic beta irradiation.

STOMACH

The best-known and, perhaps, strongest evidence of a relationship between ionizing radiation and cancer of the human stomach comes from the follow-up of studies of the Japanese atomic-bomb survivors. The most recent such analysis, using the new dosimetry on the combined cohort from Hiroshima and Nagasaki, shows that there is a highly significant radiation-related relative risk of mortality from stomach cancer; e.g., the average excess relative risk at 1 Gy is 0.23 in terms of the kerma at a survivor's location. Females have higher absolute and relative risks than males, although neither difference is statistically significant (Sh87). Overall, stomach cancer is one of the most common types of cancer seen in this cohort, and its average excess risk of 2.09 deaths/10^4 PYGy is, with the single exception of leukemia, the largest excess observed among specific cancer sites. As noted above, however, the relative risk is not large, since this is a commonly occurring tumor in the general population.

A similar association between radiation exposure and stomach cancer was observed in a recent follow-up study of patients irradiated for peptic ulcer at the University of Chicago (Gr84). The original study involved 1,457 patients treated with radiotherapy over a 1- to 2-week period between 1937 and 1955 and 763 nonirradiated patients with ulcers who served as controls (Cl74). In the majority of the radiation-treated patients, the stomach received an estimated organ dose of 16-17 Gy. Patient follow-up was continued through 1962, but no significant increase in tumors was noted. Twenty-two years later Griem and colleagues (Gr84) expanded the study population to 2,049 cases by including all patients with peptic ulcers treated with radiation through the mid-1960s. They then used the hospital identification numbers of the study population members to search the University of Chicago Tumor Registry and update their tumor histories. Their preliminary evaluation of the registry data found a radiation-related relative risk for stomach cancer of 3.7, with a corresponding estimated excess risk of 5.5×10^{-4} stomach cancers per person Gy, based on a life-table analysis.

The findings from the Japanese atomic-bomb survivors and the University of Chicago ulcer patients appear, at first, to be in marked contrast to the results of the investigation of second primary tumors among more than 82,000 women who underwent treatment for cervical cancer (Bo85). In the radiation-treated patient series of this study, an excess of 60 stomach cancers was predicted on the basis of risk estimates for radiation-induced stomach cancers prevalent in the literature at that time. However, only three excess stomach cancers were actually seen. Furthermore, the relative risk did not seem to vary to any extent either with age or time elapsed since the initiation of radiotherapy. The authors speculated that the failure to duplicate the effects observed among Japanese atomic-bomb survivors may have been due to differences in the age or sex composition of the study populations, even though the estimated average organ dose of 2 Gy for the patients with cervical cancer was somewhat higher than doses which produced significant effects in the Japanese atomic-bomb survivors. However, a further follow-up of this study population (Bo88b) suggests that an association between radiation exposure and stomach cancer is present in the patients with cervical cancer, i.e., when a subset of the original study population consisting of women with second primary cancers and their matched controls were selected for further study, it was found that an exposure of several Gy led to a significant ($p < 0.05$) relative risk of 2.1 (Bo88b).

The issue becomes complicated when one takes into account the data on patients with ankylosing spondylitis given a single treatment course with x rays. Previous analyses of the mortality data from the Court-Brown and Doll (Co65) cohort of over 14,000 patients with ankylosing spondylitis who had been treated with x rays indicated the presence of an elevated risk of stomach cancer among those patients for whom a sufficient period of time had elapsed since treatment. For example, Land indicated that 31 deaths from stomach cancer were observed, while only 20.1 were expected in those patients 9 or more years after their first exposure (La86). However, a more recent analysis of this data base (Da87) shows that, overall, 64 deaths from stomach cancer have been observed, compared with 63.2 expected. The authors reported relative risks of 1.01 (9 observed versus 8.88 expected deaths), 1.2 (44 observed versus 36.5 expected deaths), and 0.62 (11 observed versus 17.8 expected deaths) for the respective time intervals of less than 5 years, 5-24.9 years, and 25 or more years since their first treatment. Doses to the stomach were quite variable, ranging somewhat uniformly between 0 and 5 Gy (Le88). These values, which are based on different time intervals and a longer period of follow-up than many of the previous analyses, and some of the other results reported above, suggest that the relationship between radiation and stomach cancer may be more complicated than was formerly recognized.

A number of investigators have established that radiation is capable of causing adenocarcinomas in the mouse stomach (Wa86), although the stomach appears to be less susceptible to the carcinogenic effects of ionizing radiation than are many other similarly exposed organs in the laboratory mouse. Doses required to produce an effect are often quite large. For example, Hirose (Hi69) considered a single x-ray exposure of 20 Gy to be appropriate.

Summary

In human populations, as well as in laboratory animals, the risks of stomach cancer have been observed to be increased by irradiation. The existing data, although not sufficient to define the dose-incidence relationship precisely, are consistent with observations on atomic-bomb survivors, in whom the relative risk of mortality from stomach cancer is estimated to approximate 1.19 per Gy.

The Committee's risk model for cancer of the digestive system is based on the mortality of A-bomb survivors in the Life Span Study (LSS), ICD codes 150-159. Stomach cancer was the predominant cause of death in this group. The Committee's preferred risk model contains terms for sex and age at exposure as shown in Equation 4-7 which is reproduced below.

$$f(d) = \alpha_1 d$$
$$g(\beta) = \exp[\beta_1 I(S) + \sigma_E]$$

where $I(S)$ equals 1 for females and 0 for males and

$$\sigma_E = \begin{cases} 0 \text{ if } E \leq 25 \\ \beta_2(E - 25) \text{ if } 25 < E \leq 35 \\ 10\beta_2 \text{ if } E > 35 \end{cases}$$

The data indicate that females are at higher relative risk for cancer of the digestive system than males and there is a comparatively higher risk for those younger than 30 years when exposed. The Committee notes that this is not the usual pattern in which risk is highest for those exposed as children, diminishing for adolescents and young adults. Instead, the data indicate an abrupt decrease in risk for those over 30 years of age when exposed. In the Committee's judgment, this change is not simply a reflection of artifacts in the data, although a biological basis for it is unknown.

Although the relative risk diminishes for ages greater than 30, the baseline risk for digestive cancers increases rapidly with age, so that most of the excess risk occurs after middle age (Figure 5-16).

THYROID

Introduction

The incidence of thyroid carcinoma has been observed to be increased in a number of irradiated populations. Carcinoma of the thyroid was the first of the solid tumors noted to occur at increased frequency in the Japanese atomic bomb survivors (Ho63, So63). Even earlier, however, an increased incidence of thyroid cancer had been reported among persons exposed to therapeutic x rays as infants (Du50, Si55) and among individuals on the Marshall Islands who were exposed to radioactive fallout (Co80, Co84). Continuing studies of the atomic bomb survivors (Pr82, Wa83) and other irradiated populations (Co76, Ro77, Du80, NCRP85, Ro84a, Sc85, Sh84b) suggest the following generalizations:

1. Susceptibility to radiation-induced thyroid cancer is greater early in childhood than at any time later in life. In those exposed before puberty, however, the tumors usually do not become apparent until after sexual maturation.
2. Females are two to three times more susceptible than males to radiogenic as well as spontaneous thyroid cancer.

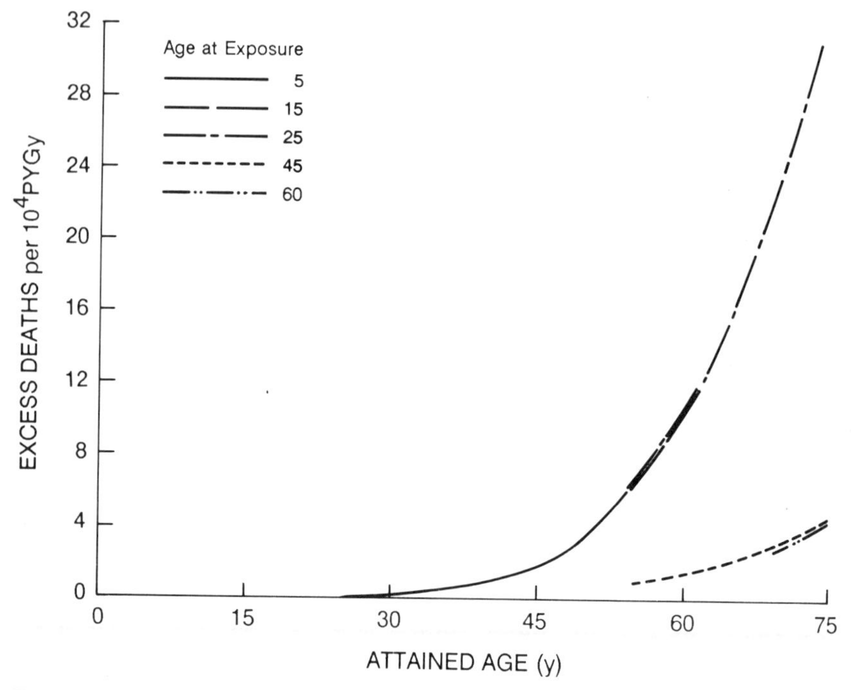

FIGURE 5-16 Excess deaths due to cancer of the digestive system in U.S. females by attained age due to exposure to low-LET radiation at various ages. Persons exposed at less than age 25 are expected to follow the same mortality pattern as those exposed at age 25.

3. Radiogenic cancer of the thyroid is frequently preceded or accompanied by benign thyroid nodules; the frequency of hypothyroidism and simple goiter also is increased in those exposed to large doses of radiation when young.

4. Radiogenic carcinomas of the human thyroid are generally papillary growths; relatively few are of follicular or mixed histopathology.

5. The development of overt cancer from thyroid epithelial cells is dependent on hormonal stimulation, with the result that any condition leading to sustained elevation of thyroid-stimulating-hormone levels increases the risk of thyroid neoplasia.

The following aspects of thyroid carcinomia are discussed below: (1) the Committee's analyses and risk estimates from two sets of subjects exposed to external radiation, (2) thyroid physiology as related to radiogenic thyroid cancer, (3) successive phases in the development of thyroid neoplasia, and (4) the special problem of radioiodide. This discussion is limited to the epithelium of the thyroid follicles; there is no evidence that radiogenic neoplasms develop from the parafollicular C cells. For reviews of radiogenic thyroid neoplasia, see references Cl86c, Co80, Du80, NCRP85.

Risk of Human Thyroid Cancer from External Radiation

The incidence of thyroid cancer was evaluated in two groups of patients who were exposed to low-LET radiations for benign conditions: children in the Israel Tinea Capitis Study (Ro84a) (10,834 irradiated subjects of ages 0-15 years when exposed and 16,226 nonirradiated subjects for comparison) and the Rochester Thymus Study (He75) (Sh85) (Sh84b) (2,652 irradiated subjects who were less than 1 year old when exposed and 4,823 nonirradiated subjects who were siblings of the irradiated subjects). The Israeli series included 39 cases of thyroid cancer among the irradiated group and 16 cases among the comparison group. In the Rochester cohort there were 37 cancer cases among the irradiated subjects and 1 in the nonirradiated group. It should be noted that the Committee's analyses are based on new individual dose estimates for individuals in the tinea study that take account of possible movement by the patients during irradiation. The mean doses remain the same as in an earlier analysis of these data which also included this feature (Ro84a). Doses in the Rochester thymus study are rough estimates of individual doses. It is likely that these dose estimates will be refined in the future.

The Committee's analyses were carried out with the use of the AMFIT program (see Annex 4C) and were restricted to cases occurring 5 years or more after exposure. Since all subjects in the cohorts described above were irradiated during childhood, and because concerns about bias in ascertainment of this often nonfatal cancer among the LSS subjects mitigated against their use in risk estimation, the Committee's risk estimates for adults are based solely on extrapolations from the childhood exposures.

Background Rates

In the Israeli study, the 16 thyroid cancer cases that occurred in the nonexposed group were used to estimate background rates. However, since the Rochester study had only one cancer case among the nonexposed, background rates were modeled on the basis of a standardized incidence ratio relative to sex-, age-, and calendar period-specific rates in the Connecticut Cancer Registry. In the final models, the Rochester background rates were estimated to be about 97% of the Connecticut rates and about one half of the rates in the Israeli cohort.

TABLE 5-3 Estimated Risk of Thyroid Cancer Incidence Under Various Modeling Assumptions

Ethnic Origin	Age[a] (yr)	Sex	Latency (yr)	Additive Risk[b]	Relative Risk[c]
Israel Tinea Study					
Israeli	0–4	f	20	15.1	23.6
Israeli	0–4	f	30	25.5	23.6
Israeli	0–4	m	20	4.7	23.6
Israeli	0–4	m	30	7.9	23.6
Israeli	5–15	f	20	7.2	8.3
Israeli	5–15	f	30	10.8	8.3
Israeli	5–15	m	20	2.3	8.3
Israeli	5–15	m	30	3.4	8.3
Non-Isr	0–4	f	20	45.1	68.7
Non-Isr	0–4	f	30	75.7	68.7
Non-Isr	0–4	m	20	14.1	68.7
Non-Isr	0–4	m	30	23.7	68.7
Non-Isr	5–15	f	20	21.4	23.1
Non-Isr	5–15	f	30	32.3	23.1
Non-Isr	5–15	m	20	6.7	23.1
Non-Isr	5–15	m	30	10.1	23.1
Rochester Thymus Study					
—	0.5	f	20	1.8	19.2
—	0.5	f	30	1.8	6.7
—	0.5	m	20	0.6	19.2
—	0.5	m	30	0.6	6.7

[a] Age at exposure.
[b] Excess is cases per 10,000 PYGy at age 40.
[c] Relative risk at 1 Gy at age 40.

NOTE: In the preferred model, constant relative risk is for Israelis exposed after age 5.

Dose-Response Relationships and Their Modification

A highly significantly elevated risk of thyroid cancer following radiation exposure during childhood was demonstrated in both the Israeli and Rochester cohorts (Table 5-3). Moreover, there was no evidence of significant nonlinearity in the shape of the dose-response function in either cohort. The level and pattern of the excess thyroid cancer incidence appears to depend on a number of the factors which are discussed briefly below.

Cohort effects: The additive cancer excess among children over 5 years old at the time of exposure in the Israeli cohort was 9 times the excess in the Rochester cohort ($p < 0.01$). A cohort indicator was thus included in the final model. The relative risk in Israeli children over the age of five at the time of exposure was about 2.7 times as large as the relative risk in the

Rochester cohort. Due in part to the small number of cases, this difference was not statistically significant.

Additive versus relative risk models: For the Israeli cohort, a constant relative risk model with age at exposure and ethnic origin effects fit the data as well as a time dependent excess risk model with sex, age at exposure, and ethnic origin effects. For the Rochester data, after allowing for a 10-year minimal latent period, the excess risk appeared to be relatively constant with time since exposure. In terms of deviance this model fit the data better than constant or simple time dependent relative risk models.

Sex: The excess absolute risk of thyroid cancer in irradiated females was approximately three times greater than that in irradiated males ($p = 0.002$). There was no significant sex difference ($p > 0.9$) in the excess relative risk. The thyroid cancer background rates among unirradiated females in both cohorts were about 3 times those among unirradiated males.

Age at exposure: In the Israeli study, age at irradiation ranged from less than 1 year to 15 years, which allowed a limited assessment of the effect of age at exposure on cancer incidence. The effect was pronounced. When the data were subdivided into three age subgroups (0-4, 5-9, and 10-15 years), the estimated number of excess cancer cases was 31/10,000 person-year-Gy (PYGy) at age 40 for children exposed under the age of 5 as compared with 10/10,000 PYGy among those exposed over the age of 5 years. It should be noted that individual doses were estimated by using a model that assumed that children exposed at younger ages received higher doses; however, an age at exposure effect was also demonstrated when the number of radiation treatments was used as an indicator of dose. A comparable analysis of the Rochester cohort was not possible, as all exposed subjects were irradiated when they were less than 1 year of age.

Time since exposure: Because it has been shown that radiation-induced thyroid cancer usually does not occur until five or more years after exposure, the results from the first five years after exposure were eliminated from the analysis. The Israeli data suggested that there is a continual increase in excess risk over the entire study period. The increase in risk was more than linear, that is, was proportional to time since exposure to the 1.4 power, but the trend was of borderline statistical significance ($p = 0.08$). In contrast, the Rochester data are well described by a constant-excess-risk model.

There was no evidence of a significant time trend in the Israeli relative-risk data, but there was a significant decrease in relative risk over time in the Rochester cohort ($p = <0.001$). This decrease in relative risk was substantially greater than linear, that is, was proportional to time since exposure to the -2.8 power. The effect of this decrease in the relative risk was to make the excess risk roughly constant. However, despite allowance for this trend the constant-excess-risk model provides a better fit to the Rochester data.

Ethnic origin: Within the Israeli cohort, thyroid cancer risks differed among different Jewish ethnic subgroups. Persons born in Israel had one-third the risk (excess risk, 3.4) of those born in Asia (mostly in the Middle East) or North Africa (excess risk, 10.2). Since the cohort of those born in Israel was restricted to those whose fathers were born in Asia or northern Africa, this difference in risk appears to originate from environmental rather than genetic bases. The westernization of the Israeli life-style, and hence of those born in Israel, may be the reason why their excess cancer risk was closer to that of the Rochester cohort.

Recommended model: The choice of a model to be used in projecting thyroid cancer risks is difficult, because both available studies are limited to childhood exposures, and the levels of risk, as well as the latency distributions, in the two studies are inconsistent. Model choice is further complicated by the fact that thyroid cancer incidence rates are highly dependent on both the method of ascertainment and the criteria for surgery and hence confirmation of diagnoses used in a given study area. These differences profoundly affect excess-risk estimates. Thus, it is unlikely that projections based on an excess-absolute-risk model can provide a reliable indication of lifetime risk when applied to different populations.

In view of these considerations, projections of lifetime thyroid cancer risks are based on the relative-risk model for Israeli-born children who were over 5 years old at the time of irradiation. This is a constant relative-risk model in which the relative risk at 1 Gy for all ages is 8.3 (Table 5-3). The distribution of this estimate is so highly skewed that the usual asymptotic standard error provides little useful information and a likelihood based confidence interval (Co74a) provides a more accurate summary of its variability. The likelihood based 95% confidence interval for this relative risk coefficient is 2 to 31 per Gy. Since there was no sex differential with the relative risk model, an equal risk is predicted for both sexes. Israeli-born children over the age of 5 were the preferred reference population because they had an intermediate risk within the Israeli cohort and their risk was relatively similar to the overall Rochester risk estimate. These lifetime risk estimates are less than those estimated from data from non-Israeli-born children exposed when they were less than age 5, but are greater than estimates based on the Rochester data.

On the basis of unpublished data on A-bomb survivors made available to the Committee by RERF, we conclude that the risk of radiation-induced thyroid cancer in adults is, at most, one-half that in children.

Thyroid Neoplasia from Internally Deposited Radionuclides

Interpretation of the effects of internally deposited ^{131}I and other iodine radionuclides has been complicated by dosimetric problems (Du80). From 73 to 96% of the energy absorption from these nuclides is from beta particles (NCRP85), and the fraction of energy absorbed within the gland depends critically on gland size and geometry and on the beta particle energy. For example, it is estimated that about 70% of the total beta energy from resident ^{131}I is absorbed in the thyroid gland of a mouse or rat, 90% in a human infant gland, and >95% is absorbed in an adult human gland. Because of geometric considerations, follicular cells near the gland surface will absorb only 35-40% of the ^{131}I beta dose delivered to cells in the center of a mouse thyroid (Du80).

The issue is further complicated by biological factors. About 90% of the iodine in the thyroid is contained in the colloid in the follicular lumena. The efficiency of iodide uptake differs among follicles, as does the efficiency of hormone secretion. Thus, relative dose distribution varies from follicle to follicle and over time. Dose is also a function of electron energy at the cellular as well as the glandular level. The low-energy electrons released by ^{125}I decay in colloidal TG deposit most of their energy at the lumenal end of the follicular cell, with little reaching the nucleus located at the basal pole. Finally, total dose depends on the resident time of the nuclide within the gland. This in turn is a function of dietary iodine content and hormonal status.

These factors, coupled with limited numbers of measurements of nuclide concentrations over time, small and variable numbers of animals, use of various goitrogens or diets as promoters, and the lack of standardization of rat age, sex, strain, and husbandry, render quantitative interpretations and comparisons of the earlier experimental data difficult (NCRP85). Estimates of the relative effectiveness per unit of radiation dose from ^{131}I varied from 1/15th to 1/2 of that from external x-irradiation in the older experimental literature (Du80, NCRP85).

The effects of internally deposited radioiodides have been investigated in three categories of human subjects: patients who received relatively large therapeutic doses of ^{131}I (Do74, Ho84a, Ho84b, Ho80a), patients who received much smaller doses of ^{131}I for diagnostic purposes (Ho80b, Ho80c, Ho88), and those on the Marshall Islands who were exposed to iodine radionuclides in fallout from a bomb test (Co80, Co84). The majority of subjects who received therapeutic ^{131}I were suffering from thyrotoxicosis. In this disease, the rate of thyroid hormone synthesis and release is accelerated, and thyroid stimulating hormone (TSH) or long-acting thyroid-stimulating protein (LATS) is generally elevated. Hence, the maximum ^{131}I uptake by the thyroid is increased and the residence time in the gland is reduced. The total radiation doses were often very large (approximately 50-100 Gy); that is, they were designed to cause extensive cell death; indeed, the probability of development of hypothyroidism within 2 years after treatment increased nearly linearly with dose over the range 0.9-8.3 megabequerels ^{131}I/gram of thyroid (Cl86a). Furthermore, the incidence of thyroid cancer among patients with thyrotoxicosis may be as much as 10 times that among the general population (NCRP85). Hence, the choice of an appropriate unirradiated control group is difficult (Ho84a). Finally, given the long latency of radiogenic cancers, the periods of follow-up have been brief (means of 8-15 years).

When the observed cancer incidence in a prospective study of 1,005 ^{131}I-treated patients with thyrotoxicosis (mean follow-up of 15 years) was compared to that of 2,141 surgically treated patients, the risk was 9.1-fold greater ($p < 0.05$) in the irradiated group (Ho84a). When compared with the Connecticut Cancer Registry, however, the relative risk from ^{131}I (3.8) was not statistically significant. After a mean follow-up of 8 years, the thyroid cancer incidence ratio of 21,714 ^{131}I-treated patients to 11,732 surgically treated patients was 2.6, and was not significant (Do74). A relative risk (insignificant) of 1.01 was found when the cancer incidence in 4,557 ^{131}I-treated patients followed for an average of 9.5 years was compared with data from the Swedish Cancer Registry (Ho84b).

An initial follow-up study of 10,133 subjects who received diagnostic doses of ^{131}I at the Radiumhemmet, Stockholm, Sweden, yielded no evidence of an increase in thyroid cancer risk (Ho80b, Ho80c). A more recent study included 35,074 Swedish subjects (28,180 women and 6,884 men) from six institutions who had survived 5 years or more after a diagnostic dose of ^{131}I (Ho88). The mean dose was 1.92 MBq of ^{131}I (range: 0.04-35.52 MBq), the mean radiation dose was approximately 0.5 Gy, the mean age at exposure was about 44 years (range: 1-74 years), the mean follow-up was 20 years during the period 1951-1984, and the data were compared with the Swedish Cause-of-Death Registry. A total of 50 thyroid cancers were found in the ^{131}I group compared with an expected number of 39.37 cases, yielding an overall standardized incidence ratio of 1.27 observed to 1.0 expected cancers (95% confidence interval, 0.94-1.67). Of the 50 observed cancers, 10 were either medullary or poorly differentiated and 1 was a sarcoma. Medullary carcinomas have not been seen to be associated with radiation exposure. Six thyroid cancers occurred among men who were 50-74 years old at the time of exposure; this subgroup yielded the only significantly increased standardized incidence ratio 3.14 (95% confidence ratio, 1.15-6.84). Sixty-eight percent of the cancers occurred among 31% of the subjects who had received a diagnostic dose of ^{131}I because of suspected thyroid cancer. Of these 34 cases, 15 cancers (44%) became clinically apparent 5-9 years after exposure, suggesting that they were occult at the time of the ^{131}I diagnostic procedure. In summary, the results of these studies do not support the conclusion that diagnostic doses of ^{131}I significantly increase the risk of thyroid cancer (Ho88).

People on the Marshall Islands were exposed to fallout from the thermonuclear BRAVO bomb test on Bikini atoll on March 1, 1954 (Co80). The atoll Bikini is approximately 95, 100, and 300 miles from Alingnae, Rongelap, and Utirik atolls, respectively. The radiation dose to the thyroid glands of the residents of the Utirik atoll was in part from external gamma rays from fallout dust (1.75, 0.69, and 0.14 Gy for those on Rongelap, Alingnae, and Utirik atolls respectively) and in part from inhaled and ingested radioiodides. Doses of the ingested radioiodides were calculated from the ^{131}I content of pooled urine samples collected 15 days after the first exposure (Co80, Co84); the dose contributions from the short lived radionuclides ^{132}I, ^{133}I and ^{135}I were assumed to be equal to 2-3 times the ^{131}I dose. Two-thirds of those on Rongelap atoll and 5% of those on Alingnae atoll suffered nausea within 48 hours. Half of the Rongelap atoll natives developed partial epilation beginning 2 weeks after exposure, indicating significant total-body and body surface doses. By 8 years after exposure, two boys who were 1 year of age when they were irradiated were diagnosed with myxedema (Co80). Nine years after exposure, the first thyroid nodule was noted in a 12-year-old girl. The seriousness of the situation was apparent by 1965, and prophylactic thyroid hormone treatment was then initiated in residents of Rongelap atoll; prophylaxis was initiated 4 years later in residents of Alingnae atoll. How this treatment has influenced the development of neoplasia is unknown. However, insofar as increased thyroid hormone levels would be expected to reduce TSH release, the rate of malignant progression of the progeny of radiation-initiated thyroid cells would be expected to have been reduced (Cl86b, Du80).

The thyroid status of the Marshall Islanders 27 years after exposure is summarized in Table 5-4. Although the dose estimation is open to question, the prevalence of hypothyroidism, thyroid nodules and proven thyroid cancer all appears to increase with dose (Co84).

These studies have recently been extended to residents of more distant atolls who were not previously considered to have been at risk (Ha87). The 14 atolls were chosen to include all northern atolls that could possibly have been in the fallout path and as many southern atolls as feasible; the latter were chosen as sources of unexposed controls. Alignae atoll was excluded, as it was uninhabited at the time of the study (1983-1985). Study subjects included 2,273 persons who were alive and residing on 1 of the 14 atolls at the time of the BRAVO test. For purposes of this study, a thyroid nodule or neoplasm was defined as a solitary discreet nodule of at least 1 cm in diameter; nodules of less than 1 cm were not considered as positive, nor were multinodular goiters, diffuse hyperplasias, or cases of Graves' disease.

TABLE 5-4 Prevalence of Thyroid Abnormalities Among Marshall Islanders 27 Years (1981) After Exposure to Fallout (Co84)

Group and Age, 1954	Number of Subjects	Dose (Gy)	Percent with Condition		
			Hypothyroid	Nodules	Cancer[a]
Rongelap					
1 yr	6	≥15	83.3	66.7	0
2–9	16	8–15	25.0	81.2	6.2
≥10	45	3.4–8	8.9	13.3	6.7
Alingnae					
<10	7	2.8–4.5	0	28.6	0
≥10	12	1.4–1.9	8.3	33.3	0
Utirik					
<10	64	0.6–1.0	0	7.8	1.6
≥10	100	0.3–0.6	1.0	12.0	2.0
Controls					
<10	229	—	0.4	2.6	0.9
≥10	371	—	0.3	7.8	0.8

[a] Values are conservative estimates; unoperated nodules were considered benign, and occult carcinomas were excluded.

Included as positive were a number of Rongelap and Utirik atoll residents who had previously been thyroidectomized for a thyroid nodule. Among the residents of the 12 atolls who had previously been considered to be unexposed, the thyroid nodule prevalence varied from 0.9 to 10.6%. The null hypothesis of no differences among these populations was rejected ($p < 0.025$). The age-adjusted prevalence of thyroid nodules bore a highly significant inverse relationship to the distance of the residence from Bikini atoll ($p < 0.002$). As suggested by the control data in Table 4-9, previous analyses had assumed a prevalence of non-radiation-related thyroid nodules of 6-7% (Co84). In the current analysis, the prevalence of spontaneous thyroid nodules was taken to be 2.45%, which was the mean prevalence of residents of the two southernmost atolls, which were far from the path of fallout (Ha87).

Logistic regression analysis showed a significant dependence of nodule prevalence on distance from Bikini atoll, age at exposure, sex, and the angle of deviation in latitude from Bikini atoll. Women were 3.7-fold more susceptible to nodule formation than men. The effect of distance in angle of deviation in latitude from Bikini atoll was attributed to shifting wind patterns that at first carried the fallout approximately due east from Bikini atoll but that later carried it southwest from Utirik atoll.

Using the age-adjusted spontaneous thyroid nodule prevalence of 2.45%, a new estimate of risk was calculated based on the nodule prevalence in persons residing on Rongelap and Utirik atolls. The absolute risk coefficient so calculated was 11 excess nodule cases/10^4 PYGy (Ha87). This estimate is approximately 33% greater than previous estimates, largely because of the difference in assumed spontaneous prevalence. These calculations are subject to the same limitations imposed by post hoc dosimetry as all previous estimates. Furthermore, there is a wide variation among reports of the spontaneous thyroid cancer incidence among different Polynesian populations. For example, reported incidences among the female Polynesians of Cook Island, New Zealand, and Hawaii were 18.6, 2.5, and 9.3/10^5 person-year Gy, respectively; however, the total numbers of cancer cases were only 4, 6, and 19 respectively (He85). The SEER report also gives a high estimate for Hawaiian Polynesian women: 19.2 cases of cancer/10^5 person-year Gy (Se81). The above analysis of the data on the Marshall Islanders should thus be interpreted with caution.

A survey of the thyroid status of school children in Lincoln Co., Nevada, and Washington Co., Utah, was performed annually from 1965 to 1971 (Ra74, Ra75). A total of 1,378 children were identified who had lived as infants in these counties during the period 1952-1955 when there was estimated to be fallout from atmospheric atomic bomb tests. Most of the dose from this fallout was assumed to be from the ingestion of ^{131}I-contaminated milk, in which the nuclide was metabolically concentrated. Cumulative radiation doses to the thyroids of children residing in southwestern Utah were estimated to average as high as 1 Gy. A cohort of 1,313 children in the same schools who had moved to the counties after the cessation of atmospheric atomic-bomb testing and another cohort of 2,140 children from a county in Arizona that was remote from the fallout path were chosen as unexposed controls. In the original reports of this study, no significance was attached to the relatively modest differences in thyroid abnormalities noted among the exposed and unexposed groups (Ra74, Ra75).

The data from the first report (Ra74) have recently been reanalyzed and are shown in Table 5-5 (Ro84b). Although there was no increase in thyroid cancer incidence in the presumably exposed populations, there was a suggestive 20-30% greater prevalence of all thyroid abnormalities in exposed versus unexposed children. It is also important to note, however, that the lower 90% confidence limits of the prevalence ratios are individually and collectively less than or equal to 1.0. It is also of importance to note that the follow-up period was approximately 14 years from the time of exposure.

In a second report of this study, the prevalence of nodular goiter among these three groups was compared (Ra75). The prevalences were 8.7, 4.6, and 4.7 per 1,000 children in the exposed group, the unexposed Utah-Nevada group, and the Arizona group, respectively (Ra75, Ro84b). The exposed/unexposed prevalence ratio was 1.9 (90% confidence limits,

TABLE 5-5 Prevalence of Thyroid Abnormalities in Fallout-Exposed and Unexposed Children (Ro84)

Children, Group, and Number	Prevalence of Thyroid Abnormalities per 1,000 Subjects					
	Cancer	Benign Neoplasm	Adolescent Goiter	Thyroiditis	Hyperthyroid, Misc.	Total
1,378 exposed Utah-Nev.	0	4.4	16.0	13.1	3.6	37.0
1,313 unexposed, Utah-Nev.	0.8	3.0	9.1	12.9	2.3	28.2
2,140 unexposed, Arizona	0.5	2.8	15.4	8.4	1.9	29.0
Ratio, exposed:unexposed,	0	1.5	1.2	1.3	1.8	1.3
and 90% conf. lim.	0, 5.4	0.6, 3.5	0.8, 1.9	0.8, 2.1	0.6, 4.8	1.0, 1.7
Ratio, excluding Arizona,	0	1.4	1.7	1.0	1.6	1.2
and 90% conf. lim.	0, 8.4	0.5, 4.5	1.0, 3.2	0.6, 1.8	0.5, 6.0	0.9, 1.9

1.0-3.5) (Ro84b). Again, the lower confidence limit was 1.0. The analyst concludes that although the data showed a weak but positive radiation effect, in the absence of better dosimetric information they revealed little about the effects of such exposure (Ro84b).

In contrast to human studies, a large scale animal experiment showed little difference between the effects of x rays and ^{131}I. In this study, oncogenic effects of ^{131}I on the thyroid were compared with those of x rays using 3,000 female rats of the Long Evans strain (Le82). The carcinogenesis experiment was preceded by detailed dosimetric studies (Le79). The thyroids of 6 week-old rats were exposed to 0, 0.94, 4.10, or 10.60 Gy of highly localized x rays or were injected intraperitoneally with 0, 0.48, 1.9, or 5.4 μCi of ^{131}I. The ^{131}I doses were chosen to yield radiation doses to the gland of 0.80, 3.30, and 8.50 Gy, respectively. Two additional groups of rats received 4.10 Gy of x rays to the pituitary alone or to the pituitary and the thyroid to yield 10 equal groups in all. A total of 2,762 animals that died or were killed from 6 months until the termination of the experiment at 24 months after exposure were included in the analysis.

The incidence of thyroid cancer increased as a positive exponential function of dose following administration of either x rays or ^{131}I; in both cases, however, the coefficient of dose in the exponent was significantly less than 1.0 (Le82). The ratios of x-ray-induced cancer to ^{131}I-induced cancer at 0.80, 3.30, and 8.50 Gy were 1.3, 1.0 and 0.9, respectively, suggesting nearly equal effectiveness per unit dose. The 24 month thyroid cancer risk per 0.01 Gy for both ^{131}I and x rays in the 0.80-0.90 Gy range was 1.9×10^{-4} which is similar to the estimated human life time risk from x rays (Le82). In contrast, the slopes of the dose-response relationship for adenoma and total tumor had upward curvature, and appeared to rise more rapidly in the x-ray-treated groups. The parameters of the dose-response relationships following administration of ^{131}I or x rays did not differ significantly, however. Irradiation of the pituitary gland did not alter the results.

The National Council on Radiation Protection and Measurements (NCRP) reviewed the data and analyses on radiation induced thyroid cancers that were available through 1985 and recommend the use of a Specific Risk Estimate (SRE) according to the following formula (NCRP85):

$$\text{SRE} = RFSAY,$$

where R is the absolute risk in excess thyroid cancer cases per 10^4 person-year-Gy; F is the dose effectiveness factor, which is assumed to be 1 for external radiation, ^{132}I, ^{133}I, and ^{135}I and 1/3 for ^{125}I and ^{131}I; S is the sex factor taken to be 4/3 for women and 2/3 for men; A is the age factor which is equal to 1 for those <18 years of age and 1/2 for those >18 years of age at time of exposure; and Y is the anticipated mean number of years at risk. The absolute risk, R, chosen for this calculation is assumed to be that for an ethnically homogeneous population of children (<18 years of age) of equal numbers of each sex who were exposed to external radiation and corrected for a minimum 5 year latency. The SRE calculated in this way is the risk of development of thyroid cancer during the rest of an individual's life. If a thyroid cancer mortality risk is desired, the SRE is multiplied by L, the lethality factor, assumed to be 1/10. The NCRP uses an absolute risk factor, R, of 2.5 thyroid cancers/10^4 PYGy for doses in the range of 0.06-15.0 Gy (NCRP85).

Parallel and combined analyses of six cohorts of children and two cohorts of adults exposed to external radiation and one cohort of children and one cohort of adults exposed to ^{131}I have been reported recently (La87). Data from the study of the large Long Evans strain of rat were also included. A constant relative potency for neoplasia induction by ^{131}I as compared with that by external x rays was assumed across ethnic and sex cohorts and ages and across species lines. The risk ratio estimate so derived for ^{131}I compared to x rays was 0.66 (95% confidence limits, 0.14-3.15) and did not differ significantly from 1.0 (La87).

Physiology of Radiogenic Thyroid Cancer

Thyroid neoplasia has been an attractive model in experimental carcinogenesis because thyroid epithelial cell proliferation and function affect susceptibility to thyroid neoplasia and can be readily manipulated, the thyroid and pituitary hormones that regulate and/or reflect thyroid cell proliferation and function are easily measured, thyroid tissue or cells are readily transplantable, and thyroid neoplasia is a significant human risk (Cl86b, Du80).

The rate of proliferation of thyroid cells is regulated by the concentration of thyroid-stimulating hormone (TSH) in blood. Synthesis and release of TSH from the anterior pituitary gland is stimulated by TSH-releasing hormone (TRH), which is synthesized in the hypothalamus and reaches the TSH-secreting cells via the hypothalamic-hypophyseal portal system. TSH levels reaching the thyroid via the general circulation cause the synthesis and release of thyroid hormone and the proliferation of thyroid follicular cells. Serum thyroid hormone reaching the hypothalamus inhibits TRH release, thus modulating hypophyseal TSH release and, in turn, maintaining thyroid hormone titers within normal levels. This long-loop feedback regulation is supplemented by neural input via the hypothalamus and by additional short-loop feedback systems which operate under special circumstances (Cl86b).

The prime functions of the thyroid follicular cells are the synthesis, storage, and release of the thyroid hormones thyroxine (T4) and 3-5-3′ triiodothyronine (T3). T4 synthesis and T3 synthesis occur in three phases: (1) uptake and concentration of inorganic iodide, (2) the preceding or concurrent synthesis of thyroglobulin (TG), and (3) iodine organification and iodothyronine formation in the TG molecule. The iodinated TG is then either hydrolyzed to release T3 and T4 for secretion or is stored in the thyroid follicular lumina as colloid (De65).

Feedback regulation of the thyroid is vulnerable to disruption by natural, therapeutic, or experimental means at virtually every step (Cl86b, Du80). The goitrogenic effect of iodide deficiency has been recognized since antiquity, and experimental hypothyroidism is readily induced by diets low in iodine. Pharmacological disruption of the iodide concentration by perchlorate and of iodide oxidation and iodothyronine synthesis by thiocarbamides and other goitrogens are common experimental techniques used to block T4 and T3 synthesis. Partial or total destruction of the thyroid epithelial cells can be induced by administration of radioiodide. A TSH-mimicking molecule, long-acting thyroid-stimulating protein (LATS), results in hyperthyroidism in some humans. These observations have been experimentally exploited in studies of thyroid radiobiology and carcinogenesis.

Phases of Thyroid Carcinogenesis

The events in thyroid carcinogenesis can be divided into three phases: (1) an acute phase, including early radiation injury, neoplastic initiation, and intracellular repair; (2) a latent phase, from the acute phase until overt tumor formation; and (3) the phase of tumor growth (Cl86b).

The Acute Phase

The first step in radiogenic thyroid cancer induction is initiation, that is, the formation of one or more heritable precancerous changes in one to many thyroid cells (Cl86b, Du80). About 1-2% of young rat thyroid epithelial cells are clonogenic, that is, they are capable of forming new clonal thyroid follicles under TSH stimulation (Cl85a); these proliferation-competent cells are presumed to be the cells of origin of thyroid neoplasms. Clonal follicular units have been used as an endpoint in a quantitative transplantation assay of the relative numbers, acute response to radiation, postirradiation repair capacity, and frequency of neoplastic initiation in the thyroid clonogens (Cl85a, Mu84). The evidence indicates that 98-99% of the thyroid epithelial cells are nonclonogenic, that is, they are capable of but a few rounds of mitosis in response to TSH (Du80).

Radiogenic cellular damage after doses in the carcinogenic range of less than 20 Gy is usually expressed during mitosis. There is little biochemical evidence of acute impairment of secretory function after radiation exposure; that is, the irradiated animals remain euthyroid for a time. Hence, TSH levels are within the normal range, mitotic activity remains low, and cellular damage is not immediately expressed (Du80).

The Latent Phase

Whether radiogenic damage is expressed in frank tumor formation—and if so, when—depends on the interaction of internal environmental factors with the initiated thyroid cells. Under normal circumstances, the rate of thyroid epithelial cell division is low, but it is not nil (Do67, Do77). In euthyroid rats, TSH in concert with other factors is normally present at concentrations that are sufficient to stimulate a small portion of grafted clonogens to follicle formation (Mu80b). In the early latent phase, the cells in the irradiated thyroid are thus subjected, a few at a time, to mitosis-triggering stimuli. TSH-triggered normal cells respond with normal mitosis. Altered but reproductively viable cells, including initiated cells, also proliferate in response to triggering stimuli. Triggered and terminally damaged cells may pass successfully through one or a few mitoses before death, or they may survive without division but with persistence of the secretory function for several months or longer. Ultimately, however, the proportion of terminally damaged cells decreases through cell death; as a result, compensatory triggering stimuli increase to bring about the replacement of lost cells. Depending on radiation dose, and hence, the fraction of the population made up of terminally damaged cells, this process of triggering and cell death continues very slowly, perhaps undetectably, over many months, or it may accelerate in a cascade fashion after a slow beginning (Cl86b).

In rats there is a threshold between 2.5 and 5.0 Gy for histologically detectable radiogenic damage (Do67). After a dose of 10 Gy, however, when only a small fraction (approximately 7%) of clonogens would be expected to retain reproductive capacity (Mu80a), although the animals may remain euthyroid, partial glandular atrophy coupled with epithelial cell hypertrophy and interstitial fibrosis occurs with time (Do67). Similar changes are observed after injection of 30 micro-Ci of ^{131}I (Do77). Higher doses of x rays (greater than 15-20 Gy) result in widespread evidence of epithelial cell damage (Do67) with cell degeneration, follicle disruption, and interstitial and vascular fibrosis. These changes are qualitatively similar following external radiation or internal radiation by ^{131}I (Do77, Ga63, Li63). They occur soon after exposure to single doses in excess of 20 Gy, but are delayed for weeks to months in smaller animals and for years in large species after a dose of about 20 Gy. If such damage is extensive, hypothyroidism develops, but neoplasia is a less common result in extensively damaged glands than in glands in which 5 to 50% of the epithelial clonogens survive (i.e., at doses of about 6-11 Gy) (Mu80a) (NCRP85).

From the standpoint of carcinogenesis, the important processes during the latent phase include amplification of the radiation-initiated clonogen population under repeated mitosis-triggering stimuli. During this process, insofar as repeated rounds of DNA synthesis and mitosis play a role in neoplastic changes in initiated cells, promotion and progression, as well as clonal expansion, occur in the initiated cells. In endocrine-responsive cell populations, progression is frequently associated with quantitative or qualitative changes in hormone responsiveness (Cl75, Fu75).

The Phase of Tumor Growth

Radiation-induced thyroid tumors first appear as localized hyperplastic nodules. They are often multifocal, suggesting that they originated from randomly distributed initiated cells. Adenomas are most common, occurring 10-16 months after exposure in rats and in increasing frequency with time thereafter (Do63). Carcinomas appear after 18-30 months in rats and are frequently found within or associated with adenomas.

The development of the thyroid cell transplantation technique has permitted studies of carcinogenesis in vivo in terms of surviving clonogenic cells, that is, thyroid carcinoma and total tumor incidence per 0 or 5 Gy of x-irradiated grafted clonogen have been investigated in thyroidectomized rats maintained on an iodine-deficient diet (Mu84, Wa88). On a cellular basis, the radiogenic initiation frequency was high. For example, cancers developed in 34% of the transplant sites, each of which were grafted with 11 surviving clonogens irradiated with a dose of 5 Gy. This corresponds to one initiated clonogen for about every 32 clonogens grafted (Wa88).

In summary, although both benign and malignant thyroid neoplasms arise from the relatively small radiation-initiated cell subpopulation, this occurs gradually over time as the result of neoplastic promotion, progression, and clonal amplification under the mitogenic stimulation of TSH. The intensity of the TSH stimulation depends in turn on the functional capacity of the entire thyroid follicular cell population, a significant fraction of which, although it retains secretory capacity, may die during mitosis from radiation injury.

Summary

Thyroid cancer is well established as a late consequence of exposure to ionizing radiation from both external and internal sources in humans and experimental animals. The histopatholoy of radiogenic thyroid cancer indicates that it appears to arise exclusively from the follicular epithelium. It is relatively indolent and causes death infrequently (mortality/incidence ratio = approximately 0.1) in comparison with more malignant medullary thyroid cancers, the incidence of which has not been found to be increased in irradiated subjects.

Analysis of two cohorts of subjects exposed to therapeutic radiation for benign conditions and a review of the literature have revealed that:

1. There are major differences in background thyroid cancer rates among unirradiated individuals of different reported cohorts. Analysis suggests that these differences are related, at least in part, to life-style, although ascertainment may also play a critical role.
2. Females are roughly 3 times as susceptible to radiogenic, as well as nonradiogenic (background), thyroid cancer as males. Hence, relative-risk estimates do not differ significantly by sex.
3. The excess risk from radiation exposure is considerably greater among children who are exposed during the first 5 years of life than in those exposed later. On the basis of an examination of data from the Japanese adult health study, not yet published by the RERF, the Committee concludes that the risk of radiation-induced thyroid cancer in adults is only one half, or less, of that in children.
4. Although the data are best fit by an excess-risk model that includes allowance for cohort effects, latency, age at exposure, and sex, a relative risk model is preferred because of the strong dependence of the estimates on the background incidence of the particular cohort under consideration. The model, which is based on the risk in Israeli-born children who were exposed when they were more than 5 years of age, yields a relative risk at 1 Gy of 8.3 for both sexes.
5. The risk ratio for ^{131}I/x rays has been estimated as 0.66, but the 95% confidence interval of the ratio is broad (0.14-3.15), since the risks from internally deposited radionuclides of iodine are not well understood.
6. The development of thyroid cancer from initiated cells is profoundly dependent on hormone balance.

ESOPHAGUS

In a recently completed registry-based study of the incidence of second primary cancers in women following radiation treatment for cervical cancer (Bo85), an overall relative risk for esophageal cancer of 1.5 ($p < 0.05$) was seen among patients with invasive cervical cancer who were treated with radiotherapy (40 observed cases versus 27 expected cases). The corresponding risks for women with invasive cervical cancer who did not receive radiation and women with *in situ* cervical cancer (the majority of whom only received surgical treatment) were 1.0 and 0.5, respectively. However, when attention was restricted to patients with invasive cervical cancer who were treated with radiation and followed for 10 or more years, the relative risk was reduced to 1.1, which was no longer statistically significant. The authors of the study included the esophagus among the organs estimated to receive small doses of radiation, that is, an estimated average dose of less than 0.5 Gy, and concluded that cancers at these sites were either not elevated or were probably increased because of other major risk factors, such as the use of cigarettes or alcohol. Esophageal cancer was not included in a recent case control analysis of these data (Bo88).

Stronger evidence in support of the relationship between esophageal cancer and exposure to ionizing radiation is provided by the analysis of the updated mortality experience of the cohort of patients with ankylosing spondylitis (Da87). The authors reported that while a highly significant ($p < 0.001$) increase was observed for all neoplasms other than leukemia or colon cancer when considered collectively, the number of deaths observed 25 years or more after treatment tended to decline, closely approaching expected values (i.e., relative risk, 1.07). The main exception to this trend was esophageal cancer, which was significantly increased ($p < 0.01$) during both the intervals of 5-24.9 years and 25 or more years posttreatment, with relative risks of 2.05 and 2.41, respectively. Overall, the relative risk for x-ray-treated patients 5 years or more after treatment was 2.20 (28 observed versus 12.73 expected esophageal cancer deaths), which was highly significant ($p < 0.001$). The estimated mean dose to the esophagus was quite high (over 4 Gy) (Le88).

The 1950-1982 follow-up data for the combined Japanese survivor populations (Pr87a), based on T65 dosimetry, continues to support the hypothesis of radiation-induced cancer of the esophagus. In this report from the Radiation Effects Research Foundation, relative risks ranging from 0.65 in the 1-9-rad dose group to 2.03 in the >400 rad dose group were observed. Preston and colleagues (Pr87a) also reported a nonsignificant ($p = 0.30$), decreasing trend in the relative risk of esophageal cancer over time and a significant ($p = 0.03$) effect of sex on the relative risk of esophageal cancer (i.e., the estimated relative risk for exposures of 1 Gy for males and females were 1.09 and 2.23, respectively). If ethnic and other differences in potential risk factors are ignored, then this latter result suggests that the lack of a more impressive and sustained esophageal effect in the cohort of patients with cervical cancer may be related, at least in part, to the relatively low organ-specific levels of exposure that they received. The latest analysis of the Japanese cohorts (Sh88) using the new DS86 dosimetry data indicates that in terms of the kerma at a survivor's location the relative risk for esophageal cancer is estimated at 1.43/Gy ($p < 0.05$), with a corresponding excess risk of $0.34/10^4$ PYGy. In terms of dose to the esophagus, the relative risk is 1.58/Gy; excess risk $0.45/10^4$ PyGy (Sh88).

Summary

Carcinoma of the esophagus has been observed to occur with increased frequency in several irradiated human populations. The available dose-incidence information is sparse, but the data from the various studies are consistent with those from the A-bomb survivors, in whom the relative risk is estimated to approximate 1.58 per Gy (organ dose).

SMALL INTESTINE

Although carcinomas of the small intestine can be induced with a high frequency in mice and rats by intensive irradiation of the ileum or jejunum, as noted below, their induction by irradiation in humans has yet to be established. In comparison with cancers of the stomach and colon, however, cancer of the small intestine occurs infrequently—its annual incidence in humans approximates only 0.8 cases/100,000 people (Yo81). In addition, little is known about the factors that affect its occurrence in the general population (Li82).

In 2,068 women treated with irradiation of the ovaries for excessive menstrual bleeding, an excess of mortality from cancer of the small intestine was observed; that is, there were 3 observed deaths, compared with only 0.4 expected deaths (Sm76). Similarly, in an international study of 82,616 women treated with radiation for carcinoma of the uterine cervix, a two-fold excess of cancer of the small intestine was observed; that is, there were 21 observed versus 9.5 expected cases (Bo85). The excess was evident, however, within the first year after treatment and did not significantly increase with time. Furthermore, a comparable excess was observed in women with invasive cervical cancer who received no radiotherapy; that is, there were 4 observed versus 0.9 expected cases. New case control analyses of these data yield a relative risk of 1 for cancer of the small intestine (Bo88). Hence, any causal relationship between the excess cases and radiation is questionable. No excess cases of the disease have been reported in Japanese A-bomb survivors, patients treated with radiation for ankylosing spondylitis, or other irradiated populations; but cancers of the small intestine have not been reported separately from cancers of the colon in most such studies (La86).

Adenocarcinomas of the small intestine have been observed in more than 20% of LAF_1 mice surviving midlethal doses (3.5 Gy) of whole-body neutron radiation, whereas such tumors are rare in nonirradiated controls or in mice surviving midlethal doses of x-irradiation (No59). Similarly, in rats, a high incidence (>50%) of such tumors has been observed after localized exposure of the ileum or jejunum to x rays (Os63, Ts73, and Co74) or deuterons (Bo52) at doses exceeding 15 Gy.

Summary

Although adenocarcinomas of the small intestine can be produced by intensive localized irradiation in laboratory animals, no carcinogenic effects of radiation on the small intestine have been evident in any of the irradiated human populations studied to date. Hence the risk of radiation carcinogenesis in the small intestine, although not quantifiable, appears to be low.

COLON AND RECTUM

Colon

Irradiation has been observed to increase the risk of colon cancer in humans and laboratory animals. The strongest evidence of the carcinogenic effects of radiation on the human colon is provided by the dose-dependent excess of colon cancers observed in Japanese A-bomb survivors. At doses of 1 Gy or higher, a total of 25 deaths from colon cancer were observed between 1950 and 1982 in members of the Life Span Study cohort population versus 14.50 expected deaths; no such excess was evident during the first 15 years after irradiation (i.e., before 1959), nor has any excess been evident at doses below about 1.0 Gy (Pr87a). The relative risk per Gy (organ dose) in the DS86 subcohort was estimated to amount to 1.85 (1.39-2.45), which corresponds to an excess of 0.81 (0.40-1.30) deaths per 10^4 PYGy (Sh88).

A comparable association between cancer of the colon and therapeutic irradiation has been observed in two series of women treated for benign gynecologic conditions. Four deaths from intestinal cancer were observed (versus one expected death) in a series of 297 women followed for an average of 16 years after irradiation of the ovaries for benign pelvic disease (Br69), and 24 deaths from colon cancer were observed (versus 13.86 expected deaths) more than 5 years after treatment in a series of 2,067 women treated with irradiation for metropathic hemorrhagica (Sm76). No significant excess deaths were observed, however, in two other series of women treated with radiation for similar disorders (Di69, Wa84). Likewise, in a large series of patients (82,616 women) treated with x rays for carcinoma of the uterine cervix, in whom the average dose to the colon was estimated to have exceeded 5 Gy, no consistent excess number of deaths was observed within the first three decades after irradiation (Bo85). The new case control analysis of these data yielded an insignificant excess risk of only 1.02 (Bo88).

In 14,106 patients who were treated with x rays to the spine for ankylosing spondylitis during 1935-1954 and who were followed until 1985, a total of 47 deaths from colon cancer were observed, versus 36.11 expected (relative risk, 1.30) (Da87); however, the relative risk in this population was higher during the first 2-5 years after irradiation (ratio of observed to expected deaths 6/2.50 = 2.40), in keeping with the known associations between ankylosing spondylitis and ulcerative colitis and between ulcerative colitis and colon cancer. In view of the confounding influence of these associations, the excess deaths in this population have not been attributed to radiation per se (e.g., NRC80, Sm82).

In laboratory rats, localized exposure of the colon to 45 Gy of collimated x rays has been observed to cause a high incidence (47%) of adenocarcinomas, with smaller increases at higher and lower dose levels (De78). Such neoplasms have also been induced in a large percentage of rats (75%) by localized beta irradiation from yttrium administered in the diet (Li47). Similarly, rats and dogs exposed to neutron beams or subjected to irradiation of the bowel by dietary polonium-210 or cerium-144 have been observed to develop benign and malignant tumors of the colon (Le73). Although whole-body gamma or x-irradiation at doses in the range of 5-10 Gy has been reported to cause only a small increase in the incidence of such tumors (5%) in rats (Br53, Wa86) and mice (Up69), a high incidence (27%) has been induced in mice (No59) by near-lethal whole-body fast neutron irradiation.

Rectum

Carcinoma of the rectum has been observed to be increased in frequency in humans (La86) and laboratory animals (Wa86) by intensive localized irradiation.

In a large series of women (82,616) who were treated with radiation for carcinoma of the uterine cervix, and who were estimated to have received an average dose of more than 50 Gy to the rectum, no excess in the number of rectal cancers was observed within the first decade after irradiation, but a growing excess was observed at later intervals, with the relative risk after 30 years reaching 4.1 ($p < 0.05$) (Bo85). A similar excess, which also arose in the second decade after treatment, was observed in a smaller series of women treated with radiation for carcinoma of the uterine cervix (ratio of observed to expected = 20/8.8) (Kl82). Suggestive evidence for an excess of rectal cancers also has been reported in women treated with radiation for benign pelvic disease (Br69) but not in the Japanese A-bomb survivors (Pr87a). As yet, however, there is no evidence of such an excess in patients treated with radiation for ankylosing spondylitis (Da87).

In ICR and CF_1 mice, the incidence of rectal carcinoma was observed to be increased by intensive x-irradiation of the pelvis, rising from zero at a dose of 20 Gy to 95% in ICR mice exposed to 60 Gy delivered in three exposures and to 70% in CF_1 mice exposed to 40 Gy delivered in two exposures (Hi77). In C57Bl mice, the induction of rectal carcinomas by intensive x-irradiation has been observed to be enhanced by the administration of the radiosensitizer midonidazole shortly before irradiation (Ro78). In rats, similar tumors have been reported to be induced by localized exposure to negative pi-mesons (B180).

Summary

The data imply that the risks of cancer of the colon and cancer of the rectum can be increased by intensive irradiation in humans and laboratory animals; however, the shapes of the dose-incidence curves and the risks per unit dose are highly uncertain. In the Japanese A-bomb survivors, the dose-dependent excess of colon cancers corresponds to a relative risk of 1.85/Gy, or 0.81 fatal cases per 10^4 PYGy, and was not evident until more than 15 years after irradiation.

LIVER

Introduction

Evidence of radiation-induced liver cancer comes mainly from observations on human populations and laboratory animals with high intrahepatic concentrations of radionuclides.

Human Studies

Follow-up studies of patients in West Germany, Portugal, and Denmark have noted the occurrence of increased numbers of liver cancers, particularly angiosarcomas, bile duct carcinomas, and hepatic cell carcinomas, many years after intravascular injection of Thorotrast, an x-ray contrast medium containing colloidal $^{232}ThO_2$. From the results of these studies, a linear lifetime risk coefficient of 300 liver cancers/10^4 person-Gy of alpha radiation was estimated by the BEIR III Committee (NRC80); however, the extent to which chemical toxicity of Thorotrast may have influenced the risk was not known.

More recently, the data from patients who received Thorotrast, including those in West Germany, Portugal, Japan, Denmark, and the United States, were extensively re-reviewed and reanalyzed by the BEIR IV Committee (NRC88). The follow-up of the West German patients was the largest of these studies, involving 5,159 Thorotrast-exposed and 5151 control subjects (Va84) (NRC88), in which 347 cases of liver cancer (primarily cholangiocarcinomas and hemangiosarcomas) were observed in the Thorotrast-exposed group and 2 cases of liver cancer in the control group. Latency ranged from 16 to more than 40 years. The average alpha dose to the liver was calculated to range from 2 to 1.5 Gy. Based on an assumed latent period of 20 years, the lifetime cancer risk from alpha irradiation of the liver was estimated to be 300 cancers/10^4 person-Gy (NRC88). Similar lifetime risks were calculated on the basis of the Japanese and Danish studies. If a 10-year instead of a 20-year latent period had been assumed, the risk estimates would have been reduced by about one-third. In the BEIR IV report, it was noted that the risk estimates applied only to intravascularly administered Thorotrast. The same radionuclide administered by different routes, or other radionuclides, could cause different patterns of dose distribution and thus different risks of liver cancer.

Experiments in animals have demonstrated that the chemical toxicity of Thorotrast contributes little to the induction of liver cancer (NRC88).

The follow-up of Japanese A-bomb survivors covering the period 1950-1982 (Pr87a) is the first in which cancers of the liver, gall bladder and bile ducts were reported separately from those of other organ sites. The dose-trend test for liver cancer was suggestive of a significant response ($p = 0.05$), there being 59 deaths due to primary cancers of the liver and intrahepatic bile ducts, 19 of which occurred in the unexposed group and 40 of which occurred in the exposed group. The estimated relative risk in terms of the T65 dosimetry was 1.35 (90% confidence interval, 0.98-2.04), and the excess risk was 0.08 deaths/10^4 PYGy (90% confidence interval, 0.00-0.20). These results are based on death certificate diagnoses for which both poor detection and poor confirmation of liver cancer have been observed. A study of a smaller number of histologically diagnosed cases of liver cancer for the period 1950-1980 found no relationship between radiation dose and the incidence of primary liver cancer for persons in either Hiroshima or Nagasaki or for both cities combined (As82).

Additional information on the occurrence of liver cancer in the Japanese A-bomb survivors for the years 1950-1985 has been reported by Shimizu et al. (Sh88) who discussed the questionable significance of an increase in mortality from liver cancer among survivors. On the basis of the 420 liver cancer deaths that were not otherwise specified, the relative risk using the DS86 dosimetry was estimated to be 1.26 (90% confidence interval, 1.05-1.53) at 1 Gy, and the excess risk 0.45/10^4 PYGy (90% confidence interval, 0.09, 0.88). Such estimates are complicated by the inclusion of metastases of other cancers to the liver. For the 77 cases of confirmed primary liver cancer, the relative risk of 1.12 was not statistically significant (90% confidence interval, 0.87-1.71).

In 14,106 patients who received a single treatment course of x rays for ankylosing spondylitis and were followed through 1982, a total of 6 liver cancers were observed more than 5 years after exposure, with 2 cases between 5 and 25 years and the other 4 cases more than 25 years after exposure; the observation of the 6 liver cancers was not significantly different from the expected number, 5.44 (Da87). The x-ray dose to the liver in this study population was estimated to be 1.63 ± 1.26 Gy (Le88).

In a study of second cancers arising after radiation treatment of the pelvic region for cancer of the uterine cervix, Boice et al. (Bo85) (Bo88) found no evidence of radiation-induced liver cancer (ratio of observed to expected cancers 19/20).

Animal Studies

Much of our knowledge of the induction of liver cancer from intrahepatic radionuclides is derived from studies in laboratory animals. As noted in Chapter 1, not all species have prolonged hepatic retention of actinide or lanthanide radionuclides. The BEIR IV report briefly discussed the prolonged retention of actinide radionuclides in the livers of Chinese hamsters, deer mice, grasshopper mice, and beagle dogs, compared with the shorter retention half-times seen in laboratory mice and rats (NRC88). Prolonged retention times increase the radiation doses received by the liver and increase the carcinogenic effects observed in some species.

In a series of life-span studies in which beagles received a single inhalation exposure to monodisperse aerosols of $^{238}PuO_2$, late-occurring cancers were prevalent findings in the skeleton, liver, and lungs (Gi88). Almost all of the cancers found in the liver and skeleton were considered to have been induced by the ^{238}Pu that was absorbed after particle fractionation in the lung. These results demonstrate that inhaled as well as injected alpha-emitting radionuclides can cause liver cancers under appropriate conditions.

In regard to low-LET irradiation, primary liver cancers, principally hemangiosarcomas, bile duct carcinomas, and hepatocellular carcinomas, were prominent long-term effects of chronic beta irradiation of the liver in dogs that had inhaled $^{144}CeCl_3$ or had been injected intravenously with $^{137}CsCl$ (Mu86). The estimated lifetime risk of liver cancer in these dogs was 90 liver cancers/10^4 dog Gy. On the basis of this value and other information on the induction of liver cancers by alpha-emitting radionuclides in humans and in dogs, Muggenburg et al. (Mu86) estimates the lifetime risk for humans exposed to internally deposited beta-emitters to be 30 liver cancers/10^4 person-Gy).

Summary

Follow-up studies of Thorotrast-exposed patients have provided conclusive evidence of carcinogenic effects on the liver from chronic alpha irradiation by internally deposited ^{232}Th and its radioactive decay products. In laboratory animals, likewise, prolonged hepatic retention of actinide and lanthanide radionuclides has produced similar carcinogenic effects on the liver, through chronic irradiation by alpha-emitters and beta-emitters. Collectively, the data indicate that the lifetime risk of liver cancer from Thorotrast is about 300 liver cancers/10^4 person Gy and that the risk from chronic beta irradiation may be about 10 times lower.

SKELETON

Human Data

Low-LET Irradiation

Among 14,106 persons given a single treatment course of x rays for ankylosing spondylitis, four bone cancers were observed at times greater than 5 years after exposure (ratio of observed to expected bone cancers, 4/1.36), corresponding to a relative risk of 2.95 ($p < 0.05$) (Da87). The mean doses received by various parts of the skeleton were estimated to be 9.44 ± 6.05 Gy to the pelvis, 4.41 ± 3.42 Gy to the ribs, 14.39 ± 9.66 Gy to the spine and 0.48 ± 0.61 Gy to other parts of the skeleton (Le88); however, the doses received by individual subjects are not available. Furthermore, each individual's total radiation dose was received in successive fractions over several weeks, with rather large dispersions in the numbers of fractions and numbers of weeks. Also, the dispersion of the dose at each site in each patient was usually quite large, as was the dispersion of doses among patients. The lack of information at this time on the doses to bone in each individual, as well as the small number of extra bone cancers seen in this study, precludes using the data to estimate a risk coefficient for bone cancer.

In a long-term follow-up study on the occurrence of second cancers following radiation treatment of women for cancer of the uterine cervix, an ostensibly radiation-related distribution of bone cancers was observed, with 55% of the bone cancers in the exposed group occurring in the pelvis, compared with 15% in a control group (Bo88). The overall relative risk was 1.3, rising threefold for bone doses greater then 10 Gy.

The possible induction of bone cancer from medically related x- or gamma-radiation was examined by Kim et al. (Ki78), who reported 27 cases of bone sarcoma that were judged to have been induced by radiation. The latent periods for the tumors that occurred in the irradiated field ranged from 4 to 27 years, with a median of 11 years. No bone sarcomas were seen after treatment doses of less than 30 Gy given over a period of 3 weeks. Similarly, Yoshizawa reviewed 262 cases of skeletal cancer attributed to therapeutic external irradiation (Yo77b); however, these cases do not lend themselves to analysis of the risks of bone cancer.

In the most recent report on the Life Span Study population of A-bomb survivors, covering the period 1950-1985, bone cancer was reported to show no statistically significant increase with dose (Sh88). Likewise, in a long-term follow-up study of 339 British radiologists who began their practice prior to 1921, no statistically significant excess of deaths from bone cancer was found (Sm81).

From the radiotherapy studies described above, it can be seen that large doses of acutely delivered x- or gamma-radiation can produce bone cancer; however, the uncertainties in dosimetry preclude the estimation of dose-response relationships for low doses of low-LET irradiation. Therefore, other approaches, including studies of the effects of internally deposited alpha emitters in human populations or studies of the comparative carcinogenic effects of alpha irradiation and beta irradiation in laboratory animals, are required.

High-LET Irradiation

With internally deposited ^{224}Ra, ^{226}Ra or ^{228}Ra, the main long-term biological effect has been observed to be the induction of bone cancer, primarily osteosarcoma (NRC80, NRC88, UN86, Va86). The internal deposition of ^{226}Ra in dial painters, chemists, and medical patients has resulted in life-long alpha irradiation of the bone volume, whereas with ^{224}Ra administered medically the alpha irradiation has been of relatively short duration (because of the short half-life of ^{224}Ra) and delivered mainly to endosteal bone surfaces. Risk coefficients for alpha-radiation-induced bone cancer given in the BEIR III report (NRC80) included a linear function based on the ^{224}Ra data, (i.e., 27 $\times$ 10^{-6} sarcomas/person rad) and a dose squared function based on the ^{226}Ra and ^{228}Ra data (i.e, 3.7 $\times$ 10^{-8} sarcoma/person-rad^2).

Recognizing the lack of available human data from which to estimate a risk coefficient for low-LET radiation-induced bone cancer, the BEIR III Committee divided the risk factors for high-LET radiation, by an estimated relative effectiveness factor of 20, to estimate the risk coefficient for low-LET irradiation of the skeleton. In this way, a lifetime linear risk coefficient of 1.4 $\times$ 10^{-6} bone sarcomas/person-rad and a dose-squared risk coefficient of 9.2 $\times$ 10^{-11} bone sarcoma/person rad^2 were derived. No direct evidence for the derivation of this relative effectiveness factor was cited, except that it corresponded to the ratio of currently used quality factors for alpha emissions, as compared with beta emissions, from internally deposited radionuclides. In discussing these risk coefficients, the BEIR III Committee noted that the shapes of the dose-response relationships for radiation-induced bone cancer were uncertain and that a quadratic dose-response function might be more appropriate than a linear function for low-LET radiation because of the sparsely ionizing nature of the radiation.

The long-term follow up studies of persons with elevated body burdens of ^{224}Ra, ^{226}Ra, or ^{228}Ra were examined again in detail in the BEIR IV report (NRC88). Because of the short (3.62-day) radioactive half-life of ^{224}Ra, alpha radiation is confined primarily to the sites of initial deposition on bone surfaces. In ^{224}Ra-injected subjects, bone cancers have been seen at times ranging from 3.5 to 25 years after initial exposure, with a peak occurrence at about 8 years. Several different dose-response functions for internally deposited ^{224}Ra, and their associated uncertainties, were discussed by the BEIR IV Committee, and the lifetime risk of osteosarcoma was estimated to be about 2×10^{-2}/person-Gy for a well-protracted exposure (NRC88).

A range of intake-response or dose-response functions for internally deposited ^{226}Ra and ^{228}Ra was also examined by the BEIR IV Committee (NRC88). The lifelong presence of ^{226}Ra in the skeleton after deposition affects both the doses that are received and the biological responses that are observed. In contrast to the results for ^{224}Ra, the alpha dose from ^{226}Ra continues to accumulate throughout life, and bone cancers have occurred over a much longer period of time after initial deposition of ^{226}Ra (up to 63 years after the first exposure). Because of the long-continued alpha irradiation of the skeleton, the ongoing biological processes of remodeling of bone tissues, and the associated nonuniform local deposition and redeposition of ^{226}Ra, the BEIR IV Committee recommended the use of intake-response instead of dose-response functions. No estimate of the lifetime risk as a function of the dose in person-Gy comparable to the estimate given above for ^{224}Ra was given for ^{226}Ra.

Studies in Laboratory Animals

Because of the sparseness of human data for the risks of bone cancer from low-LET irradiation, studies with laboratory animals provide another means of estimating these risks. Most of the currently available data on the long-term biological effects of low-LET irradiation in laboratory animals have been derived from chronic beta irradiation by internally deposited beta-emitting radionuclides such as ^{32}P, ^{45}Ca, and ^{90}Sr (Go86a). Of these radionuclides, ^{90}Sr has been studied in the greatest detail because of its relatively long-term persistence in fission product mixtures that may be released into the environment.

A broad range of species of laboratory animals has been used in these studies. The most extensive of these have been the life-span studies of dogs exposed to ^{90}Sr in relatively soluble form by intravenous injection, at the

University of Utah, Salt Lake City; by ingestion, at the University of California at Davis; and by inhalation, at the Lovelace Inhalation Toxicology Research Institute, Albuquerque, New Mexico. Parallel studies of ^{226}Ra injected 1 or 8 times into young adult beagle dogs have been conducted at the University of Utah and the University of California at Davis, respectively, to provide a direct link between the biological responses seen in dogs and the human data base (Go86a).

Mays (Ma80) examined the relative effectiveness of chronic alpha and chronic beta radiations by comparing the average absorbed doses of alpha and beta radiations required to produce equal incidences of bone cancer in dogs. When the incidence of bone cancer induced by ^{226}Ra was plotted against the average absorbed dose of alpha radiation, an approximately linear relationship was obtained, whereas the plot of ^{90}Sr-induced tumor incidence was concave upward at higher doses. It was observed that the effectiveness of alpha irradiation relative to that of beta irradiation increased as the dose decreased, reaching a value of 26 at an incidence of 8.7%. A similar comparison of mice injected with ^{226}Ra and with ^{90}Sr gave a relative effectiveness factor of 25 at an incidence of 7.7%. The increase in relative effectiveness resulted primarily from the decreased response seen at lower doses in ^{90}Sr-exposed dogs. No bone tumors were seen in dogs that received an average skeletal dose of 6 Gy or less from ^{90}Sr.

Experiments designed to study the long-term effects of ^{90}Sr ingested daily in food, as compared with the effects of eight fortnightly injections of ^{226}Ra (Go86b), also demonstrated the reduced carcinogenic response of the dog skeleton to chronic beta irradiation as compared with chronic alpha irradiation. Raabe et al. (Ra83), using a log-normal dose-response model, reported that the relative effectiveness of these two chronic exposure modalities was of about equal potency when the average skeletal dose rate was about 0.1 Gy/day, but that the relative effectiveness of ^{90}Sr at lower dose rates decreased in comparison with that of ^{226}Ra, eventually reaching a point at which ^{90}Sr was only 1/30 as effective as ^{226}Ra. The nonparallel nature of the dose-response relationships seen for ^{90}Sr and ^{226}Ra was consistent with similar observations made at the University of Utah (Ma80).

A proportional hazards model also has been used to compare the life-span carcinogenic response from ^{90}Sr inhaled by beagle dogs (Mc86, Gi87) with the carcinogenic responses from inhalation of $^{238}PuO_2$ or intravenous injection of ^{90}Sr, ^{226}Ra or ^{239}Pu (Me86). The results of this comparison show that the relative risk coefficients for bone cancer from ^{90}Sr are the same in dogs that received one exposure to inhaled ^{90}Sr or injected ^{90}Sr and were about 5, 48, and 30 times lower than the relative risk coefficients for injected ^{226}Ra, injected ^{239}Pu, or inhaled $^{238}PuO_2$, respectively. Collectively, these studies demonstrate that the risks per Gy of bone cancer from internally deposited ^{90}Sr are appreciably less than those from internally deposited ^{226}Ra (NRC88).

Summary

1. Currently available information on persons who have received x ray or gamma radiation delivered therapeutically for medical purposes indicates that large doses of low-LET radiation can produce bone cancer. Skeletal dosimetry in these cases is too uncertain to provide precise information about the dose-reponse relationship.

2. The data currently available from the study of Japanese A-bomb survivors provide no evidence of an excess of bone cancer resulting from low-LET irradiation at levels in the 0 to 4 Gy range.

3. The most definitive dose-response relationships for radiation-induced bone cancer come from studies of persons with elevated body burdens of the alpha-emitting radionuclides ^{224}Ra and ^{226}Ra, in whom the lifetime risk of bone cancer from internally deposited ^{224}Ra has been estimated to be about 2×10^{-2}/person Gy.

4. Studies of the carcinogenic response of the skeleton to internally deposited ^{90}Sr in beagle dogs have demonstrated a nonlinear, concave upward, dose-response relationship for chronic beta irradiation of the dog skeleton. Parallel studies with ^{226}Ra in beagles dogs have demonstrated chronic beta irradiation from ^{90}Sr to be less effective than chronic alpha irradiation from ^{226}Ra, by a factor of up to 25, the carcinogenic effectiveness of chronic beta irradiation being greatly reduced at low doses and low dose rates.

BRAIN AND NERVOUS SYSTEM

Radiation has been observed to increase the incidence of tumors of the nervous system in humans and laboratory animals. The human data are derived from studies of populations exposed prenatally to diagnostic x radiation and populations exposed postnatally to therapeutic x radiation or A-bomb radiation (La86, Ku87).

In a 1% sample of 734,000 children exposed to diagnostic x radiation in utero, MacMahon (Ma62) observed an excess mortality from cancer of thc central nervous system, amounting to approximately 6.3 deaths/10^4 PYGy (80% confidence limits, 1.1-17.2) after adjustment for birth order, religion, maternal age, sex, and pay status of parents. A similar risk estimate (6.1 excess deaths/10^4 PYGy) was subsequently reported by Bithell and Stewart (Bi75), based on their finding of a history of antenatal irradiation in 1,332 British children dying of malignant central nervous system tumors before the age of 15. Although later studies of children exposed prenatally to x rays (Di73, Mo84) or A-bomb radiation (Ja70) failed to confirm an excess number of central nervous system tumors, the results of such studies were not statistically inconsistent with the previous risk estimates (La86).

A smaller excess has been reported in persons given radiotherapy to the scalp for tinea capitis in childhood. In one series of such persons, including 2,215 patients who were followed for an average of 25 years after a dose to the brain that was estimated to average 1.4 Gy, 8 brain tumors (3 malignant) were observed, versus 1.4 expected (none were observed in 1,413 controls), corresponding to an excess of 1.0 $\pm$ 0.4 cases/10^4 PYGy, or an excess relative risk of 3.4 $\pm$ 1.5%/cGy (Sh76, La86). The tumors included glioblastomas as well as meningiomas. In another series, which included 10,842 persons followed for an average of 22.6 years after receiving an estimated brain dose of 1.21-1.39 Gy, 21 brain tumors (10 malignant) were observed, versus 6 (4 malignant) in an equal number of controls; 9 other central nervous system tumors (2 malignant) were also observed, versus none in the controls (Ro84a). From these observations, the excess of brain tumors has been estimated to approximate 0.71 $\pm$ 0.20 cases/10^4 PYGy, and the total excess of all central nervous system tumors has been estimated to approximate 1.09 $\pm$ 0.24 cases/10^4 PYGy (La86).

Other patients in whom the risk of central nervous system tumors has been observed to be increased after therapeutic irradiation include a series of 3,108 persons who were followed for an average of 22 years after x-ray treatment of the head and neck during childhood, in whom 14 intracranial tumors (6 malignant) were observed (Co78). On the basis of an estimated average midbrain dose of 0.8 Gy, a latent period of 5 years, and an expectation of about 1.6 intracranial tumors, the excess of intracranial tumors in this series has been calculated to approximate 2.9 $\pm$ 0.9 cases/10^4 PYGy, or 9.7 ($\pm$ 2.9)%/cGy (La86). Also among 592 children treated with irradiation of the cranium for acute lymphatic leukemia (ALL), the relative risk of subsequent brain tumor was reported by Rimm et al. (Ri87) to approximate 20 (0.25 cases/10^4 person-year); and in a comparable series of 468 children, the relative risk was reported by Albo et al. (Al85) to approximate 226 (ratio of observed to expected cases, 9/0.0398). Similarly, 3 of 904 patients treated with radium implants in the nasopharynx and followed for an average of 25 years after treatment were observed to develop brain tumors, versus none in 2,021 controls (Sa82); on the basis of an estimated average dose to the brain of 0.15-0.4 Gy and an expectation of 0.57 brain tumors, the excess in this series has been calculated to range from 3.4 $\pm$ 2.4 to 9.0 $\pm$ 6.4 cases/10^4 PYGy for doses of 0.4 Gy and 0.15 Gy, respectively (La86). Likewise, an excess of intracranial tumors (22 observed after a latency of 5 years, versus 14.03 expected) has been observed in a series of 14,106 patients treated with spinal irradiation for ankylosing spondylitis and followed for up to 48 years after treatment (Da87); if the average dose to the affected part of the brain in such patients is assumed to have been less than 0.15 Gy, the excess of intracranial tumors can be calculated to exceed $5.8/10^4$ PYGy. In atomic-bomb survivors no excess of intracranial or other central nervous system tumors has been evident thus far (La86, Pr87a).

In pioneer radiologists who entered practice in the United States during the 1920s, mortality from brain cancer was about 3 times higher than that in other medical specialists (Ma75); however, the numbers that were exposed and the doses that they received are not known. Hence, the magnitude of the risk per unit dose cannot be estimated (La86).

An association between intracranial meningiomas and previous medical or dental radiography has been suggested by the results of a case-control study of such tumors diagnosed during 1972-1975 in women of Los Angeles County (Pr80). The strongest association (relative risk, 4.0, $p < 0.01$) was with a history of exposure to full-mouth dental x-ray examinations at a young age (<20 years). Radiation dose estimates were not reported. Other series of patients in whom an excess of meningiomas occurred after previous localized irradiation have been reported by Soffer et al. (So83) and Rubinstein et al. (Ru84).

In monkeys exposed acutely to x rays, neutrons, or protons, the incidence of glioblastoma multiforme has been observed to be increased at high doses; for example, two of four rhesus monkeys exposed to 15 Gy of x radiation delivered in a single exposure to the head alone developed such tumors during the subsequent decade (Wa82). Similarly, the incidence of such tumors was greatly increased in *Macaca mulatto* monkeys exposed to whole-body fission neutron irradiation (Br81) or whole-body 55-MeV proton irradiation delivered in a single exposure (Yo85). In the latter, the incidence rose from zero at a surface dose of 2 Gy or less to about 30% at a surface dose of 6 Gy and 33% at a surface dose of 8 Gy. In addition to glioblastomas, other types of intracranial tumors (ependymomas, meningiomas, and pituitary adenomas) also occurred with increased frequency, the total excess of intracranial tumors greatly exceeding that of all other radiation-induced neoplasms combined. The incidence of intracranial tumors in rats has been observed to be increased after acute whole-body x-irradiation at doses in the range of 1-2 Gy (Kn82). A brain tumor (oligodendroglioma) was also observed in 1 of 11 dogs exposed to 1.33 Gy of fast neutrons delivered to the head alone (Zo80).

Summary

Radiation has been observed to increase the incidence of nervous system tumors in human populations and laboratory animals. The tumors include malignant as well as benign growths of the brain, meninges, and peripheral nerves. The increase has been evident after irradiation in childhood at doses of less than 1-2 Gy. Although the dose-incidence relation is uncertain, the data indicate the brain to be relatively sensitive to the carcinogenic effects of radiation.

OVARY

There is a wealth of experimental data on the induction of ovarian tumors in mice by ionizing radiation. It has been demonstrated that all nonreproductive cells (i.e., all cells other than oocytes) in the ovary are susceptible to the carcinogenic effects of radiation; that even low levels of acute exposure (e.g., as little as 50 rad) can cause a significant increase in the rate of tumor induction; and that the dose-response curve for radiation-induced mouse ovarian tumors is generally S-shaped, depending on the total dose, dose rate, LET of the radiation, and the strain and age of the mice exposed (Up70 and Ul79). The induction of these growths is ascribed to abnormal gonadotrophic stimulation incident to the cessation of ovulatory cycles resulting from radiation-induced sterilization (Cl59).

The strongest evidence of an association between radiation and cancer of the human ovary comes from the combined cohort of Japanese atomic bomb survivors in whom the relative risk at 1 Gy (using the new dosimetry) was estimated to be 2.33/Gy (Sh88)

In women with cancer of the uterine cervix who were treated with radiation (Bo85), the relative risk of ovarian cancer increased significantly over time, reaching a (nonsignificant) high of 3.4 (3 observed versus 0.9 expected cases) in those whose first treatment had occurred 30 or more years previously. Overall, the risk of ovarian cancer was significantly decreased ($p < 0.001$) (relative risk, 0.7); however, when only those women who were followed for at least 10 years after radiotherapy were considered, the risk was similar to that seen in the general population (i.e., 70 cases observed versus 76 expected cases). The significance of the increasing trend over time is likely to be due in large part to the marked reduction in risk seen the first 10 years after radiotherapy, which, in turn, may have arisen because of the frequent incorporation of bilateral oophorectomies in the treatment regimen.

The results from the latest update (Da87) of the Court-Brown and Doll cohort (Co65) of patients with ankylosing spondylitis offer little additional support for the presumed relationship between ovarian cancer and exposure to ionizing radiation. The overall mortality risk among patients 5 years or more after radiotherapy is slightly below the expectation based on the experience of the national population (i.e., 5 observed versus 5.37 expected deaths). As was noted above in the discussion of uterine cancer deaths in this cohort, the results need to be interpreted cautiously since they are based on such a small number of cases.

UTERUS

Epidemiologic evidence of an association between irradiation and uterine cancer has been based in the past primarily on the Tumor Registry data from A-bomb survivors of Hiroshima (but not those of Nagasaki), which gave some indication of a linear trend in cervical cancer cases in groups exposed to 10+ and 50+ rads (i.e., p values of 0.06 and 0.09, respectively), and on data from the Smith and Doll (Sm76) cohort of 2,068 women who had undergone pelvic irradiation for benign uterine bleeding. In the latter study, 16 deaths from uterine cancer were observed among the patients 5 or more years after treatment, while only 10.3 deaths were expected (p = 0.08).

The most recent analysis of the mortality experience of the combined cohorts of A-bomb survivors from Hiroshima and Nagasaki (Sh87) provides the strongest evidence of a potential relationship between radiation exposure and uterine cancer. Using the new dosimetry calculations, Shimizu and coworkers estimated that the relative risk for cancer of the cervix uteri and uterus was 1.22, which represented a suggestive but not significantly increased risk (i.e., $p = 0.07$).

In a study of the occurrence of second primary tumors among 182,040 women treated for cervical cancer (Bo85), an overall deficit in the number of cancers of the uterine corpus was observed among patients who received radiotherapy, in comparison with the expectated number based on general population rates. A total of 133 cases of cancer were observed in this subgroup, versus an expected value of 215, giving a significantly reduced ($p < 0.001$) relative risk of 0.6. However, when attention was restricted to those radiation-treated patients who were followed for 10 years or more, the relative risk returned to a nearly normal or baseline value (84 observed versus 86 expected cases) and was no longer significant. Indeed, the trend in the relative risk over 5-year intervals since the administration of treatment was significantly increased ($p < 0.001$), primarily because of the significant reduction in risk seen in the first 10 years after radiation exposure. Since many of the radiation-treated patients also had a hysterectomy as part of their treatment regimen, the authors speculated that use of general population rates to predict the expected number of uterine cancers may actually lead to an overestimation of risk. As a result, they concluded that ". . . an RR of nearly one in all women followed for more than 10 years probably corresponds to a substantial excess in those women with intact uteri and is likely associated with the prior radiotherapy." No excess was observed in recent case control analysis of these data (Bo88).

Data from the recently updated study of long-term mortality in patients treated for ankylosing spondylitis with a single treatment course of x rays (Da87) provide little additional insight into the potential relationship between radiation exposure and uterine cancer. Among patients who died in less than 5 years, 5.0-24.9 years, and 25 or more years, respectively, after the first treatment, uterine cancer relative risks of 0.00 (0 observed versus 1.24 expected deaths), 1.15 (5 observed versus 4.35 expected deaths), and 0.65 (1 observed versus 1.54 expected deaths) were seen. The study's authors qualified these negative findings to some extent by noting that their cohort of patients with ankylosing spondylitis provided little information on radiation-based cancer of the uterus, since so few women were included in the study population.

Summary

Taken collectively, the new data that have been accumulated since the BEIR III report (NRC80) still do not resolve the question of the potential association between radiation exposure and uterine cancer.

TESTIS

Relatively little information on the possible association between ionizing radiation and testicular cancer is available, particularly with reference to potential human risk. In the BEIR III report (NRC80), testes were included as a site or tissue in which radiation-induced cancer has not been observed. Darby et al. (Da85) specifically addressed the question of radiation-induced testicular cancer in a study involving a parallel analysis of cancer mortality among the Japanese A-bomb survivors and patients with ankylosing spondylitis in the United Kingdom who had been given x-ray therapy. A comparison of observed and expected testicular cancer deaths among these two study populations revealed no observed deaths in either group. In another recent analysis (Sh88), of the Japanese A-bomb survivors, testicular cancer was not among the site-specific cancers that showed a significant increase in occurrence with dose (DS86 system). No excess cases of testicular cancer have been identified in other epidemiologic studies to date. Thus, the limited data available for humans suggest that the human testes may be relatively resistant to cancer induction by exposure to ionizing radiation.

There are a number of studies in the experimental literature that indicate that there is some association between whole-body or site-specific radiation exposure and the induction of testicular cancer, particularly interstitial cell tumors in rats (see, e.g., Wa86). It has been postulated that these tumors occur in part because of a probable hormonal imbalance from radiation damage to the testes. At present, however, the experimental data have not been paralleled by epidemiologic evidence.

Summary

The exisiting data imply that the human testis is relatively insensitive to the carcinogenic effects of radiation.

PROSTATE

Introduction

In the 1980 BEIR report (NRC80), the prostate was recorded as an organ with little or no sensitivity to the induction of cancer by radiation, since no epidemiologic evidence suggesting radiogenic prostate cancer was available at that time. For the same reason, the 1985 National Institutes of Health report (NIH85) on the probability of causation did not include prostate cancer as a radiogenic neoplasm. In the interim since these reports, data suggesting a weak association between prostatic cancer and radiation have been reported, as summarized below.

Human Studies

Japanese A-Bomb Survivors

In the 1987 RERF report on the cancer experience of the Japanese A-bomb survivors (Pr87a), prostate cancer was considered separately for the first time. In the survivor population, mortality from carcinoma of the prostate was uncommon; only 51 deaths were reported. A causal association with radiation dose was not statistically significant, the average relative risk under the T65 dosimetry being 1.27 at 1 Gy. However, the combination of a moderate excess risk of 0.14 excess prostate cancer deaths/10^4 PYGy together with low background mortality rates yielded an attributable risk of 11.7%. The data suggested that the relative risk of death from prostate cancer may have increased with time. However, the time trend was not statistically significant, even though the average increase in the excess relative risk of 37.5%/year was one of the largest estimated among the cancers considered. Additional uncertainties were introduced due to inaccuracies of death certificate diagnoses of prostate cancer; confirmation rates were 39%, while the detection rate was only 21%.

In the most recent Life Span Study cohort report (Sh88), prostate cancer mortality in the years 1950-85 showed no significant increase with increasing dose. In the DS86 subcohort, which included 75,991 exposed persons, there were 52 deaths from cancer of the prostate; the estimated relative risk at 1 Gy (shielded kerma) was 1.05; the absolute risk was 0.03 excess cancer deaths/10^4 PYGy; and the attributable risk was 1.95%.

Ankylosing Spondylitis Patients

Darby et al. (Da85) included prostate cancer patients in the category of those who were treated with x-irradiation to heavily irradiated sites. In the most recent follow-up on cancer mortality, to January 1, 1983, covering 11,772 men with ankylosing spondylitis given a single course of x ray treatment during the period 1935-1954 (152,979 person-years at risk), there were 21 observed versus 18.15 expected prostate cancers (ratio of observed to expected cases, 1.16) during the period $\geq$ 5 years after the first treatment (Da87). An early excess (ratio of observed to expected cancers 4.0/1.31 = 3.04, $p < 0.5$) was limited to the first 5 years after treatment. Thus, the relative risk for cancer of the prostate was the third highest in this series, and the risk was significantly increased; but the authors noted that this disease is often confused with ankylosing spondylitis, since it is frequently present with pain in the back due to metastases to the spine (Da87). The mean dose to the prostate was calculated to be 24 rad (0.24 Gy) (Le88). Based on this average organ dose, estimated by using the Monte Carlo method (Le88), the increase in relative risk of prostate cancer for the period 5.0-24.9 years after exposure (ratio of observed to expected cancers, 1.24) was estimated to be 0.66%/rad at 0.01 Gy (1 rad).

Nuclear Workers

Beral et al. (Be85) have recently reported an increase in the relative risk of mortality from prostate cancer among British nuclear workers, as discussed in Chapter 7 (SMR = 145 for those with 10 or more years of employment). In other groups of nuclear workers also, a nonsignificant elevation of risk has been reported (Sm86).

United States Radiologists

The early U.S. radiologists are estimated to have had lifetime exposures of 2 to 20 Gy. Their cancer mortality experience has been analyzed by Matanoski et al. (Ma84), who reported standardized mortality ratios (SMRs) for selected cancers among members of the Radiological Society of North America (RSNA), the American College of Physicians (ACP), and the American Academy of Ophthalmology and Otolaryngology (AAOO). During the period 1920-1939, SMRs for prostate cancer were 1.24 for members of RSNA, 1.03 for members of ACP, and 0.81 for members of AAOO. The excess of prostate cancer mortality in radiologists was not statistically significant ($p < 0.05$). During the period 1940-1969, the SMRs were 1.01 for members of RSNA and ACP, 0.98 for members of AAOO and OTOL, and 1.40 for members of OPH, values which were not significantly different from unity.

Patients Receiving Iodine-131 Therapy

In a study (Ho84b) of the incidence of malignant tumors in 4,557 Swedish patients treated with ^{131}I for hyperthyroidism, prostate cancer was analyzed in 726 men, with 6,400 person-years at risk. With doses of less than 370 megabecquerels (MBq) (10 mCi), there were 11 observed, and 10.3 expected cases (relative risk, 1.07). For doses of greater than or equal to 370 MBq (10 mCi), there were 5 observed and 8.0 expected cases (relative risk, 0.63). For the overall doses combined, there were 16 observed, and 18.4 expected cases (relative risk, 0.87). Thus, none of the relative risks were considered significantly different from unity.

Animal Data

In rats, intensive x-irradiation has been observed to induce carcinoma of the prostate, but only at doses of 10 Gy or more (Wa86).

Summary

From the studies available thus far, the relative risk of radiation-induced prostate cancer appears to be small. Hence, although the data suggest that there may be a weak association between prostate cancer and radiation, the sensitivity of the prostate to the induction of cancer by irradiation appears to be comparatively low.

URINARY TRACT

In the 1980 BEIR report (NRC80) the urinary organs, particularly the kidney and urinary bladder, were included among those tissues with definite but low sensitivity to radiation carcinogenesis. Since then substantial new information in support of this conclusion has become available.

Japanese A-Bomb Survivors

The Radiation Effects Research Foundation (RERF) Life Span Study Report 10 on cancer mortality among A-bomb survivors in Hiroshima and Nagasaki (1950-1982), which was based on T65D dosimetry, indicated a significant dose-related increase in the number of cases of urinary bladder cancer (Pr87a). Among 91,231 exposed survivors with T65D dose estimates, there were 6,270 cancer deaths during 1950-1982. Death certificate diagnoses for cancers of the urinary tract are moderately accurate. Of the 131 deaths caused by cancers of the urinary tract (bladder, kidney, and other unspecified urinary organs), 95 were urinary bladder, 33 were kidney, and 3 were ureter. For all sites combined, the relative risk at 1 Gy (T65D) was 1.55. The highly significant radiation dose response ($p = 0.006$) occurred primarily among those who died from bladder cancer ($p = 0.003$); for those who died from kidney cancer, the positive radiation dose response was slight ($p = 0.3$).

An analysis of radiation-related cancer mortality during 1950-1985 according to the new DS86 dosimetry systems has recently been made on data from 75,991 Hiroshima and Nagasaki survivors, a subgroup of the RERF Life Span Study cohort designated the DS86 subcohort (Sh87, Sh88). In terms of the kerma at the survivor's location, the overall relative risk of death from cancers of the bladder and other urinary organs based on 133 cases was 2.06/Gy (90% confidence interval: 1.46-2.90). This relative risk is 1.49 times the risk calculated according to the T65D dosimetry system on the same cohort (Sh87).

There was a suggestive trend ($p < 0.10$) of an increase in relative risk with time from exposure (Sh88). For bladder and kidney cancer mortality individually, the relative risk values were 2.13 (90% confidence interval, 1.40-3.28) and 1.58 (90% confidence interval, 0.91-2.94) at 1 Gy kerma, respectively (Sh88). The corresponding absolute risks were 0.41 (90% confidence interval, 0.16-0.70) and 0.09 (90% confidence interval, −0.02-0.26)/10^4 PYGy, respectively. Sex had little effect on the relative risk of bladder cancer mortality; the male relative risk/female relative risk was 0.9. However, the absolute risk in males was nearly twice that in females; this reflects the higher incidence of bladder cancer in Japanese males than in females that is unrelated to radiation exposure (Sh88). The dose-response relationship had a strong linear component. In terms of organ dose rather than kerma, the relative risk at 1 Gy was 2.27 for bladder cancer mortality (90% confidence interval, 1.53-3.37).

Ankylosing Spondylitis Series

Darby et al. (Da85) compared radiation-induced cancer mortality among the Japanese A-bomb survivors who received doses of ≥ 1 Gy with that of British patients who had received x-ray therapy for ankylosing spondylitis. The estimated mean organ doses to the bladder were 1.02 Gy, based on T65D dosimetry, in the Japanese studies and 0.31 Gy in the spondylitics (NRC80). Eleven cases of bladder cancer were observed in each group. The risk estimates were given, with 90% confidence interval given in parentheses. The relative risk for A-bomb survivors was 3.0

(1.7-5.2); for those with ankylosing spondylitis it was 1.6 (0.9-2.7). The excess risk/10^5 person-years for A-bomb survivors was 5.1 (1.3-8.8), and for spondylitics it was 4.1 (0.6-11.2).

Land's review (La86) of the analysis of patients with ankylosing spondylitis indicated that urinary tract cancers occurred in excess numbers among those patients given one course of x-ray treatment. Eight deaths as a result of bladder cancer were seen while 5.1 were expected ($p = 0.14$). During the first 9 years the mortality was 3 deaths from bladder cancer versus 2.6 expected. The observed and expected values more than 9 years after the beginning of treatment corresponded to an absolute risk of 1.7 excess deaths from bladder cancer/10^4 PYGy and a relative risk of 1.9/Gy.

Radiotherapy for Benign Uterine Bleeding

Smith and Doll (Sm76, Sm77) reviewed the experience of English women treated by x radiation for metropathia hemorrhagica (benign uterine bleeding); 3 deaths from bladder cancer were observed 5 years or more after treatment, and 2.15 deaths from bladder cancer were expected (13.5 years of mean follow-up). Wagoner (Wa84) found 10 cases of bladder cancer versus 5.1 expected ($p = 0.026$) among women who had received radiation therapy for benign gynecologic disorders in Connecticut between 1935 and 1966.

Cervical Cancer Series

An international study was recently performed on 150,000 women with uterine cervical cancer who had been treated at 1 of 20 oncologic clinics and/or who were reported to 1 of 19 population-based registries. A sample of 4,188 of these women who had received radiation therapy for cervical cancer and who had second cancers and 6,880 matched controls were selected for detailed study (Bo88). A dose of 30 to 60 Gy to the bladder was associated with a relative risk for bladder cancer of 4.0. Women who were <55 years of age at the time of treatment were at especially high risk (relative risk of 16.0). Risk increased with time after exposure; the relative risk was 8.7 among those who survived ≥ 20 years. Over the range of 30-60 Gy, the relative risk increased significantly with dose ($p < 0.001$). The effects of smoking and radiation exposure were independent; controlling for smoking did not appreciably alter the radiation risk estimates. Overall bladder cancer risks were greater in the U.S. registry data (Bo88).

An analysis of 148 cases of kidney cancer and 285 matched controls revealed an overall radiation-related relative risk of 1.2 (Bo88). The relative risk increased to 3.5 (90% confidence interval, 1.3-9.2) among those who survived $\geq$15 years after exposure. Women in U.S. registry areas and those exposed when they were <55 years of age had the greatest radiogenic kidney cancer risks (Bo88). Finally, a relative risk of 2.9 was found for cancers of the renal pelvis (23 cases) and ureter (4 cases); the epithelia of these structures are similar to those of the bladder.

Cancer after Iodine-131 Therapy

A recent report on iodine-131 treatment for hyperthyroidism in Sweden (1951-1975) indicated no increase in bladder and kidney cancers during the 24 years after therapy despite the administration of relatively high doses to the urinary tract (Ho84b). The values for kidney and bladder cancers combined were as follows: (1) males, 9 observed versus 7.5 expected (relative risk, 1.20); (2) females, 17 observed versus 19.0 expected (relative risk 0.89); (3) both sexes, 26 observed, versus 26.5 expected (relative risk 0.98).

Conclusions

The epidemiologic evidence shows that radiation can cause cancer of the bladder and, to a lesser extent, of the kidneys and other urinary organs. For such effects, the observed dose-response relationship is consistent with a linear nonthreshold function over a broad range of doses, from a few Gy to 60 Gy. Women less than 55 years of age at the time of exposure appear to be at greater risk than older women, and risk appears to increase with time after exposure. The most recent analysis of the A-bomb survivor data using the new DS86 dosimetry indicates a relative risk of 2.3 (90% confidence interval, 1.5-2.4) urinary tract cancer deaths/Gy of absorbed DS86 dose, and an absolute risk of 0.7 urinary tract cancer deaths/10^4 PYGy (Sh88). If the incidence of urinary tract cancer is assumed to be 3-4 times the rate of mortality from such cancer, the relative risk can be estimated to be ~6.8-9.1/Gy of absorbed dose.

PARATHYROID GLANDS

By the time of publication of the BEIR III report, (NRC80), the parathyroid glands were included among those tissues that are susceptible to radiogenic neoplasia. Of the 64 women and 36 men examined more than 25 years after radiotherapy for cervical tubercular adenitis, a total of 11 were found to have parathyroid abnormalities, including 7 with adenomas and 4 with diffuse hyperplasias (Ti77). Of these 11, 7 were hypercalcemic; that is, they had hyperparathyroidism (HPT). None of the 27 subjects who had received <3 Gy had parathyroid dysfunction, while 3/39 (8%) and 4/28

(14%) who had received 3-6 Gy and 6-12 Gy, respectively, had HPT. Four of six (67%) of those who had received >12 Gy had parathyroid disease. The mean time from exposure to diagnosis was 38 years.

In a health survey done in Stockholm, Sweden, 15,903 subjects were screened, of whom 58 (44 women, 14 men) had HPT with parathyroid adenomas. None of 58 matched eucalcemic control subjects had had radiotherapy to the parathyroid region, whereas 8 (14%) of the patients with HPT had been irradiated at a mean age of 8.1 years. The dose range was 2-5 Gy, and the mean time to diagnosis was 47 years (Ch78).

A total of 17% of 130 patients with HPT at the Henry Ford Hospital had a history of radiation exposure at a mean age of 16 years, whereas only 3% of 400 ambulatory eucalcemic patients had been irradiated ($p < 0.025$) (Ra80). Similarly, 8/73 (11%) of patients with HPT in a Dutch study had received radiotherapy for benign disease (Ne83). The mean time to diagnosis of HPT was 36 years among those in the Henry Ford Hospital study (Ra80) and 34 years among those in The Netherlands study. Among 200 subjects in the Henry Ford Hospital and other series who were known to have a history of irradiation received during childhood, the prevalence of HPT was 5%, ≥30 fold the prevalence in the general population ($p < 0.025$) (Ra80).

HPT is not always associated with parathyroid hyperplasia or adenoma. In a study of 23 patients who received surgery for nodular thyroid disease and who had no known HPT, five women and three men (35%) had either parathyroid adenoma or hyperplasia (Pr81). All 23 patients had received radiotherapy when they were an average of 16 years of age; the time from irradiation to surgery averaged 33 years.

The incidence of thyroid disease among 42 patients with HPT who had a history of receiving irradiation was compared with that in 162 patients with HPT who had not been exposed to radiation (Ka83). Seventy-nine percent of the irradiated patients with HPT had thyroid abnormalities, including 38% with thyroid adenomas and 29% with cancer; in contrast, 43% of patients with HPT with no history of irradiation had thyroid disease and only 10% each had thyroid adenoma and carcinoma (Ka83). Although thyroid disease was not reported to accompany HPT or parathyroid tumors in the Stockholm study (Ch78), an association between radiation-induced diseases of these two glands is a common finding in other series. All 11 patients with parathyroid disease (7 with adenomas, 4 with hyperplasias) of 100 irradiated subjects had thyroid abnormalities, including 2 with thyroid carcinomas and 1 with adenoma (Ti77). Of 73 patients with HPT, 8 were found to have a history of irradiation; 5 had parathyroid adenomas and 3 had hyperplasias. Of these eight patients, six had concurrent nodular goiters and one had struma lymphocytica (Ne83).

Possible radiation-related carcinomas of the parathyroid have been reported sporadically (Ir85). The low frequency of overt cancer may be related in part to the often very long latency of symptomatic parathyroid disease. Mean times from exposure to diagnosis of HPT varied from 30 to 50 years in the different series. In addition, 90% of parathyroid adenomas were accompanied by clinically important HPT and a high percentage of those with radiogenic parathyroid hyperplasia or adenoma had concurrent thyroid disorders. Both conditions commonly require surgical intervention and, hence, removal of the possibly premalignant parathyroid tissue.

A review of experimental studies, including those involving mice, rats, guinea pigs, dogs, and monkeys, illustrates that the parathyroid glands do not acutely express radiation damage at the histological level at x-ray doses below 5 Gy (Be72). Doses between 5 and 25 Gy cause modest edema and hyperemia, and higher doses cause severe damage. Late changes include hyperplasia, cyst formation, adenomatous nodules, gross adenomas, and carcinomas. In one experiment, a cumulative total of 12 parathyroid tumors were noted among 80 rats of each sex that were exposed to 250-kVp x rays at 5 Gy when they were 100 days of age; no tumors occurred in 160 unirradiated controls. Ten animals of each sex in the irradiated and control groups were necropsied every 3 months for 24 months after exposure, so the cumulative reported incidence of 8% parathyroid tumors is an underestimate of the true 24-month (or life span) incidence (Be67).

Parathyroid neoplasia was also observed to follow irradiation with radioiodide. In one experiment, 185 or 370 kBq of ^{131}I was administered to neonatal rats. Sixty-one percent (28/46) of such animals that survived 15 months were found to have parathyroid adenomas (Tr77); adenomas were found in 31 untreated control rats. Some of the animals had HPT as evidenced by elevated serum calcium.

In summary, both experimental and human studies confirm that HPT and parathyroid hyperplasia, parathyroid adenoma, and less frequently, parathyroid carcinoma are late sequelae of radiation exposure. Most parathyroid neoplasms are hyperfunctional, and radiogenic HPT is frequently accompanied by thyroid dysfunction, neoplasia, or both. In humans, the time from irradiation until the time of diagnosis is most commonly at least 30 years. Although the incidence of HPT and neoplasia appears to increase with dose (Ti77), the data are inadequate for quantitative risk estimation. It is clear, however, that parathyroid neoplasia may eventually follow doses in the range of 1 to 5 Gy after exposures that cause little or no acute histopathologic evidence of damage in the glands. The possibility of HPT and parathyroid neoplasia should be considered in those individuals with a history of irradiation of the head and neck, and particularly those with thyroid dysfunction or thyroid nodules.

NASAL CAVITY AND SINUSES

In the BEIR III report (NRC80), the induction of cancer of the paranasal sinuses and mastoid air cells by internally deposited ^{226}Ra was described, but no information was presented on the induction of such cancer by low-LET irradiation. In the latest report from the Life Span Study cohort of Japanese A-bomb survivors, Shimizu et al. (Sh88) reported that a total of 44 cases of nasal cancer had been seen during the interval 1950-1985, without any evidence of a dose-response relationship. Similarly, no radiation-induced excess has been evident in the 14,106 patients who received a single course of x-ray treatment for ankylosing spondylitis (Da87), although the mean dose received by the nasal region in such patients was estimated to be 0.47 ± 0.44 Gy (Le88). These results imply that the nasal cavity is not highly sensitive to low-LET radiation.

Conversely, carcinomas of the paranasal sinuses and mastoid air cells have been observed in radium-dial painters and other people exposed to internally deposited ^{226}Ra. The occurrence of these cancers and the underlying radiation etiology are discussed in detail in the BEIR IV report (NRC88). The carcinomas are thought to arise as a result of alpha irradiation of the epithelium from ^{222}Rn gas and radon progeny in the air above the epithelium and from emissions, primarily beta and gamma radiations, from ^{226}Ra and its progeny in the underlying bone.

Thirty-five carcinomas of the paranasal sinuses and mastoid air cells have occurred in the 4,775 ^{226,228}Ra-exposed subjects, of which there has been at least one determination of vital status (NRC88). The observed latent periods for these cancers have been quite long, ranging from 19 to 52 years (NRC80). In the BEIR III report (NRC80), the lifetime risk of ^{226}Ra-induced paranasal sinus and mastoid carcinomas was estimated to be 64 carcinomas/10^6 person-rad.

The response of the paranasal sinuses to radiation was also demonstrated in patients who received Thorotrast (^{232}Th), an x-ray contrast medium, by antral injection into their sinuses. Fabrikant (Fa64) et al. reported on 10 patients with maxillary sinus carcinomas after maxillary sinus instillations, and Rankow et al. (Ra74b) reported that 13 of 14 patients who received Thorotrast by this route developed cancers of the maxillary and adjacent sinuses.

Studies of inhaled or intravenously injected beta-emitting radionuclides in beagle dogs have shown that relatively high local dose rates can occur in the nasal cavity because of patterns of radionuclide deposition and retention (Be79, Bo86). These local accumulations, which result from deposition and retention of the inhaled material or from subsequent translocation of radionuclide to the underlying bone, persist for long periods of time, resulting in the accumulation of high local doses. Such irradiation from inhaled $^{144}CeCl_3$, $^{91}YCl_3$, or $^{90}SrCl_2$ or from intravenously injected $^{137}CsCl$ has been observed to induce nasal cavity cancers in dogs (Be79, Bo86a).

In summary, there are currently no human dose-response data on cancers of the nasal cavity or cranial sinuses from low-LET irradiation. The only data on the induction of such tumors in human populations pertain to the internally deposited alpha-emitters ^{226}Ra or ^{232}Th and their decay products. The latency for such cancers has been at least 10 years. The induction of nasal cavity cancers in dogs by intensive irradiation from beta-emitting radionuclides implies that such a response might occur in humans under appropriate conditions of low-LET irradiation, but also that the risk would be vanishingly small.

SKIN

In pioneer radiation workers, carcinomas of the epidermis arising in areas of chronic radiodermatitis were the first radiation-induced neoplasms to be recognized as such (Br36, Ca48, He50). The early literature, consisting largely of case reports, affords no adequate basis for assessing the dose-incidence relationship (Al86). Although some epidemiologic studies of irradiated cohorts have provided dose-incidence data in recent years, such studies have been complicated by the fact that skin cancer—unlike cancer of other sites—carries a low mortality and is grossly underreported. The result is that its ascertainment is difficult and uncertain. These limitations notwithstanding, the results of several studies (summarized below) imply that the skin has a higher susceptibility to radiation carcinogenesis than has generally been suspected.

Perhaps the most extensive study of radiation-induced skin cancer is an investigation of 2,226 persons who were treated in childhood with epilating doses of 100 kVp x rays to the scalp for tinea capitis and who have since been followed for an average of more than 25 years (Sh84a, Ha83b). The absorbed dose to the scalp in such persons averaged 4.5 Gy (3.3-6.0 Gy), while the dose at the margins of the scalp averaged 2.4 Gy and the dose to the face and neck averaged 0.1-0.5 Gy. In 41 of the 1,680 white members of the cohort, 80 basal cell carcinomas of the skin had appeared, whereas none had appeared in the 546 nonwhite members and only 3 have appeared in a control group of 1,387 nonirradiated white tinea cases (Sh84a). The tumors began to appear about 20 years after exposure and were not limited to the most heavily irradiated parts of the scalp but tended to occur more commonly at the margins of the scalp and in neighboring areas of skin that were not covered by hair or clothing; an excess has been detected even on the cheek and the neck, where the dose is estimated to have been only 0.12 and 0.09 Gy, respectively (Ha83b). The distribution of tumors in relation to the dose suggests, therefore, that the carcinogenic effects of x-irradiation were enhanced by exposure to ultraviolet (UV) radiation (Ha83b). The cumulative excess increased with the dose of x rays in a manner consistent with linearity, amounting to about 3.3 $\times$ 10^{-5} cases/cm^2 Person Gy in areas that were exposed to both x radiation and UV radiation, as compared with about 7.1 $\times$ 10^{-6} cases/cm^2 Person Gy in areas that were exposed to x rays alone. The average follow-up period for these observations was 25.7 years (Sh84a).

Other populations for which risk estimates have been derived include 2,653 persons given x-ray therapy to the chest for enlargement of the thymus gland in infancy, in whom 8 skin cancers were observed to develop later in the irradiated area, versus 3 skin cancers in the corresponding area among 4,791 controls; the 8 cancers included 6 basal cell carcinomas and 2 malignant melanomas (E. Woodward and L. Hempelman, personal communication). The average dose to the irradiated skin was estimated to approximate 3.3 Gy, and the excess relative risk of cancer in the irradiated area between 10 and 49 years postirradiation was interpreted to amount to 4.8 (Al86), giving an average excess relative risk of about 1.5/Gy. The absolute risk has been estimated to range from 0.66/10^4 PYGy at doses of less than 4 Gy to 0.32/10^4 PYGy at doses exceeding 4 Gy (Al86).

An excess of skin cancer, primarily basal cell carcinomas of the face, has been observed also in Czechoslovakian uranium miners (Se78). On the basis of an estimated relative risk of 4.5 in this population and a cumulative dose to the affected skin from alpha radiation of approximately 1-2 Gy (20-40 Sv), the relative risk may be calculated to approximate 15%/Sv and the absolute risk 0.95/10^4 PYSv (Al86). As has been noted previously, however, the excess may not be attributable entirely to radiation, in view of the possible causal contribution that may have been made by arsenic in the uranium ore dust. No excess in numbers of skin cancers has been observed thus far in a number of other irradiated populations that have been studied epidemiologically (Al86). The failure to detect an excess in such studies may be attributable, however, to underascertainment of skin cancers, for the reasons cited above.

Radiation carcinogenesis in the skin has been studied experimentally in several species of laboratory animals (UN77). In the rat, a variety of different types of skin tumors occur in response to irradiation, including tumors of hair follicles; in total numbers, the tumors induced by a given dose in the rat exceed those induced by the same dose in the mouse, a species in which the tumors are composed predominantly of squamous cell carcinomas (Bu86). The incidence of tumors in the rat increases as a linear-quadratic function of the dose and reaches a peak at 20-30 Gy of low-LET radiation or 9-10 Gy of high-LET (125 keV/μ) radiation. For maximal tumorigenic effectiveness per unit dose, the full thickness of the epidermis must be irradiated in the rat, including the entire hair follicle

(Bu76). Fractionation or protraction of the dose to the skin of rat reduces the cumulative incidence per unit dose with low-LET radiation, but not with high-LET radiation (Bu80). For a given total dose, moreover, the yield of tumors may be increased by exposure to ultraviolet radiation, tumor-promoting agents, or other factors, depending on the particular experimental conditions in question (UN82, Fr86).

Summary

The risks of basal cell and squamous cell carcinomas of the skin have been observed to be increased by occupational and therapeutic radiation exposure. Although the data do not suffice to define the dose-incidence relationship precisely, the cumulative 30-year excess of basal cell carcinomas in fair-skinned persons treated with x rays to the scalp for tinea capitis in childhood has been observed to increase over a 25-year period with dose in a manner consistent with linearity, corresponding to 7.1×10^{-6} excess cases/cm^2 Person Gy in areas of skin not exposed to sunlight and 3.3×10^{-5} excess cases/cm^2 Person Gy in areas of skin exposed to sunlight as well as x rays.

LYMPHOMA AND MULTIPLE MYELOMA

An increase in the frequency of some forms of lymphoma has been associated with irradiation in humans and laboratory animals (UN77). In humans, the forms include multiple myeloma, in which the tumor cells proliferate primarily in the bone marrow, and non-Hodgkin's lymphoma, in which the tumor cells proliferate primarily in the lymph nodes. Multiple myeloma and non-Hodgkin's lymphoma, like chronic lymphocytic leukemia, are malignancies of B lymphocytes. Of the three diseases, however, only multiple myeloma and non-Hodgkin's lymphoma have been observed to increase in frequency after irradiation in humans.

Multiple Myeloma

Multiple myeloma has been observed to be increased in frequency by irradiation more consistently than that of any other human lymphoma. A review of the literature by Cuzick (Cu81) showed such an increase in 12 of the 17 irradiated populations analyzed (Table 5-6). In the cohorts tabulated, the pooled excess corresponded to a relative risk of 2.25 (ratio of observed to expected, 50/22.21), with the largest excesses occurring in those exposed to internal-emitters (14 observed versus 3.24 expected cases); however, a deficit of multiple myeloma was reported (3 observed versus 10.17 expected

TABLE 5-6 Occurrence of Multiple Myeloma, as Compared with Leukemia, in Some Irradiated Human Populations

	No. with Multiple Myeloma[a]		No. with Leukemia		
Cohort	Observed	Expected	Observed	Expected	Reference
Atomic-bomb survivors, exposure ≥100 rad (T65D kerma)	5	1.59	58	8.24	Ic79
Spondylitic patients:					
United Kingdom	3	1.87	52[b]	5.48	Co65
West Germany	0	0.25	2	0.84	Sp58
Fluoroscopy of chest	0	0.39	2	1.20	Bo79
U.S. radiologists	11	7.91	12	4.02	Le63
British radiologists	0	1.04	4	0.65	Sm81
Windscale workers	4	1.00			Do76
Hanford workers	3	1.10	1	0.70	Gi79
Radium-dial painters	6	0.86	3[c]	1.41	Po78
Thorotrast patients					
Denmark	4	0.77	11	<4.70	Fa78
West Germany	2	0.93	17	1.0[d]	Va78
Portugal	1	0.16	12	0.00[d]	Da78
Metropathia hemorrhagica	5	2.09	7	2.69	Sm76
Benign gynecologic disorders, Connecticut	5	1.98	12	9.50	Sm76
Uranium millers and miners	1	0.27	8	8.00	Wa64
TOTAL	50	22.21	201	<48.40	

[a] Myeloma data from Cu81.
[b] In addition, 15 men died of aplastic anemia (aleukemic leukemia?), versus 0.52 expected.
[c] Additional cases may have been misdiagnosed as aplastic anemia before surveillance began.
[d] Number observed in the comparison group.

SOURCE: Miller and Beebe (Mi86). Copyright © 1986 by Elsevier Publishing Co.

cases) in two large cohorts of women receiving radiation for uterine cancer (Bo88, Cu81).

In A-bomb survivors, mortality from multiple myeloma has been observed at doses as low as 0.5-0.99 Gy (Pr87a), and the relative risk at 1 Gy is estimated to approximate 3.29 (1.67-6.31), corresponding to 0.26 excess fatal cases/10^4 PYGy (Sh87). For persons exposed to radiation in both Hiroshima and Nagasaki, the relative risk increased with dose in males and females aged 20-59 at the time of bombing but did not become evident until 20 years after exposure (Ic79). As noted elsewhere (Mi86), the data from A-bomb survivors, as well as from other populations, imply that for multiple myeloma the minimal latent period is appreciably longer, the relative risk smaller, and the age distribution later than for leukemia.

Although mortality from multiple myeloma has been observed to be comparably increased (ratio of observed to expected mortality, 9/4.6 = 1.72) in 14,106 patients who were followed for up to 25+ years after radiation therapy for ankylosing spondylitis (Da87), no excess has been evident in 150,000 women who were followed for more than 15 years after radiation therapy for carcinoma of the uterine cervix (relative risk, 0.26) (Bo88).

In Hanford nuclear plant workers, mortality from multiple myeloma was observed to be elevated in the 1970s (Gi79) and has been found to remain elevated in a more recent, expanded analysis of the same population (To83). A similar excess has since been reported in workers at two other nuclear installations (Be85, Sm86). No excess, however, has been evident in an early cohort of 27,011 Chinese x-ray workers, in whom the ratio of observed to expected cases is 0/0.5 (relative risk, 0) (Wa88b).

Malignant Lymphoma

For Hodgkin's disease, the data are reasonably consistent in showing no excess in irradiated populations. For other lymphomas, however, the data are inconsistent.

As concerns non-Hodgkin's lymphoma, mortality from this disease has not been increased in A-bomb survivors (Sh87), notwithstanding a previous suggestion to the contrary (An64). Patients treated with radiation for ankylosing spondylitis, however, continue to show increased mortality from the disease (ratio of observed to expected mortality 16/7.14 = 2.24) (Da87). An excess of the disease has also been observed in women who were treated with radiation for benign gynecologic disorders (Wa84) and in women who were treated with radiation for carcinoma of the uterine cervix (relative risk, 2.51; 90% confidence interval, 0.8-7.6) (Bo88).

Mortality from lymphosarcoma has been observed to be increased in American radiologists who entered practice in the 1920s and 1930s, when the average occupational radiation levels were higher than they are today. Although such early cohorts showed an increased standardized mortality ratio (2.73) for lymphosarcoma, no excess of this disease or of other lymphomas has been evident in American radiologists of more recent cohorts (Ma81b) or in pioneer Chinese x-ray workers (Wa88b).

In laboratory animals, a variety of lymphoid neoplasms can be induced by irradiation (UN77). The best studied of these growths is the thymic lymphoma of the mouse, which, as discussed above (see the Section on parathyroid glands), often terminates as a lymphatic leukemia (Yo86). The dose-incidence curves for experimentally induced lymphomas vary markedly with the lymphoma in question, as well as with species, sex, age at exposure, conditions of irradiation, and other variables (UN86). Paradoxically, the incidence of one such neoplasm, a reticulum cell sarcoma of the mouse, typically decreases with increasing dose of whole-body radiation (UN77, UN86).

Summary

The incidence of multiple myeloma has been observed to be elevated after widespread irradiation of the bone marrow in the majority of populations studied to date. In A-bomb survivors, although the excess did not become detectable until 20 years after irradiation, it is now evident at doses as low as 0.05-0.99 Gy and corresponds to a relative risk of 3.29/Gy or to 0.26 fatal cases/10^4 PYGy. No other form of lymphoma has been consistently observed to be increased in frequency in irradiated human populations.

PHARYNX, HYPOPHARYNX, AND LARYNX

The review of radiation-induced cancers of the pharynx, hypopharynx and larynx by the BEIR III Committee (NRC80) was based primarily on several small studies of the late effects of therapeutic irradiation of adjacent tissues, such as the esophagus, larynx, thyroid and spine. In the cases of cancer reported, the mean latent periods ranged from 23 to 27 years, and the radiation doses involved were high (fractionated doses of 3,000-6,000 rad delivered over 3 to 6 weeks).

In patients with ankylosing spondylitis treated with radiation and observed through January 1, 1970, 3 deaths from cancer of the larynx were observed versus 1.29 expected (ratio of observed to expected cancer deaths, 2.33) and 3 deaths from cancer of the larynx were observed versus 2.25 expected (ratio of observed to expected cancer deaths, 1.33) (Sm82). Neither the excess for the pharynx nor that for the larynx was significant at $p < 0.05$. Similarly, Darby and colleagues later found no significant excess in deaths from cancer of the pharynx or larynx in ankylosing spondylitis patients and in Japanese A-bomb survivors (Da85).

In an update of the Japanese A-bomb survivor data for the period 1950-1985, Shimizu et al. likewise found no excess mortality from cancer of the pharynx, hypopharynx, or larynx, reporting a total of 23 cancers of the pharynx and 46 cancers of the larynx (Sh88).

Summary

Although cancers of the pharynx and larynx have been observed to arise as a late complication of therapeutic irradiation, after doses in the range of 30-60 Gy, no significant excess of such cancers has been found in the Japanese A-bomb survivors or other populations exposed to doses in the range below 1 Gy. The sensitivity of the pharynx, hypopharynx, and larynx to radiation carcinogenesis thus appears to be relatively low.

SALIVARY GLANDS

The incidence of salivary gland tumors has been observed to be increased in patients treated with irradiation for diseases of the head and neck, in Japanese A-bomb survivors, and in persons exposed to diagnostic x radiation.

The therapeutically irradiated populations fall primarily into three groups: (1) those treated with x rays to the head and neck during childhood or infancy, in whom the dose to the salivary glands has usually exceeded 1 Gy (Sa60, Ja71, He75, Sc78, Ma81); (2) those treated with x rays to the scalp for tinea capitis in childhood, in whom the dose to the salivary gland is estimated to have averaged about 0.4 Gy (Mo74, Sh76); and (3) women treated with iodine-131 during later middle age, in whom the dose to the thyroid gland is estimated to have averaged about 5.3 Gy (Ho82, La86). The data from the three types of studies are remarkably consistent in yielding an average excess of 0.26 ± 0.06 malignant tumors/10^4 PYGy, or an average increase in relative risk of 6.9 ($\pm$5.5)%/rad, excluding the first 5 years after irradiation (chi-square of 12.7 on 12 degrees of freedom; $p = 0.39$). For benign tumors, an average excess of $0.44 \pm 0.11/10^4$ PYGy, or 3.6 ($\pm$2.1)%/rad (chi-square of 12.9 on 10 degrees of freedom; $p = 0.23$) (La86).

In Japanese A-bomb survivors, although mortality from salivary gland tumors has not been detectably affected, the incidence of such tumors has shown a dose-dependent increase (Table 5-7). The increase is smaller than that in radiotherapy patients, however, possibly because of differences in ascertainment or case reporting. No marked variation of susceptibility with age at the time of irradiation has been evident in the A-bomb survivors.

Persons exposed to diagnostic x radiation of the head and neck also have been reported to show an increase in the risk of cancer of the parotid gland, the risk being highest in those receiving full-mouth or panoramic dental radiography or some other type of major diagnostic examination of the head before the age of 20 (Pr88b).

In laboratory animals, irradiation has been observed to induce cancer of the salivary gland infrequently (Gl62, Ta75), indicating that the susceptibility of the salivary gland to radiation carcinogenesis is relatively low in comparison with that of other organs.

TABLE 5-7 Incidence of Salivary Gland Tumors Among A-Bomb Survivors, Hiroshima, 1953–1971, Open City Population

	Exposure Distance (m)				Trend Test, 0–5,000 m (*P* value)	No. of Cases/10^6 P rad	Estimated Risk (%/rad)[a]
	<1,500	1,501–2,000	2,001–5,000	>5,000			
Person years	322,768	366,887	800,486	4,333,814			
Average kerma (rad)[a]	124.3	9.9	0	0			
No. malignant tumors	7	4	6	5	0.030	0.056 ± 0.036	0.69 ± 0.57
No. benign tumors	7	2	5	25	0.010	0.063 ± 0.035	1.10 ± 0.86
Total no. tumors	14	6	11	30	0.002	0.120 ± 0.050	0.86 ± 0.49

[a]Based on T65D dose estimates (see La86).

SOURCE: From La86, based on data from Ta76 and Oh78.

Summary

The incidence of salivary gland tumors has been observed to be increased by irradiation in A-bomb survivors, patients treated with x rays to the head and neck in childhood, and women treated with iodine-131 in middle age. The excess relative risk of salivary gland cancer averages 550% per Gy, or 0.26 cases/10^4 PYGy. In patients treated for tinea capitis, the excess was evident at an estimated average dose of only 0.4 Gy, indicating that the susceptibility of the salivary gland to radiation carcinogenesis is relatively high.

PANCREAS

Cancer of the pancreas is the fourth leading type of fatal cancer in the United States (Yo81), although it is difficult to diagnose clinically and is verified histologically in only a small percentage of cases (Ma82). Excess mortality from the disease has been observed inconsistently in irradiated human populations and has borne no clear relationship to dose or time after irradiation.

One of the first populations in which an excess of the disease was observed is the well-studied series of 14,106 patients who were treated with radiation to the spine for ankylosing spondylitis, in whom 27 deaths from the disease have been reported versus 22.39 expected (Da87); however, the relative risk in this population was increased significantly only within the first 5 years after treatment (6 observed versus 1.85 expected deaths). Because cancer of the pancreas frequently causes pain in the back and is thus prone to be confused with ankylosing spondylitis, it is conceivable that the disease was present before irradiation in some of the observed cases (Da87).

Other therapeutically irradiated patients in whom an excess has been reported include a series of men and women treated for lymphoma (Jo76) and a series of 82,616 women treated for cervical cancer (Bo85). In the latter—as in the patients with ankylosing spondylitis—the relative risk was increased maximally soon (1-4 years) after irradiation and not consistently thereafter. Furthermore, a comparable excess (ratio of observed to expected cases = 34/25 = 1.4) was observed in a companion series of women with in situ carcinoma of the cervix who received no therapeutic radiation (Bo85). A case control analysis of these data also yielded a null result (Bo88).

In Japanese A-bomb survivors, no dose- or time-dependent excess in mortality from cancer of the pancreas has been observed; the relative risk at 1 Gy (T65DR shielded kerma) is estimated to approximate 0.9974 $\pm$ 0.1069 (Pr87a). Although data from the Nagasaki Tumor Registry for 1959-1978 suggested an increase with dose (P value for trend test, 0.0740), corresponding to an excess of 1.15 ± 0.66 cases/10^4 PYGy (Wa83, La86), no dose-dependent excess was evident in the concurrent data (then incomplete) from the Hiroshima Tumor Registry (Be78, La86).

Among occupationally exposed persons, an excess in the number of deaths from the disease was reported among British radiologists who entered the practice of radiology before 1921 (6 deaths versus 1.9 expected through 1976) but was not evident in later cohorts (Sm81) nor in U.S. radiologists who entered practice after 1920 (Ma75). Among radiation workers at the Hanford Plant, a dose-related excess number of deaths from pancreatic cancer was reported a number of years ago (Ma77, Ma78b, Gi79), but the excess has not been confirmed by more recent follow-up (To83).

Summary

An association between cancer of the pancreas and previous irradiation, suggested by several reports in the past, has not been confirmed in more recent and thorough studies of irradiated human populations. The pancreas appears, therefore, to be relatively insensitive to radiation carcinogenesis.

REFERENCES

Al85 Albo, V., D. Miller, S. Leiken, N. Satlok, and D. Hammond. 1985. Nine brain tumors (BT) as a late effect in children "cured" of acute lymphoblastic leukemia (ALL) from a single protocol study. Proc. Am. Soc. Clin. Oncol. 4:172.

Al86 Albert, R. E., and R. E. Shore. 1986. Carcinogenic effects of radiation on the human skin. Pp. 335-345 in Radiation Carcinogenesis, A. C. Upton, R. E. Albert, F. J. Burns, and R. E. Shore, eds. New York: Elsevier.

An64 Anderson, R. E., and K. Ishida. 1964. Malignant lymphoma in survivors of the atomic bomb in Hiroshima. Am. Inst. Med. 61:853-862.

As82 Asano, M., K. Yoshimoto, S. Seyama, H. Itakura, T. Hamada, and S. Iijuma. 1982. Primary liver carcinoma and liver cirrhosis in atomic bomb survivors, Hiroshima and Nagasaki, 1961-75 with special reference to hepatitis B surface antigen. J. Natl. Cancer Inst. 69:1221-1227.

Ba78 Barendsen, G. W. 1978. Fundamental aspects of cancer induction in relation to the effectiveness of small doses of radiation. Pp. 263-276 in Late Biological Effects of Ionizing Radiation, Vol. II. Vienna: International Atomic Energy Agency.

Ba86 Bair, W. J. 1986. Experimental carcinogenesis in the respiratory tract. Pp. 151-167 in Radiation Carcinogenesis, A. C. Upton, R. E. Albert, F. J. Burns, and R. E. Shore, eds. New York: Elsevier.

Be67 Berdjis, C. C. 1967. Pathogenesis of radiation-induced endocrine tumors. Oncology 21:49-60.

Be78 Beebe, G. W., H. Kato, and C. E. Land. 1978. Studies of the mortality of A-bomb survivors. 6. Mortality and radiation dose, 1950-74. Radiat. Res. 75:138-201.

Be79 Benjamin, S. A., B. B. Boecker, R. G. Cuddihy, and R. O. McClellan. 1979. Nasal carcinomas in beagles after inhalation of relatively soluble forms of beta-emitting radionuclides. J. Natl. Cancer Inst. 63:133-139.

Be85 Beral, V., H. Inskip, P. Fraser, et al. 1985. Mortality of employees of the United Kingdom Atomic Energy Authority, 1946-79. Br. Med. J. 29:440.

Be72 Berdjis, C. C. 1972. Parathyroid diseases and irradiation. Strahlentherapie 143:48-62.

Bi75 Bithell, J. F., and A. M. Stewart. 1975. Pre-natal irradiation and childhood malignancy: A review of British data from the Oxford survey. Br. J. Cancer 31:271-287.

Bl80 Black, W. C., L. S. Gomez, J. M. Yuhas, and M. M. Kligerman. 1980. Quantitation of the late effects of x-radiation on the large intestine. Cancer 45:444-451.

Bl84 Blot, W. J., S. Akiba, and H. Katon. Ionizing radiation and lung cancer: A review including preliminary results from a case-control study among a-bomb survivors. In Atomic Bomb Survivor Data: Utilization and Analysis, R. L. Prentice, and D. J. Thompson, eds. Philadelphia: Society for Industrial and Applied Mathematics.

Bo52 Bond, V. P., M. N. Swift, C. A. Tobias, and G. Brecher. 1952. Bowel lesions following single deuteron irradiation. Fed. Proc. 11:408-409.

Bo60 Bond, V. P., E. P. Cronkite, S. W. Lippincott, and C. J. Shellabarger. 1960. Studies on radiation-induced mammary gland neoplasms in the rat. III. Relation of the neoplastic response to dose of total body radiation. Radiat. Res. 12:276-285.

Bo79 Boice, J. D., Jr. 1979. Multiple chest fluoroscopies and the risk of breast cancer. Pp. 147-156 in Advances in Medical Oncology Research and Education, Vol. 1. Oxford: Pergamon.

Bo84 Boice, J., and J. Fraumeni, Jr., eds. 1984. Radiation Carcinogenesis: Epidemiology and Biological Significance. New York: Raven Press.

Bo85 Boice, J. D., Jr., N. E. Day, A. Andersen, L. A. Brinton, R. Brown, N. W. Choi, E. A. Clarke, M. P. Coleman, R. E. Curtis, J. T. Flannery, M. Hakama, T. Hakulinen, G. R. Howe, O. M. Jensen, R. A. Kleinerman, D. Magnin, K. Magnus, K. Makela, B. Malker, A. B. Miller, N. Nelson, C. C. Patternson, F. Petterssen, V. Pompe-Kirn, M. Primic-Zakelj, P. Prior, R. Ravnihar, R. G. Skeet, J. E. Skjerven, P. G. Smith, M. Sok, R. F. Spengler, H. H. Storm, M. Stovall, G. H. O. Thomkins, and C. Wall. 1985. Second Cancers Following Radiation Treatment for Cervical Cancer. An international collaboration among cancer registries. J. Natl. Cancer Inst. 74:955-975.

Bo86 Boice, J. D., Jr., and R. H. Kleinerman. 1986. Meeting highlights: Radiation studies of women treated for benign gynecologic disease. J. Natl. Cancer Inst. 76:549-551.

Bo86a Boecker, B. B., F. F. Hahn, R. G. Cuddihy, M. B. Snipes, and R. O. McClellan. 1986. Is the human nasal cavity at risk from inhaled radionuclides? Pp. 564-576 in Life-Span Radiation Effects Studies in Animals: What Can They Tell Us? U.S. Department of Energy Report CONF 830951. Springfield, Va.: National Technical Information Service.

Bo87 Boice, J. D., Jr., M. Blettner, R. A. Kleinerman, M. Stovall, W. C. Moloney, G. Engholm, D. F. Austin, A. Bosch, D. L. Cookfair, E. T. Krementz, H. B. Latourette, L. J. Peters, M. D. Schulz, M. Lundell, F. Pattersson, H. H. Storm, C. M. J. Bell, M. P. Coleman, P. Fraser, M. Palmer, P. Prior, N. W. Chol, T. G. Hislop, M. Koch, D. Robb, D. Robson, R. F. Sprengler, D. von Fournler,

R. Frischkorn, H. Lochmuller, V. Pompe-Kirn, A. Rimpels, K. Kjorstad, M. H. Pejovic, K. Sigurdsson, P. Pisani, H. Kucera, and G. B. Hutchison. 1987. Radiation Dose and Leukemia Risk in Patients Treated for Cancer of the Cervix. JNCI 79:1295-1311

Bo88 Boice, J. D., Jr., G. Enghohm, R. A. Kleinerman, M. Blettner, M. Stovall, H. Lisco, W. C. Moloney, D. F. Austin, A. Bosch, D. L. Cookfair, E. T. Krementz, H. B. Latourette, J. A. Merrill, L. J. Peters, M. D. Schulz, H. H. Storm, E. Bjorkholm, F. Pettersson, C. M. J. Bell, M. P. Coleman, P. Fraser, F. E. Neal, P. Prior, N. W. Choi, T. G. Hislop, M. Koch, N. Kreiger, D. Robb, D. Tobson, D. H. Thomson, H. Lochmuller, D. V. Fournier, R. Frischkorn, K. E. Kjorstad, A. Rimpela, M. H. Pejovic, V. P. Kirn, H. Stankusova, F. Berrino, K. Soigurdsson, G. B. Hutchison, and B. MacMahon. 1988. Radiation dose and second cancer risk in patients treated for cancer of the cervix. Radiat. Res. 116:3-55.

Bo88b Boecker, B. B., F. F. Hahn, B. A. Muggenburg, R. A. Guilmette, W. C. Griffith, and R. O. McClellan. 1988. The relative effectiveness of inhaled alpha- and beta-emitting radionuclides in producing lung cancer. Pp. 1059-1062 in Radiation Protection Practice. Sydney: Pergamon Press.

Br53 Brecher, G., E. P. Cronkite, and J. H. Peers. 1953. Neoplasms in rats protected against lethal doses of irradiation by parabiosis or para aminopropriophenone. J. Natl. Cancer Inst. 14:159-175.

Br69 Brinkley, D., and J. L. Haybittle. 1969. The late effects of artificial menopause by x-radiation. Br. J. Radiol. 42:519-521.

Br81 Broerse, J.J., C. F. Hollander, and M. J. Van Zwieten. 1981. Tumor induction in Rhesus monkeys after total body irradiation with X-rays and fission neutrons. Int. J. Radiat. Biol. 40:671-676.

Br85 Broerse, J. J., L. A. Hermen, and M. J. van Zwieten. 1985. Radiation carcinogenesis in experimental animals and its implications for radiation protection. Int. J. Radiat. Biol. 48:167-187.

Br36 Brown, P. 1936. American Martyrs to Science through the Roentgen Ray. Springfield, Ill.: Charles C Thomas.

Bu80 Burns, F. J., and R. E. Albert. 1980. Dose-response for rat skin tumors induced by single and split doses of argon ions. In Biological and Medical Research with Accelerated Heavy Ions at the Bevalac. Berkeley: University of California.

Bu86 Burns, F. J., and R. E. Albert. 1986. Radiation carcinogenesis in rat skin. Pp. 199-214 in Radiation Carcinogenesis, A. C. Upton, R. E. Albert, F. J. Burns, and R. E. Shore, eds. New York: Elsevier.

Bu76 Burns, F. J., I. P. Sinclair, R. E. Albert, and M. Vanderlaan. 1976. Tumor induction and hair follicle damage for different electron penetrations in rat skin. Radiat. Res. 67:474-481.

By85 Byers, T., S. Graham, T. Rzepka, and J. Marshall. 1985. Lactation and breast cancer: Evidence for a negative association in premenopausal women. Am. J. Epidemiol. 121:664-674.

Ca48 Cade, S. 1948. Malignant Disease and Its Treatment by Radium, Vol. 1. Baltimore: Williams & Wilkins.

Ch78 Christensson, T. 1978. Hyperparathyroidism and radiation therapy. Ann. Intern. Med. 89:216-217.

Cl74 Clayman, C. B., W. H. Kruskal, J. W. J. Carpenter, and W. L. Palmer. 1974. The neoplastic potential of gastric irradiation. In Gastric Irradiation in Peptic Ulcer, W. L. Palmer, ed.

Cl59 Clifton, K. H. 1959. Problems in experimental tumorigenesis of the pituitary gland, gonads, adrenal cortices, and mammary glands: a review. Cancer Res. 19:2-22.

Cl75 Clifton, K. H., and B. N. Sridharan. 1975. Endocrine factors and tumor growth. Pp. 249-285 in Cancer: A Comprehensive Treatise, Vol. 3. F. F. Becker, ed. New York: Plenum.

Cl77 Clifton, K. H. 1977. The physiology of endocrine therapy. Pp. 573-597 in Cancer: A Comprehensive Treatise, Vol. 5. New York: Plenum Press.

Cl78 Clifton, K. H., and J. Crowley. 1978. Effects of radiation type and role of glucocorticoids, gonadectomy and thyroidectomy in mammary tumor induction in MtT-grafted rats. Cancer Res. 38:1507-1513.

Cl79 Clifton, K. H. 1979. Animal models of breast cancer. Pp. 1-20 in Endocrinology of Cancer, D. P. Rose, ed. Boca Raton, Fla.: CRC Press.

Cl85a Clifton, K. H., and M. N. Gould. 1985. Clonogen transplantation assay of mammary and thyroid epithelial cells. Pp. 128-138 in Cell Clones: Manual of Mammalian Cell Techniques, C. S. Potten and J. H. Hendry, eds. Edinburgh: Churchill Livingstone.

Cl85b Clifton, K. H., K. Kamiya, R. T. Mulcahy, and M. N. Gould. 1985. Radiogenic neoplasia in the thyroid and mammary clonogens: Progress, problems and possibilities. Pp. 329-342 in Assessment of Risk From Low-Level Exposure to Radiation and Chemicals: A Critical Overview. A. D. Woodhead, C. J. Shellabarger, V. Pond, and A. Hollaender, eds. New York: Plenum.

Cl85c Clifton, K. H., J. Yasukawa-Barnes, M. A. Tanner, and R. V. Haning, Jr. 1985. Irradiation and prolactin effects on rat mammary carcinogenesis: Intrasplenic pituitary and estrone capsule implants. J. Natl. Cancer Inst. 75:167-175.

Cl86a Clifton, K. H., M. A. Tanner, and M. N. Gould. 1986. Assessment of radiogenic cancer initiation frequency per clonogenic rat mammary cell *in vivo*. Cancer Res. 46:2390-2395.

Cl86b Clifton, K. H. 1986. Thyroid cancer: Reevaluation of an experimental model for radiogenic carcinogenesis. Pp. 181-198 in Radiation Carcinogenesis, A. C. Upton, R. E. Albert, F. J. Burns, and R. E. Shore, eds. New York: Elsevier-North-Holland.

Cl86c Clifton, K. H. 1986. Thyroid and mammary radiobiology: Radiogenic damage to glandular tissue. Br. J. Cancer 53(Suppl. VII):237-250.

Co85 Coggle, J. E., D. M. Peel, and J. D. Tarling. 1985. Lung tumor induction in mice after uniform and nonuniform external thoracic x-irradiation. Int. J. Radiat. Biol. 48:95-106.

Co76 Colman, M., L. R. Simpson, L. K. Patterson, and L. Cohen. 1976. Thyroid cancer associated with radiation exposure. Pp. 285-288 in Biological and Environmental Effects of Low Level Radiation, Vol. II. Vienna: International Atomic Energy Agency.

Co78 Colman, M., M. Kirsch, and M. Creditor. Radiation induced tumors. Pp. 167-180 in Late Biological Effects of Ionizing Radiation, Vol. 1. Vienna: International Atomic Energy Agency.

Co80 Conard, R. A., et al. 1980. Review of Medical Findings in a Marshallese Population Twenty-six Years after Accidental Exposure to Radioactive Fallout. Report BNL 5126 (Biology and Medical TID-4500). Upton, N.Y.: Brookhaven National Laboratory.

Co84 Conard, R. A. 1984. Late radiation effects in Marshall Islanders exposed to fallout 28 years ago. Pp. 57-71 in Radiation Carcinogenesis: Epidemiology and Biological Significance, J. D. Boice, Jr. and J. F. Fraumeni, Jr., eds. New York: Raven.

Co74 Coop, K. L., J. G. Sharp, J. W. Osborne, and G. R. Zimmerman. 1974. An animal model for the study of small-bowel tumors. Cancer Res. 34:1487-1494.

Co58 Court, W. M., and R. Doll. 1958. Expectation of life and mortality from cancer among British radiologists. Br. Med. J. 181-187.

Co65 Court-Brown, W. M., and R. Doll. 1965. Mortality from cancer and other causes after radiotherapy for ankylosing spondylitis. Br. Med. J. 2:1327-1332.

Co74a Cox, D. R., and D. V. Hinkley. 1974. Theoretical Statistics. London: Chapman and Hall.

Cu81 Cuzick, J. 1981. Radiation-induced myelomatosis. N. Engl. J. Med. 304:204-210.

Cu84 Curtis, R., B. Hankey, M. Myers, et al. 1984. Risk of leukemia associated with a first course of cancer treatment: An analysis of Surveillance, Epidemiology and End Results Program experience. J. Natl. Cancer Inst. 72:531-544.

Da85 Darby, S. C., E. Nakashima, and H. Kato. 1985. A parallel analysis of cancer mortality among atomic bomb survivors and patients with ankylosing spondylitis. J. Natl. Cancer Inst. 75:1-21.

Da87 Darby, S. C., R. Doll, S. K. Gill, and P. G. Smith. 1987. Long-term mortality after a single treatment course with X-rays in patients treated for ankylosing spondylitis. Br. J. Cancer 55:179-190.

Da78 da Silva Horta, J., M. E. da Silva Horta, L. C. da Motta, and M. H. Tavares. 1978. Malignancies in Portuguese Thorotrast Patients. Health Phys. 35:137-151.

De65 DeGroot, L. J. 1965. Current views of formation of thyroid hormones. N. Engl. J. Med. 272:243-250, 297-303, 355-362.

De78 Denman, D. L., F. R. Kirchner, and J. W. Osborne. 1978. Induction of colonic adenocarcinoma in the rat by x-irradiation. Cancer Res. 38:1899-1905.

Di73 Diamond, E. L., H. Schmerler, and A. M. Lilienfeld. 1973. The relationship of intrauterine radiation to subsequent mortality and development of leukemia in children. Am. J. Epidemiol. 97:283-313.

Di69 Dickson, R. J. 1969. The late results of radium treatment for benign uterine hemorrhage. Br. J. Radiol. 42:582-594.

Do74 Dobyns, B. M., G. E. Sheline, J. B. Workman, E. A. Tompkins, W. M. McConahey, and D. V. Becker. 1974. Malignant and benign neoplasms of the thyroid in patients treated for hyperthyroidism: A report of the cooperative thyrotoxicosis therapy follow-up study. J. Clin. Endocrinol. 38:976-998.

Do76 Dolphin, G. W. 1976. A comparison of the observed and the expected cancers of the haematopoietic and lymphatic systems among workers at Windscale. London: Her Majesty's Stationery Office.

Do63 Doniach, I. 1963. Effects including carcinogenesis of ^{131}I and x-rays on the thyroid of experimental animals: A review. Health Phys. 9:1357-1362.

Do67 Doniach, I. 1967. Damaging effect of x-irradiation of less than 1000 rads on goitrogenic capacity of rat thyroid gland. In Thyroid Neoplasia, S. Young and D. R. Inman, eds. New York: Academic Press.

Do77 Doniach, I. 1977. Pathology of irradiation thyroid damage. In Radiation-Associated Thyroid Carcinoma, L. J. DeGroot, L. A. Grohman, E. L. Kaplan, and S. Refetoff, eds. New York: Grune and Stratton.

Du50 Duffy, B. J., Jr., and P. J. Fitzgerald. 1950. Cancer of the thyroid in children: A report of 28 cases. J. Clin. Endocrinol. 10:1296-1308.

Du80 Dumont, J. E., J. F. Malone, and A. J. Van Herle. 1980. Irradiation and Thyroid Disease: Dosimetric, Clinical and Carcinogenic Aspects. Report EUR 6713ER. Luxembourg: Commission of the European Communities.

Ev86 Evans, J. S., J. E. Wennberg, and B. J. McNeil. 1986. The influence of diagnostic radiography on the incidence of breast cancer and leukemia. N. Engl. J. Med. 315:800-815.

Fa78 Faber, M. 1978. Malignancies in Danish Thorotrast patients. Health Phys. 35:153-158.

Fa64 Fabrikant, J. I., R. J. Dickson, and B. F. Fetter. 1964. Mechanism of radiation carcinogenesis at the clinical level. Br. J. Cancer 18:459-477.

Fe87 Feola, J. M., Y. Maruyama, A. Pattarasumunt, and R. M. Kryscio. 1987. Cf-252 leukemogenesis in the C57BL mouse. Inst. J. Radiat. Oncol. Biol. Phys. 13:69-74.

Fr73 Fritz, T. E., W. P. Norris, and D. V. Tolle. 1973. Myelogenous leukemia and related myeloproliferative disorders in beagles continuously exposed to 60Co y-radiation. Pp. 170-188 in Unifying Concepts of Leukemia, R. M. Dutcher and L. Chieco-Bianchi, eds. Bibliography of Haematology, No. 39. Basel: Karger.

Fr86 Fry, R. J. M., J. B. Storer, and F. J. Burns. 1986. Radiation induction of cancer of the skin. Br. J. Radiol. 19(Suppl.):58-60.

Fu36a Furth, J., and J. S. Butterworth. 1936. Neoplastic diseases occurring among mice subjected to general irradiation with x-rays. II. Ovarian tumors and associated lesions. Am. J. Cancer 28:66-95.

Fu36b Furth, J., and O. B. Furth. 1936. Neoplastic diseases produced in mice by general irradiation with x-rays. I. Incidence and types of neoplasms. Am. J. Cancer 28:54-65.

Fu75 Furth, J. 1975. Hormones as etiological agents in neoplasia. In Cancer: A Comprehensive Treatise, Vol. 1. F. F. Becker, ed. New York: Plenum.

Ga63 Garner, R. J. 1963. Comparative early and late effects of single and prolonged exposure to radioiodine in young and adults of various animal species—a review. Health Phys. 9:1333-1339.

Gi72 Gibson, R., S. Graham, A. Lilienfeld, et al. 1972. Irradiation in the epidemiology of leukemia among adults. J. Natl. Cancer Inst. 48:301.

Gi79 Gilbert, E. S., and S. Marks. 1979. An analysis of the mortality of workers in a nuclear facility. Radiat. Res. 79:122-128.

Gi87 Gillett, N. A., B. A. Muggenburg, B. B. Boecker, W. C. Griffith, F. F. Hahn, and R. O. McClellan. 1987. Single inhalation exposure to $^{90}SrCl_2$ in the beagle dog: Late biological effects. J. Natl. Cancer Inst. 79:359-376.

Gi88 Gillett, N. A., B. A. Muggenburg, J. A. Mewhinney, F. F. Hahn, F. A. Seiler, B. B. Boecker, and R. O. McClellan. Primary liver tumors in beagle dogs exposed by inhalation to aerosols of plutonium-238 dioxide. Am. J. Pathol. (submitted).

Gl62 Glucksman, A., and C. P. Cherry. 1962. The induction of adenoma by the irradiation of salivary glands of rats. Radiat. Res. 17:186-202.

Go86a Goldman, M. 1986. Experimental carcinogenesis in the skeleton. Pp. 215-331 in Radiation Carcinogensis, A. C. Upton, R. E. Albert, F. J. Burns, and R. E. Shore, eds. New York: Elsevier.

Go86b Goldman, M., L. S. Rosenblatt, and S. A. Book. 1986. Lifetime radiation effects research in animals: An overview of the status and philosophy of studies at University of California-Davis Laboratory for Energy Related Health Research. Pp. 53-65 in Life-Span Radiation Effects Studies in Animals: What Can They Tell Us?, R. C. Thompson and J. A. Mahaffey, eds. U.S. Department of Energy Report CONF-830951. Springfield, Va.: National Technical Information Service.

Gr65 Gray, L. H. 1965. Radiation biology and cancer. Pp. 18-19 in Cellular Radiation Biology. Baltimore: Williams & Wilkins.

Gr84 Griem, M. L., J. Justman, and L. Weiss. 1984. The neoplastic potential of gastric irradiation. IV. Am. J. Clin. Oncol. 7:675-677.

Gr87 Griffith, W. C., B. B. Boecker, R. C. Cuddihy, R. A. Guilmette, F. F. Hahn, R. O. McClellan, B. A. Muggenburg, and M. B. Snipes. 1987. Preliminary Radiation Risk Estimates of Carcinoma Incidence in the Lung as a Function of Cumulative Radiation Dose Using Proportional Tumor Incidence Rates. Pp. 196-204 in 1986-87 Inhalation Toxicology Research Institute Report, J. D. Sun and J. A. Mewhinney, eds. U.S. Department of Energy Report LMF-120. Springfield, Va.: National Technical Information Service.

Gu64 Gunz, F., and H. Atkinson. 1964. Medical radiation and leukemia: A retrospective survey. Br. Med. J. 1:389.

Ha83a Hahn, F. F., B. B. Boecker, R. G. Cuddihy, C. H. Hobbs, R. O. McCellan, and M. B. Snipes. 1983. Influence of Radiation Dose Patterns on Lung Tumor Incidence in Dogs That Inhaled Beta Emitters: A Preliminary Report. Radiat. Res. 96:505-517.

Ha83b Harley, N., A. B. Kolber, R. E. Shore, R. E. Albert, S. M. Altman, and B. Pasternack. 1983. The skin dose and response for the head and neck in patients irradiated with x-ray for tinea capitis: implications for environmental radioactivity. Pp. 125-142 in Epidemiology Related to Health Physics. Proceedings of the 16th Midyear Topical Meeting of the Health Physics Society. CONF-83011. Springfield, Va.: National Technical Information Service.

Ha87 Hamilton, T. E., G. van Belle, and J. P. Lo Gerfo. 1987. Thyroid neoplasia in Marshall Islanders exposed to nuclear fallout. J. Am. Med. Assoc. 258:629-636.

He75 Hempelmann, L. H., W. J. Hall, M. Phillips, R. A. Cooper, and W. R. Ames. 1975. Neoplasms in persons treated with x-rays in infancy: Fourth survey in 20 years. J. Natl. Cancer Inst. 55:519-530.

He85 Henderson, B. E., L. N. Kolonel, R. Dworsky, D. Kerford, E. Mori, K. Singh, and H. Thevenot. 1985. Cancer incidence in the islands of the Pacific. Natl. Cancer Inst. Monogr. 69:73-81.

He88 Henderson, B.E., Ross, R. and Bernstein, L. 1988. Estrogens as a cause of human cancer Cancer Res. 48:246-253.

He50 Henry, S. A. 1950. Cutaneous cancer in relation to occupation. Ann. R. Coll. Surg. Engl. 7:425.

Hi69 Hirose, F. 1969. Experimental induction of carcinoma in the glandular stomach by localized x-irradiation of gastric region, Pp. 75-113 in Experimental Carcinoma of the Glandular Stomach. Japanese Cancer Association GANN Monograph No. 8. Tokyo.

Hi77 Hirose, F., K. Fukazawa, H. Watanabe, Y. Terada, I. Fujii, and S. Ootuska. 1977. Induction of rectal carcinoma in mice by local x-irradiation. GANN 68:669-680.

Ho83 Hoel, D. G., T. Wakabayashi, and M. C. Pike. 1983. Secular trends in the distributions of the breast cancer risk factors—menarche, first birth, menopause and weight—in Hiroshima and Nagasaki, Japan. Am. J. Epidemiol. 118:78-89.

Ho82 Hoffman, D. A., W. M. McConahey, and L. T. Kurland. 1982. Cancer incidence following treatment for hyperthyroidism. Int. J. Epidemiol. 11:218-224.

Ho84a Hoffman, D. A. Late effects of I-131 therapy in the United States. 1984. Pp. 273-280 in Radiation Carcinogenesis: Epidemiology and Biological Significance, J. D. Boice, Jr., and J. F. Fraumeni, Jr., eds. New York: Raven.

Ho63 Hollingsworth, D. R., H. B. Hamilton, H. Tamagaki, and G. W. Beebe. 1963. Thyroid disease: A study in Hiroshima, Japan. Medicine 42:47-71.

Ho80a Holm, L.-E., I. Dahlqvist, A. Israelsson, and G. Lundell. 1980. Malignant thyroid tumors after iodine-131 therapy. N. Engl. J. Med. 303:188-191.

Ho80b Holm, L.-E., G. Eklund, and G. Lundell. 1980. Incidence of malignant thyroid tumors in humans after exposure to diagnostic doses of iodine-131. II. Estimation of thyroid gland size, thyroid radiation dose, and predicted versus observed number of malignant thyroid tumors. J. Natl. Cancer Inst. 65:1221-1224.

Ho80c Holm, L.-E., G. Lundell, and G. Walinder. 1980. Incidence of malignant thyroid tumors in humans after exposure to diagnostic doses of iodine-131. I. Retrospective cohort study. J. Natl. Cancer Inst. 64:1055-1059.

Ho84b Holm, L.-E. 1984. Malignant disease following iodine-131 therapy in Sweden. Pp. 263-271 in Radiation Carcinogenesis: Epidemiology and Biological Significance. J. D. Boice, Jr., and J. F. Fraumeni, Jr., eds. New York: Raven Press.

Ho88 Holm, L.-E., K. E. Wicklund, G. E. Lundell, J. D. Boice, N. A. Bergman, G. Bjelkengren, E. S. Cederquist, U.-B. C. Ericsson, L.-G. Larsson, M. E. Lidberg, R. S. Lindberg, and H. V. Wicklund. 1988. Thyroid cancer after diagnostic doses of iodine-131: A retrospective study. J. Natl. Cancer Inst. 80:1132-1136.

Ho70 Howard, E. B., and W. J. Clarke. 1970. Strontium-90-induced hematopoietic neoplasms in miniature swine. Pp. 379-401 in Myeloproliferative Disorders of Animals and Man. U.S.A.E.C. Div. Tech. Info., W. J. Clarke, E. B. Howard, and P. L. Hackett, eds.

Hr89 Hrubec, Z., J. Boice, R. Monson, and M. Rosenstein. 1989. Breast cancer after multiple chest fluoroscopies: Second follow-up of Massachusetts women with tuberculosis. Cancer Res. 49:229-234.

Hu63 Huggins, C., and R. Fukunishi. 1963. Cancer in the rat after single exposures to irradiation or hydrocarbons. Radiat. Res. 20:493-503.

Hu87 Humphreys, E. R., J. F. Loutit, and V. A. Stones. 1987. The induction by 239 Pu of myeloid leukemia and outsosarcoma in female CBA mice. Int. J. Radiat. Biol. 51:331-339.

Ic79 Ichimaru, M., T. Ishimaru, M. Mikami, and M. Matsunaga. 1979. Multiple myeloma among atomic bomb survivors, Hiroshima and Nagasaki, 1950-1976. Technical Report 9-79. Hiroshima: Radiation Effects Research Foundation.

ICRP80 International Commission on Radiological Protection (ICRP). 1980. Pp. 1-108 in Biological Effects of Inhaled Radionuclides. ICRP Publication 31. Oxford: Pergamon.

ICRP87 International Commission on Radiological Protection (ICRP). 1987. Pp. 1-60 in Lung Cancer Risk from Indoor Exposures to Radon Daughters. ICRP Publication 50. Oxford: Pergamon.

Ir85 Ireland, J. P., S. J. Fleming, D. A. Levison, W. R. Cattell, and L. R. I. Baker. 1985. Parathyroid carcinoma associated with chronic renal failure and previous radiotherapy to the neck. J. Clin. Pathol. 38:1114-1118.

Ja70 Jablon, S., and H. Kato. 1970. Childhood cancer in relation to prenatal exposure to atomic-bomb radiation. Lancet ii:1000-1003.

Ja71 Janower, M. L., and O. S. Miettinen. Neoplasms after childhood irradiation of the thymus gland. J. Am. Med. Assoc. 215:753-756.

Ka82 Kato, H., and W. J. Schull. 1982. Studies of the mortality of a-bomb survivors. 7. Mortality 1950-1978: part 1. Cancer mortality. Radiat. Res. 90: 395-432.

Ka83 Katz, A., and G. D. Braunstein. 1983. Clinical, biochemical, and pathologic features of radiation-associated hyperparathyroidism. Arch. Intern. Med. 143:79-82.

Ka85 Kamiya, K., A. Inoh, Y. Fujii, K. Kanda, T. Kobayashi, and K. Yokoro. 1985. High mammary carcinogenicity of neutron irradiation in rats and its promotion by prolactin. Jpn. J. Cancer Res. 76:449-456.

Ke78 Kennedy, A. R., and J. B. Little. 1978. Radiation carcinogenesis in the respiratory tract. Pp. 189-261 in Pathogenesis and Therapy of Lung Cancer, C. C. Harris, ed. A Monograph in the series "Lung Biology in Health and Disease." C. Lenfant, ed. New York: Marcel Dekker.

Ki78 Kim, J. H., F. C. Chu, H.Q. Woodward, M. R. Melamed, A. Huvos, and J. Cantin. 1978. Radiation-Induced Soft-Tissue and Bone Sarcoma.

Kl87 Kleinberg, D. L. 1987. Prolactin and breast cancer (editorial). N. Engl. J. Med. 316:269-271.

Kl82 Kleinerman, R. A., R. E., Curtis, J. D. Boice, Jr., J. T. Flannery, and J. F. Fraumeni, Jr. 1982. Second cancers following radiotherapy for cervical cancer. J. Natl. Cancer Inst. 69:1027-1033.

Kn82 Knowles, J. F. 1982. Radiation-induced nervous system tumours in the rat. Int. J. Radiat. Biol. 41:79-84.

Ko86 Kopecky, K. J., E. Nakashima, T. Yamamoto, and H. Kato. 1986. Lung Cancer, Radiation and Smoking Among A-Bomb Survivors. RERF TR 13-86.

Ku87 Kumar, P. R., R. R. Good, F. M. Skultety, L. G. Liebrock, and G. S. Severson. 1987. Radiation-induced neoplasms of the brain. Cancer 59:1274-1282.

La87 Laird, N. M. 1987. Thyroid cancer risk from exposure to ionizing radiation: A case study in the comparative potency model. Risk Anal. 7:299-309.

La89 Lafuma, J., D. Chmelevsky, J. Chameaud, M. Morin, R. Masse, and A. M. Kellerer. 1989. Lung carcinomas in Sprague-Dawley rats after exposure to low doses of radon daughters, fission neutrons, or gamma rays. Radiat. Res. 118:230-245.

La80 Land, C. E., J. D. Boice, Jr., R. E. Shore, J. E. Norman, and M. Tokunaga. 1980. Breast cancer risk from low-dose exposures to ionizing radiation: Results of parallel analysis of three exposed populations of women. J. Natl. Cancer Inst. 65:353-365.

La86 Land, C. E. 1986. Carcinogenic effects of radiation on the human digestive tract and other organs. Pp. 347-378 in Radiation Carcinogenesis, A. C. Upton, R. E. Albert, F. J. Burns, and R. E. Shore, eds. New York: Elsevier.

Le63 Lewis, E. B. 1963. Leukemia, multiple myeloma, and aplastic anemia in American radiologists. Science 142:1492-1494.

Le73 Lebedeva, G. A. 1973. Intestinal polyps arising under the influence of various kinds of ionizing radiations. Vop. Onkol. 19:47-51.

Le79 Lee, W., B. Schlein, N. C. Telles, and R. P. Chiacchierini. 1979. An accurate method of ^{131}I dosimetry in the rat thyroid. Radiat. Res. 79:55-62.

Le82 Lee, W., R. P. Chiacchierini, B. Shlein, and N. C. Telles. 1982. Thyroid tumors following I-131 or localized x-irradiation to the thyroid and the pituitary glands in rats. Radiat. Res. 92:307-319.

Le88 Lewis, C. A., P. G. Smith, I. M. Stratton, S. C. Darby, and R. Doll. 1988. Estimated Radiation Doses to Different Organs Among Patient Treated for Ankylosing Spondylitis with a Single Course of X-rays. Br. J. Radiol. 61:212-220.

Li82 Lightdale, C. J., T. D. Koepsell, and P. Sherlock. 1982. Small intestine. In Cancer Epidemiology and Prevention, D. Schottenfeld and J. F. Fraumeni, Jr., eds. Philadelphia: W. B. Saunders.

Li63 Lindsay, S., and I. L. Chaikoff. 1963. The effects of irradiation on the thyroid gland with particular reference to the induction of thyroid neoplasms: A review. Cancer Res. 24:1099-1107.

Li80 Linos, A., J. Gray, and A. Orvis. 1980. Low dose radiation and leukemia. N. Engl. J. Med. 302:1101.

Li47 Lisco, H., M. P. Finkel, and A. M. Brues. 1947. Carcinogenic properties of radioactive fission products and of plutonium. Radiology 49:61-63.

Lu87a Lundgren, D. L., F. F. Hahn, W. C. Griffin, R. G. Cuddihy, P. J. Haley, and B. B. Boecker. 1987. Effects of relatively low-level exposure of rats to inhaled $^{144}CeO_2$. III. Pp. 308-312 in 1986-87 Inhalation Toxicology Research Institute Annual Report, J. D. Sun, and J. A. Mewhinney, eds. U.S. Department of Energy Report LMF-120. Springfield, Va.: National Technical Information Service.

Lu87b Ludgren, D. L., F. F. Hahn, W. C. Griffin, R. G. Cuddihy, F. A. Seiler and B. B. Boecker. 1987. Effects of relatively low-level thoracic or whole-body exposure of rats to x-rays. I. Pp. 313-317 in 1986-87 Inhalation Toxicology Research Institute Annual Report, J. D. Sun, and J. A. Mewhinney, eds. U.S. Department of Energy Report LMF-120. Springfield, Va.: National Information Service.

Ma82 Mack, T. M. 1982. Pancreas. In Cancer Epidemiology and Prevention, D. Schottenfeld and J. F. Fraumeni, Jr., eds. Philadelphia: W. B. Saunders.

Ma65 Mackenzie, I. 1965. Breast cancer following multiple fluoroscopies. Br. J. Cancer 19:1-9.

Ma62 MacMahon, B. 1962. Prenatal x-ray exposure and childhood cancer. J. Natl. Cancer Inst. 28:1173-1191.

Ma73 MacMahon, B., P. Cole, and J. Brown. 1973. Etiology of human breast cancer: A review. J. Natl. Cancer Inst. 50:21-42.

Ma78 Major, I. R., and R. H. Mole. 1978. Myeloid Leukemia in X-ray Irradiated CBA Mice. Nature 272, 455-456.

Ma77 Mancuso, R. F., A. Stewart, and G. Kneale. 1977. Radiation exposures of Hanford workers dying from cancer and other causes. Health Phys. 33:369-385.

Ma78b Marks, S., E. S. Gilbert, and B. D. Breitenstein. 1978. Cancer mortality in Hanford workers. Pp. 369-386 in Late Biological Effects of Ionizing Radiation, Vol. I. Vienna: International Atomic Energy Agency.

Ma75 Matanoski, G. M. 1975. The current mortality rates of radiologists and other physician specialists: Specific causes of death. Am. J. Epidemiol. 101:199-210.

Ma81b Matanoski, G. M. 1981. Risk of cancer associated with occupational exposure in radiologists and other radiation workers. Pp. 241-254 in Cancer Achievements, Challenges and Prospects for the 1980's, Vol. 1, J. H. Burchenal and H. F. Oettgen, eds. New York: Grune & Stratton.

Ma84 Matanoski, G. M., P. Sartwell, E. Elliott, J. Tonascia, A. Sternberg. 1984. Cancer risks in radiologists and radiation workers. Pp. 83-96 in Radiation Carcinogenesis: Epidemiology and Biological Significance, J. D. Boice, Jr., J. F. Fraumeni, Jr., eds. New York: Raven.

Ma81 Maxon, H. R., E. L. Saenger, C. R. Buncher, J. G. Kereiakes, S. R. Thomas, M. L. Shafer, and C. A. McLaughlin. 1981. Radiation-associated carcinoma of the salivary glands. A controlled study. Ann. Otol. 90:107-108.

Ma80 Mays, C. W., and M. P. Finkel. 1980. RBE of Alpha-Particles vs Beta-Particles in Bone Sarcoma Induction. Pp. 401-405 in Proceedings of the 6th Congress of the International Radiation Protection Association. Berlin: International Radiation Protection Association.

Ma88 Mays, C. W. Personal communication.

Mc86 McClellan, R. O., B. B. Boecker, F. F. Hahn, and B. A. Muggenburg. 1986. Lovelace ITRI Studies on the Toxicity of Inhaled Radionuclides in Beagle Dogs. Pp. 74-96 in Life-Span Radiation Effects Studies in Animals: What Can They Tell Us?, R. E. Thompson and J. A. Mahaffey, eds. U.S. Department of Energy Report CONF-830951.

Mc86a McTiernan, A., and D. B. Thomas. 1986. Evidence for a protective effect of lactation on risk of breast cancer in young women. Am. J. Epidemiol.

Me88 Metivier, H., R. Masse, G. Rateau, D. Nolibe, and J. Lafuma. 1988. New Data on the Toxicity of $^{239}PuO_2$ in Baboons. To be published in the Proceedings of the CEC/CEA/DOE-Sponsored Workshop on Biological Assessment of Occupational Exposure to Actinides, Versailles, France, May 30-June 2.

Me86 Mewhinney, J. A., F. F. Hahn, M. B. Snipes, W. C. Griffith, B. B. Boecker, and R. O. McClellan. 1986. Incidence of $^{90}SrCl_2$ or $^{238}PuO_2$; Implications for Estimation of Risk in Humans. Pp. 535-555 in Life Span Radiation Effects Studies in Animals: What Can They Tell Us?, R. C. Thompson and J. A. Mahaffey, eds. U.S. Department of Energy Report CONF-83051. Springfield, Va.: National Technical Information Service.

Mi86 Miller, R. W., and G. W. Beebe. 1986. Leukemia, lymphoma, and multiple myeloma. Pp. 245-260 in Radiation Carcinogenesis, A. C. Upton, R. E. Albert, F. Burns, and R. E. Shore, eds. New York: Elsevier.

Mi89 Miller, A. B., G. R. Howe, G. J. Sherman, J. P. Lindsay, M. J. Yaffe, P. Dinner, H. A. Risch, and D. C. Preston. 1989. Breast cancer mortality following irradiation in a cohort of Canadian tuberculosis patients. N. Engl. J. Med. (in press).

Mo74 Modan, B., D. Baidatz, H. Mart, R. Steinitz, and S. G. Levin. 1974. Radiation-induced head and neck tumours. Lancet i:277-279.

Mo82 Mole, R. H., and J. A. G. Davids. 1982. Induction of myeloid leukemia and other tumors in mice by irradiation with fission neutrons. Pp. 31-43 in Neutron Carcinogenesis, J. J. Broerse and G. B. Gerber, eds. Luxembourg: Commission of the European Communities.

Mo83a Mole, R. H., and J. R. Major. 1983. Myeloid leukemia frequency after protraced exposure to ionizing radiation: experimental confirmation of the flat dose-response found in ankylosing spondylitis after a single treatment course with x-rays. Leukemia Res. 7:295-300.

Mo83b Mole, R. H., D. G. Papworth, and M. J. Corp. 1983. The dose response for x-ray induction of myeloid leukemia in male CBA.H mice. Br. J. Cancer 47:285-291.

Mo84 Monson, R. R., and B. MacMahon. 1984. Prenatal x-ray exposure and cancers in children. Pp. 97-105 in Radiation Carcinogenesis: Epidemiology and Biological Significance, J. D. Boice, Jr., and J. F. Fraumeni, Jr., eds. New York: Raven.

Mo77 Montour, J. L., R. C. Hard, Jr., and R. E. Flora. 1977. Mammary neoplasia in the rat following high-energy neutron irradiation. Cancer Res. 37:2619-3623.

MRC56 Medical Research Council. 1956. The Hazards to Man of Nuclear and Allied Radiations. London: Her Majesty's Stationery Office.

Mu86 Muggenburg, B. A., B. B. Boecker, F. F. Hahn, W. C. Griffith, and R. O. McClellan. 1986. The Risk of Liver Tumors in Dogs and Man from Radioactive Aerosols. Pp. 556-563 in Life-Span Radiation Effects Studies in Animals: What Can They Tell Us?, R. C. Thompson and J. A. Mahaffey, eds. U.S. Department of Energy Report CONF-830951. Springfield, Va: National Technical Information Service.

Mu87 Muirhead, C. R., and S. C. Darby. 1987. Modelling the relative and absolute risks of radiation-induced cancers. J. R. Statist. Soc. A 150(part 2):83-118.

Mu80 Mulcahy, R. T., M. N. Gould, and K. H. Clifton. 1980. The survival of thyroid cells: *In vivo* irradiation and *in situ* repair. Radiat. Res. 84:523-528.

Mu80a Mulcahy, R. T., M. N. Gould, and K. H. Clifton. 1980 The survival of thyroid cells: *In vivo* irradiation *in situ* repair. Radiat. Res. 84:523-528.

Mu80b Mulcahy, R. T., D. P. Rose, J. M. Mitchen, and K. H. Clifton. 1980. Hormonal effects on the quantitative transplantation of monodispersed rat thyroid cells. Endocrinology 106:1769-1775.

Mu84 Mulcahy, R. T., M. N. Gould, and K. H. Clifton. 1984. Radiation initiation of thyroid cancer: A common cellular event. Int. J. Radiat. Biol. 45:419-426.

NRC80 National Research Council, Committee on the Biological Effects of Ionizing Radiations. The Effects on Populations of Exposure to Low Levels of Ionizing Radiation (BEIR III). Washington, D.C.: National Academy Press. Pp 524.

NRC88 National Research Council, Committee on the Biological Effects of Ionizing Radiations. Health Risks of Radon and Other Internally Deposited Alpha-Emitters (BEIR IV). Washington, D.C.: National Academy Press. Pp 602.

NCI81 National Cancer Institute. 1981. Surveillance, Epidemiology and End Results: Incidence and Mortality Data, 1973-77 (SEER Report). Natl. Cancer Inst. Monogr. 57:794-797.

NCRP84 National Council on Radiation Protection and Measurements (NCRP). 1984. Evaluation of Occupational and Environmental Exposures to Radon and Radon Daughters in the United States. Report No. 78. Bethesda, Md.: National Council on Radiation Protection and Measurements.

NCRP85 National Council on Radiation Protection and Measurements (NCRP). 1985. Induction of Thyroid Cancer By Ionizing Radiation. NCRP Report No. 80. Bethesda, Md.: National Council on Radiation Protection and Measurement.

Ne83 Netelenbos, C., P. Lips, and C. van der Meer. 1983. Hyperparathyroidism following irradiation of benign diseases of the head and neck. Cancer. 52:458-461.

NIH85 Report of the National Institutes of Health Ad Hoc Working Group to Develop Radioepidemiological Tables. 1985. NIH publication 85-2748. Washington, D.C.: U.S. Government Printing Office.

No59 Nowell, P. C., and L. G. Cole. 1959. Late effects of fast neutrons versus x-rays in mice: Nephrosclerosis, tumors, longevity. Radiat. Res. 11:545-556.

Oh78 Ohkita, T., H. Takebashi, N. Takeichi, and F. Hirose. 1978. Prevalence of leukemia and salivary gland tumors among Hiroshima atomic bomb survivors. Pp. 71-81 in Late Biological Effects of Ionizing Radiation, Vol. 1. Vienna: International Atomic Energy Agency.

Os63 Osborne, J. W., D. P. Nicholson, and K. N. Prasad. 1963. Induction of intestinal carcinoma in the rat by x-irradiation of the small intestine. Radiat. Res. 18:76-85.

Pa86 Park, J. F., G. E. Dangle, H. A. Ragan, R. E. Weller, and D. L. Stevens. 1986. Current Status of Life-Span Studies with Inhaled Plutonium in Beagles at Pacific Northwest Laboratory. Pp. 445-470 in Life-Span Radiation Effects Studies in Animals: What Can They Tell Us?, R. E. Thompson and J. A. Mahaffey, eds. U.S. Department of Energy Report CONF-830951. Springfield, Va.: National Technical Information Service.

Pi83 Pike, M. C., M. D. Krailo, B. E. Henderson, J. T. Casagrande, and D. G. Hoel. 1983. "Hormonal" risk factors, "breast tissue age" and the age-incidence of breast cancer. Nature 303:767-770.

Po78 Polednak, A. P., A. F. Stehney, and R. E. Rowland. 1978. Mortality among women first employed before 1930 in the U.S. radium dial painting industry. A group ascertained from employment lists. Am. J. Epidemiol. 107:179-195.

Pr80 Preston-Martin S., A. Paganini-Hill, B. E. Henderson, M. Pike, and C. Wood 1980. Case control study of interacranial meningomas in women in Los Angeles County, California. J. Natl. Cancer Inst. 65:67-73.

Pr81 Prinz, R. A., E. Paloyan, A. M. Lawrence, A. L. Barbato, S. S. Braithwaite, and M. H. Brooks. 1981. Unexpected parathyroid disease discovered at thyroidectomy in irradiated patients. Am. J. Surg. 142:355-357.

Pr82 Prentice, R. L., H. Kato, K. Yoshimoto, and M. Mason. 1982. Radiation exposure and thyroid cancer incidence among Hiroshima and Nagasaki residents. Natl. Cancer Inst. Monogr. 62:207-212.

Pr83 Prentice, R. L., Y. Yoshimoto, and M. W. Mason. 1983. Relationship of cigarette smoking and radiation exposure to cancer mortality in Hiroshima and Nagasaki. J. Natl. Cancer Inst. 70:611-622.

Pr87a Preston, D. L., H. Kato, K. J. Kopecky, and S. Fugita. 1987. Life Span Study Report 10. Part 1. Cancer Mortality among A-Bomb Survivors in Hiroshima and Nagasaki, 1950-82. Technical Report RERF TR 1-86. Hiroshima: Radiation Effects Research Foundation.

Pr87b Preston, D. L., H. Kato, K. J. Kopecky, and S. Fujita. 1987. Life Span Study Report 10. Part 1. Cancer mortality among a-bomb survivors in Hiroshima and Nagasaki, 1950-82. Radiat. Res. 111:151-178.

Pr88 Preston D., and D. Pierce, 1988. The effect of changes in dosimetry on cancer mortality risk estimates in atomic bomb survivors. Radiat Res. 114:437-466.

Pr88b Preston-Martin S., D.C. Thomas, S. C. White, D. Cohen. 1988. Prior exposure to medical and dental x-rays related to tumors of the parotid gland. J. Natl. Cancer Inst. 80:943-949.

Ra83 Raabe, O. G., S. A. Book, and N. J. Parks. 1983. Lifetime bone cancer dose-response relationships in beagles and people from skeletal burdens of ^{226}Ra and ^{90}Sr. Health Phys. 44(Suppl. 1):33-48.

Ra74 Rallison, M. L., B. M. Dobyns, F. R. Keating, Jr., J. E. Rall, and F. H. Tyler. 1974. Thyroid disease in children. A survey of subjects potentially exposed to fallout radiation. Am. J. Med. 56:457-463.

Ra75 Rallison, M. L., B. M. Dobyns, F. R. Keating, Jr., J. E. Rall, and F. H. Tyler. 1975. Thyroid nodularity in children. J. Am. Med. Assoc. 233:1069-1072.

Ra74b Rankow, R. M., J. Conley, and P. Fodor. 1974. Carcinoma of the maxillary sinus following Thorotrast instillation. J. Max. Fac. Surg. 2:119-126.

Ra80 Rao, S. D., B. Frame, M. J. Miller, M. Kleerekoper, M. A. Block, and A. M. Parfitt. 1980. Hyperparathyroidism following head and neck irradiation. Arch. Intern. Med. 140:201-207.

Ri87 Rimm, I. J., F. C. Li, N. J. Tarbell, K. R. Winston, and S. E. Sallan. 1987. Brain tumors after cranial irradiation for childhood acute lymphoblastic leukemia. Cancer 59:1506-1508.

Ro78b Robinson, C. V., and A. C. Upton. 1978. Competing risk analysis of leukemia and nonleukemia mortality in x-irradiated, male RF mice. J. Natl. Cancer Inst. 60:995-1007.

Ro87 Roesch, W. C., ed. 1987. U.S.-Japan Joint Reassessment of Atomic Bomb Radiation Dosimetry in Hiroshima and Nagasaki, Vol. 1. Hiroshima: Radiation Effects Research Foundation.

Ro84a Ron, E., and B. Modan. 1984. Thyroid and other neoplasms following childhood scalp irradiation. Pp. 139-151 in Radiation Carcinogenesis: Epidemiology and Biological Significance, J. D. Boice, Jr., and J. F. Fraumeni, Jr., eds. New York: Raven.

Ro79 Rose, D. P. 1979. Endogenous hormones in the etiology and clinical course of breast cancer. Pp. 21-60 in Endocrinology of Cancer, Vol. I, D. P. Rose, ed. Boca Raton, Fla.: CRC Press.

Ro78 Rostom, A. Y., S. L. Kauffman, and G. G. Steel. 1978. Influence of misonidazole on the incidence of radiation-induced intestinal tumors in mice. Br. J. Cancer 38:530-536.

Ro84b Rothman, R. J. 1984. Significance of studies of low-dose radiation fallout in the western United States. Pp. 73-82 in Radiation Carcinogenesis: Epidemiology and Biological Significance, J. D. Boice, Jr., and J. F. Fraumeni, Jr., eds. New York: Raven.

Ro77 Roudebush, C. P., and L. J. DeGroot. 1977. The Natural History of Radiation-Associated Thyroid Cancer. L. J. DeGroot, L. A. Grohman, E. L. Kaplan, and S. Refetoff, eds. New York: Grune and Stratton.

Ru84 Rubinstein, A. B., M. N. Shalit, M. L. Cohen, U. Zandbank, and E. Reichenthal. 1984. Radiation-induced cerebral meningioma: A recognizable entity. J. Neurosurg. 61:966-971.

Ru82 Russo, J., L. K. Tay, and L. H. Russo. 1982. Differentiation of the mammary gland and susceptibility to carcinogenesis. Breast Cancer Res. Treat. 2:5-73.

Sa60 Saenger, E. L., F. N. Silverman, T. D. Sterling, and M. E. Turner. 1960. Neoplasia following therapeutic irradiation for benign conditions in childhood. Radiology 74:889-904.

Sa82 Sandler, D. P., G. N. Cornstock, and G. M. Matanoski. 1982. Neoplasms following childhood irradiation of the nasopharynx. J. Natl. Cancer Inst. 68:3-8.

Sa88 Sanders, C. L., K. E. Lauhala, J. A. Mahaffey, and K. E. McDonald. 1988. Low-Level $^{239}PuO_2$ Lifespan Studies. Pp. 31-34 in Pacific Northwest Laboratory Annual Report for 1987 to the DOE Office of Energy Research, Part 1, Biomedical Sciences. U.S. Department of Energy PNL-6500 Pt. 1. Springfield, Va.: National Technical Information Service.

Sc78 Schneider, A. B., M. J. Favus, M. E. Stachura, M. J. Arnold, and L. A. Frohman. 1978. Salivary gland neoplasms as a late consequence of head and neck irradiation. Ann. Int. Med. 87:160-164.

Sc85 Schneider, A. B., E. Shore-Freedman, U. Y. Ryo, C. Bekerman, M. Favus, and S. Pinsky. 1985. Radiation-induced tumors of the head and neck following childhood irradiation. Medicine 64:1-15.

Se78 Sevcoca, M., J. Sevc, and J. Thomas. 1978. Alpha irradiation of the skin and the possibility of late effects. Health Phys. 35:803-806.

Sh57 Shellabarger, C. J., E. P. Cronkite, V. P. Bond, and S. W. Lippincott. 1957. The occurrence of mammary tumors in the rat after sublethal whole-body irradiation. Radiat. Res. 6:501-512.

Sh66 Shellabarger, C. J., V. P. Bond, G. E. Aponte, and E. P. Cronkite. 1966. Results of fractionation and protraction of total-body radiation on rat mammary neoplasia. Cancer Res. 26:509-513.

Sh71 Shellabarger, C. J. 1971. Induction of mammary neoplasia after *in vitro* exposure to x-rays. Proc. Soc. Exp. Biol. Med. 136:1103-1106.

Sh80 Shellabarger, C. J., D. Chmelevsky, and A. M. Kellerer. 1980. Induction of mammary neoplasms in the Sprague-Dawley rat by 430-keV neutrons and x-rays. J. Natl. Cancer Inst. 64:821-833.

Sh82 Shellabarger, C. J., D. Chmelevsky, A. M. Kellerer, J. P. Stone, and S.

Holtzman. 1982. Induction of mammary neoplasms in the ACI rat by 430-keV neutrons, x-rays and diethylstilbestrol. J. Natl. Cancer Inst. 69:1135-1146.

Sh86a Shellabarger, C. J., J. P. Stone, and S. Holtzman. 1986. Experimental carcinogenesis in the breast. Pp. 169-180 in Radiation Carcinogenesis, A. C. Upton, R. E. Albert, F. B. Burns, and R. E. Shore, eds. New York: Elsevier.

Sh87 Shimizu, Y., H. Kato, W. J. Schull, D. L. Preston, S. Fujita, and D. A. Pierce. Life Span Study Report 11. Part 1. Comparison of Risk Coefficients for Site-Specific Cancer Mortality Based on the DS86 and T65DR Shielded Kerma and Organ Doses. Technical Report RERF TR 12-87. Hiroshima: Radiation Effects Research Foundation.

Sh88 Shimizu, Y., H. Kato, and W. J. Schull. Life Span Study Report 11. Part II. Cancer Mortality in the Years 1950-85 Based on the Recently Revised Doses. Technical Report. Hiroshima: Radiation Effects Research Foundation.

Sh76 Shore, R. E., R. E. Albert, and B. S. Pasternack. 1976. Follow-up study of patients treated by x-ray epilation for tinea capitis: Resurvey of post-treatment illness and mortality experience. Arch. Environ. Health 31:17-24.

Sh84a Shore, R. E., R. E. Albert, M. Reed, N. Harley, and B. S. Pasternack. 1984. Skin cancer incidence among children irradiated for ringworm of the scalp. Rad. Res. 100:192-204.

Sh84b Shore, R. E., E. D. Woodward, and L. H. Hempelmann. 1984. Radiation-induced thyroid cancer. Pp. 131-138 in Radiation Carcinogenesis: Epidemiology and Biological Significance, J. D. Boice, Jr., and J. F. Fraumeni, Jr., eds. New York: Raven.

Sh85 Shore, R. E., E. Woodward, N. Hildreth, P. Dvoretsky, L. Hempelmann, and B. Pasternack. 1985. Thyroid tumors following thymus irradiation. J. Natl. Cancer Inst. 74:1177-1184.

Sh86b Shore, R., N. Hildreth, E. Woodward, P. Dvoretsky, L. Hempelmann, and B. Pasternack. 1986. Breast cancer among women given x-ray therapy for acute postpartum mastitis. J. Natl. Cancer Inst. 77:689-696.

Si8l Sikov, M. R. 1981. Carcinogenesis following prenatal exposure to radiation. Biol. Res. Pregnancy 2:159-167.

Si55 Simpson, C. L., L. H. Hempelmann, and L. M. Fuller. 1955. Neoplasia in children treated with x-rays in infancy for thymic enlargement. Radiology 64:840-845.

Sm76 Smith, P.G., and R. Doll. 1976. Late effects of x-irradiation in patients treated for metropathia haemorrhagica. Br. J. Radiol. 49:224-232.

Sm77 Smith, P. G. 1977. Leukemia and other cancers following radiation treatment of pelvic disease. Cancer 39:1901-1905.

Sm81 Smith, P.G., and R. Doll. 1981. Mortality from cancer and all causes among British radiologists. Br. J. Radiol. 54:187-194.

Sm82 Smith, P. G., and R. Doll. 1982. Mortality among patients with ankylosing spondylitis after a single treatment course with X-rays. Br. Med. J. 284:449-460.

Sm86 Smith, P. G., and A. J. Douglas. 1986. Mortality of workers at the Sellafield plant of British Nuclear Fuels. Br. Med. J. 293:845-854.

So63 Socolow, E. L., A. Hashizume, S. Neriishi, and R. Niitani. 1963. Thyroid carcinoma in man after exposure to ionizing radiation. N. Engl. J. Med. 268:406-410.

So83 Soffer, D., S. Pittaluga, M. Feiner, and A. J. Beller. 1983. Intracranial meningiomas following low-dose irradiation to the head. J. Neurosurg. 59:1048-1053.

Sp58 Spiess, H., A. Gerspach, and C. W. Mays. 1958. Soft-tissue effects following Ra injections into humans. Health Phys. 35:61-81.

St73 Stewart, A. 1973. The carcinogenic effects of low level radiation. A re-appraisal of epidemiologic methods and observations. Health Phys. 24:223-240.

St62 Stewart, A., W. Pennypacker, and R. Barber. 1962. Adult leukemia and diagnostic x-rays. Br. Med. J. 2:882.

Ta75 Takeichi, N. 1975. Induction of salivary gland tumors following x-ray examination. II. Development of salivary gland tumors in long-term experiments. Med. J. Hiroshima Univ. 23:391-411.

Ta76 Takeichi, N., F. Hirose, and H. Yamamoto. 1976. Salivary gland tumors in atomic bomb survivors, Hiroshima, Japan. I. Epidemiology observations. Cancer 38:2462-2468.

Th86 Thompson, R. C., and J. A. Mahaffey, eds. 1986. Life-Span Radiation Effects Studies in Animals: What Can They Tell Us? U.S. Department of Energy Report No. CONF-830951. Springfield, Va.: National Technical Information Service.

Ti77 Tisell, L. E., G. Hansson, S. Lindberg, and I. Ragnhult. 1977. Hyperparathroidism in persons treated with x-rays for tuberculous cervical adenitis. Cancer 40:846-854.

To87 Tokunaga, M., C. E. Land, T. Yamamoto, M. Asano, S. Tokuoka, H. Ezaki, and I. Nishimori. 1987. Incidence of female breast cancer among atomic bomb survivors, Hiroshima and Nagasaki, 1950-1980. Radiat. Res. 112:243-272.

To83 Tolley, H. D., S. Marks, J. A. Buchanan, and E. S. Gilbert. 1983. A further update of the analysis of mortality of workers in a nuclear facility. Radiat. Res. 95:211-213.

Tr77 Triggs, S. M., and E. D. Williams. 1977. Irradiation of the thyroid as a cause of parathyroid adenoma. Lancet i:293-294.

Ts73 Tsubouchi, S., and T. Matsuzawa. 1973. Nodular formations in rat small intestine after local abdominal x-irradiation. Cancer Res. 33:3155-3158.

Ul79 Ullrich, R. L., and J. B. Storer. 1979. Influence of irradiation on the development of neoplastic disease in mice. II. Solid tumors. Radiat. Res. 80:317-324.

Ul84 Ullrich, R. L. 1984. Tumor induction in BALB/c mice after fractionated or protracted exposures to fission spectrum neutrons. Radiat. Res. 97:587-597.

Ul86 Ullrich, R. L. 1986. The rate of progression of radiation-transformed mammary epithelial cells is enhanced after low-dose-rate neutron irradiation. Radiat. Res. 105:68-75.

Ul87 Ullrich, R. L., and R. J. Preston. 1987. Myeloid leukemia incidence in male RFM mice following irradiation with x-rays or fission spectrum neutrons. Radiat. Res. 109:165-170.

Ul87b Ullrich, R. L., M. C. Jernigan, L. C. Satterfield, and N. D. Bowles. 1987. Radiation carcinogenesis: Time-dose relationships. Radiat. Res. 111:179-184.

UN77 United Nations Scientific Committee on the Effects of Atomic Radiation (UNSCEAR). 1977. Sources and Effects of Ionizing Radiation. Report E. 77. IX. 1. New York: United Nations. Pp. 725.

UN82 United Nations Scientific Committee on the Effects of Atomic Radiation (UNSCEAR). 1982. Ionizing Radiation: Sources and Biological Effects. Report E, 82, IX, 8. New York: United Nations. Pp. 773.

UN86 United Nations Scientific Committee on the Effects of Atomic Radiation (UNSCEAR). 1986. Genetic and Somatic Effects of Ionizing Radiation, Report. New York: United Nations. Pp. 366.

UN88 United Nations Scientific Committee on the Effects of Atomic Radiation (NSCEAR). 1988. Sources, Effects and Risks of Ionizing Radiation. Report E.88.IX.7. New York: United Nations.

Up64 Upton, A. C., V. K. Jenkins, and J. W. Conklin. 1964. Myeloid leukemia in the mouse. Ann. N.Y. Acad. Sci. 114:189-201.

Up66 Upton, A. C., V. K. Jenkins, H. E. Walburg, Jr., R. L. Tyndall, J. W. Conklin, and N. Wald. 1966. Observations on viral, chemical, and radiation-induced myeloid and lymphoid leukemias in RF mice. Natl. Cancer Inst. Monogr. 22:329-347.

Up70 Upton, A. C., M. L. Randolph, and J. W. Conklin (with the collaboration of M. A. Kastenbaum, M. Slater, G. S. Melville, Jr., F. P. Conte, and J. A. Sproul, Jr.). 1970. Late effects of fast neutrons and gamma-rays in mice as influenced by the dose rate of irradiation: Induction of neoplasia. Radiat. Res. 41: 467-491.

Va78 van Kaick, G., D. Lorenz, H. Muth, and A. Kaul. 1978. Malignancies in German thorotrast patients and estimated tissue dose. Health Phys. 35:127-136.

Va84 van Kaick, G. H. Muth, and A. Kaul. 1984. The German Thorotrast Study. Results of Epidemiological, Clinical and Biophysical Examinations on Radiation-Induced Late Effects in Man Caused by Incorporated Colloidal Thorium Dioxide (Thorotrast). Report No. EUR 9504 EN. Luxembourg: Commission of the European Communities.

Va86 Vaughn, J. Carcinogenic effects of radiation on the human skeleton and supporting tissues. Pp. 311-334 in Radiation Carcinogenesis, A. C. Upton, R. E. Albert, F. J. Burns, and R. E. Shore, eds. New York: Elsevier.

Vo72 Vogel, H. H., and R. Valdivar. Neutron-induced mammary neoplasms in the rat. Cancer Res. 32:933-938.

Wa64 Wagoner, J .K., V. E. Archer, V. E. Carroll, D. A. Holaday, and P. A. Lawrence. 1964. Cancer mortality patterns among U.S. uranium miners and millers, 1950 through 1962. J. Natl. Cancer Inst. 32:787-801.

Wa84 Wagoner, J. K. 1984. Leukemia and other malignancies following radiation therapy for gynecological disorders. Pp. 153-159 in Radiation Carcinogenesis: Epidemiology and Biological Significance, J. D. Boice, Jr., and J. F. Fraumeni, Jr., eds. New York: Raven.

Wa82 Wakisaka, S., T. L. Kemper, H. Nakagaki, and R. R. O'Neill. 1982. Brain tumors induced by radiation in Rhesus monkeys. Fukuoka Acta Med. 73:585.

Wa83 Wakabayashi, T., H. Kato, T. Ikeda, and W. J. Schull. Studies of the mortality of A-bomb survivors. Report 7, Part III. Incidence of cancer in 1959-1978, based on the tumor registry, Nagasaki. Radiat. Res. 93:112-146.

Wa68 Wanebo, C. K., K. G. Johnson, K. Sato, and T. W. Thorslund. 1968. Breast cancer after exposure to the atomic bombings of Hiroshima and Nagasaki. N. Engl. J. Med. 279:667-671.

Wa88b Wang, J.-X., J. D. Boice, Jr., B.-X. Li, J.-Y. Zhang, and J. F. Fraumeni Jr.. 1988. Cancer among medical diagnostic x-ray workers in China. J. Natl. Cancer Res. 80:344-350.

Wa86 Watanabe, H., A. Ito, and F. Hirose. 1986. Experimental carcinogenesis in the digestive and genitourinary tracts. Pp. 233-244 in Radiation Carcinogenesis, A. C. Upton, R. E. Albert, F. J. Burns, and R. E. Shore, eds. New York: Elsevier.

Wa88 Watanabe, H., M. A. Tanner, F. E. Domann, M. N. Gould, and K. H. Clifton. 1988. Inhibition of carcinoma formation and of vascular invasion in grafts of radiation-initiated thyroid clonogens by unirradiated thyroid cells. Carcinogenesis 9:1329-1335.

Yo85 Yochmowitz, M.G., D. H. Wood, and Y. L. Salmon. 1985. Seventeen-year mortality experience of proton radiation in Macaca mulatta. Radiat. Res. 102:14.

Yo77 Yokoro, K., M. Nakano, A. Ito, K. Nagao, and Y. Kodama. 1977. Role of prolactin in rat mammary carcinogenesis: Detection of carcinogenicity of low-dose carcinogens and of persisting dormant cancer cells. J. Natl. Cancer Inst. 58:1777-1783.

Yo78 Yokoro, K., C. Sumi, A. Ito, K. Hamada, K. Kanda, and T. Kobayashi. 1978. Mammary carcinogenic effect of low-dose fission radiation in Wistar/Furth rats and its dependency on prolactin. J. Natl. Cancer Inst. 64:1459-1466.

Yo86 Yokoro, K. 1986. Experimental radiation leukemogenesis in mice. Pp. 137-150 in Radiation Carcinogenesis, A. C. Upton, R. E. Albert, F. J. Burns, and R. E. Shore, eds. New York: Elsevier.

Yo77b Yoshizawa, Y., T. Kusama, and K. Morimoto. 1977. Search for the lowest irradiation dose from literature on radiation-induced bone tumors. Nippon Acta Radiol. 37:377-386.

Yo81 Young, J. L., C. L. Percy, and A. J. Asire. 1981. Surveillance, Epidemiology, and End Results. Incidence and Mortality Data. 1973-1977. National Cancer Institute Monograph 57. NIH Publication No. 81-2330. Washington, D.C.: U.S. Government Printing Office.

Yu88 Yuan, J. M., M. C. Yu, R. K. Ross, Y. T. Gao, et al. 1988. Risk factors for breast cancer in Chinese women in Shaghai. Cancer Res. 48:1949-1953.

Zo80 Zook, B. C., E. W. Bradley, G. W. Casarett, and C. C. Rogers. 1980. Pathologic findings in canine brain irradiated with fractionated fast neutrons or photons. Radiat. Res. 84:562-578.

6

Other Somatic and Fetal Effects

CANCER IN CHILDHOOD FOLLOWING EXPOSURE IN UTERO

Human Epidemiologic Studies

Preliminary results of the Oxford Survey of Childhood Cancers, published over 30 years ago, suggested an association between the risk of cancer, primarily leukemia, in childhood (within 15 years of birth) and prenatal exposure to diagnostic x rays in utero (St56, 58). A subsequent survey of 734,243 children born in New England supported this suggestion (Ma62). The initial results of follow-up of prenatally irradiated atomic-bomb survivors during the first 10 years of life had failed to support the suggestion (Ja70). However, in a more recent, 1950-1984, follow-up based on DS86 dosimetry (Yo88), two cases of childhood cancer have been observed among 1,630 in utero-exposed survivors during the first 14 years of life, both of which occurred in persons who had been heavily exposed (1.39 and 0.56 Gy). The occurrence of these two cases corresponds to an upper bound risk estimate (95% confidence level) of 279 cases/10^4 PGy, an estimate consistent with Bithell and Stiller's estimate on reanalysis of the Oxford survey data (Bi88).

An extension of the New England survey to include cancer deaths in 1,429,400 children born between 1940 and 1960 in 42 hospitals in New England and the mid-Atlantic states (Mo84) also showed an excess of cancers among those exposed to diagnostic x rays in utero. In this study, cases were compared with age- and sex-matched nonirradiated controls. For leukemia and other cancers, the relative risks were 1.52 and 1.27, respectively, with no evidence that the excess was attributable to risk factors other than radiation or was limited to a particular subpopulation (Mo84).

To explore the possibility that both prenatal x-ray examination and childhood cancer might be attributable to a separate, common risk factor, and since radiographic examination of women who are pregnant with twins has usually been performed because of the twin pregnancy rather than because of other diagnostic concerns, the incidence of cancer has been investigated in irradiated twins. The first such study, conducted in the United Kingdom, found the relative risks of childhood leukemia and other cancers in irradiated twins (versus nonirradiated twins) to be 2.0 and 1.7, respectively. It also found as many excess cases of cancer in irradiated dizygotic and monozygotic twins as in irradiated singleton births (Mo74). The second study, conducted on twins in Connecticut, likewise found the relative risks of childhood leukemia and other cancers in irradiated twins versus those in nonirradiated twins, especially at ages 10-14 years, to be 1.6 (90% C.I. 0.4, 6.8) and 3.2 (0.9, 10.7), respectively (Ha85); however, the excess was restricted largely to children of mothers with a history of previous pregnancy loss, in whom the overall relative risk of cancer was 7.8 (1.2, 50.4), compared with 1.4 (0.5, 4.3) in irradiated twins born to mothers without a history of pregnancy loss (Ha85).

Because of the comparatively small magnitude of the average radiation dose to the fetus from diagnostic radiography, which has been estimated as 5-50 mGy, the data imply that susceptibility to radiation carcinogenesis is relatively high during prenatal life (NRC72, NRC80, UN77, Mo84). Such an interpretation is complicated, however, by the fact that little increase in susceptibility has been evident in prenatally x-irradiated experimental animals and there is no known biological basis for such an increase in susceptibility or for the suggested equivalence in magnitude of the leukemia excess with that of other childhood cancers (Mi86). These complications notwithstanding, the concordance of the studies of twins with the studies of prenatally irradiated singleton births prompts the tentative conclusion that susceptibility to the carcinogenic effects of irradiation is high during prenatal life.

Although, mortality from cancer now appears to be increased in prenatally exposed atomic-bomb survivors more than four decades after they were irradiated (Yo88), it remains to be established that the risk of cancer in adult life is increased by prenatal irradiation. During the observation period 1950-1984, however, the relative risk of fatal cancer at a dose of 1 Gy to the mother's uterus (DS86 organ dose), among a total of 1,630 in utero-exposed A-bomb survivors has been estimated as 3.77 (90% C.I. 1.14, 13.48), corresponding to an absolute risk of 6.57 (90% C.I. 0.47, 14.49) per 10^4 PYGy and an attributable risk of 40.9% per Gy (90% C.I. 2.9%,

90.2%). Thus, these results also suggest that susceptibility to radiation-induced cancer is higher in prenatally exposed survivors than in postnatally exposed survivors (Yo88). Comparable late-occurring carcinogenic effects from prenatal irradiation have been observed in laboratory mice (Co84).

Summary

Based on the limited epidemiologic data available through the early 1970s, the 1977 UNSCEAR committee (UN77) estimated the risk per unit absorbed dose to be about 200 to 250 excess cancer deaths/10^4 person Gy in the first 10 years of life, with one-half of these malignancies being leukemias and one-quarter tumors of the nervous system. Bithell and Stiller's (Bi88) recent estimate from the Oxford survey, 217 cases/10^4 person Gy, falls within this range. The epidemiologic studies also suggest that an association exists between in utero exposure to diagnostic x rays and carcinogenic effects in adult life; however, the magnitude of the risk remains uncertain.

EFFECTS ON GROWTH AND DEVELOPMENT

Animal Studies

The effects of prenatal irradiation on the growth and development of the mammalian embryo and fetus, mediated through direct radiation injury of developing tissues (Br87), include gross structural malformations, growth retardation, embryo lethality, sterility, and central nervous system abnormalities (UN77). Major anatomical malformations have been produced in all mammalian species by irradiation of the embryo during early organogenesis; however, the time of maximal susceptibility is sharply circumscribed, and the evidence suggests that there may be a threshold for many, if not most, major malformations (NRC80). Retardation of postnatal growth also has been observed to be produced over a broad range of mammalian gestational ages in experimental animals and humans (NRC80).

The developing central nervous system exhibits a particular sensitivity to ionizing radiation (ICRP87). In experimental animals, the central nervous system malformations most likely to be produced by irradiation during early organogenesis include hydrocephaly, anencephaly, encephalocele, and spina bifida. In rats, mice, and monkeys, radiation has been shown to induce functional and behavioral effects too, including motor defects (Ya62), emotionality (Fu58), impairment of nervous reflexes and hyperactivity (Ma66), and deficits in learning (Le62). In rodents, disturbances of conditional reflexes, impairment of learning ability, and locomotor damage also have been reported after doses that were large enough to cause gross structural damage (UN86).

Human Studies

The most definitive human data concerning the effects of prenatal irradiation are those relating to brain development (UN86).

Severe Mental Retardation

Injurious effects of ionizing radiation on the developing human brain have been documented in Japanese A-bomb survivors who were exposed in utero (Bl73, Bl75, IC86, Mi76, Ot83, Sc86a, Sc86b, UN86, Wo67), in whom the prevalence of mental retardation and small head size increases with increasing exposure. In recent studies based on a cohort of 1,598 such individuals, all of the 30 children who were found to have severe mental retardation were diagnosed before the age of 17. Nine of the mentally retarded individuals, only 3 of whom had doses greater than 0.5 Gy, also had other health problems, presumably not related to radiation, which might account for their severe mental retardation. Two individuals had conditions unlikely to be casual for mental retardation, neonatal jaundice and, possibly, neurofibromatosis. Three have or have had Down syndrome, one a retarded sibling and another Japanese encephalitis during infancy.

Dosimetry: Estimates of the dose received by the children as fetuses are not yet available from the DS86 system, but the intrauterine doses received by their mothers should provide a useful approximation. DS86 organ dose estimates for the uterus have been computed for most of the exposed mothers who were within 1600 m of the hypocenter in Hiroshima and 2000 m in Nagasaki (Ot87, Ro87). Organ doses were modeled individually to take account of house shielding and the orientation and posture of the exposed individuals. For exposed individuals with incomplete shielding histories, the calculated free-in-air (FIA) kerma was adjusted by means of average house and body transmission factors to obtain an average organ dose. Under the DS86 dose system, neutrons are not a significant contributor to most fetal exposures; the DS86 FIA neutron kerma in Hiroshima at 2000 meters was only 0.0004 Gy and in Nagasaki, 0.0003 Gy (Ro87).

Gestational Age: Gestational age is an important factor in determining the nature of the radiation injury to the developing brain of the embryo or fetus (Bl73, Bl75, Mi76, Ot83, Ot86, Ot87, Sc86a, Sc86b). Gestational ages have been grouped to reflect the known phases in normal brain development. The four categories measured from the time of conception were 0-7, 8-15, 16-25, and $\geq$26 weeks. During the first period (0-7 weeks), the precursors of the neurons and neuroglia emerge and are mitotically active (Ma82). During the second period (8-15 weeks), a rapid increase in

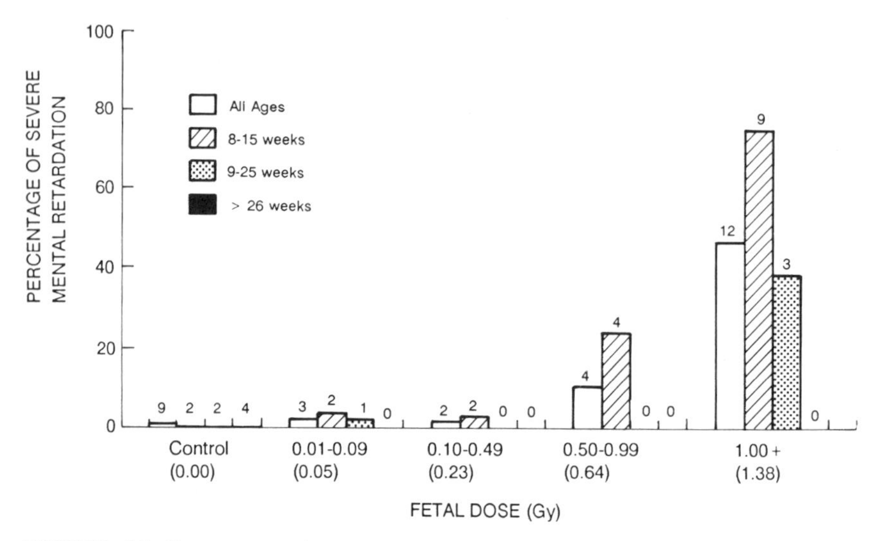

FIGURE 6-1 Percentages of severe mental retardation at various fetal doses in the combined Hiroshima and Nagasaki data. The number of cases is given at the top of the histogram (Ot87).

the number of neurons occurs; they migrate to their developmental sites and lose their capacity to divide, (Ra75, Ra78). During the third period (16-25 weeks), differentiation in situ accelerates, synaptogenesis that began at about week 8 increases, and the definitive cytoarchitecture of the brain results. The fourth period (≥26 weeks) is one of continued architectural and cellular differentiation and synaptogenesis of the cerebrum with, at the same time, accelerated growth and development of the cerebellum.

Among atomic-bomb survivors exposed in utero, a dose-dependent increase in the incidence of severe mental retardation occurred in the gestational age group 8-15 weeks after conception and, to a lesser extent, in the gestational age group 16-25 weeks after conception (Figure 6-1). No subjects exposed to radiation at less than 8 weeks or ≥26 weeks of gestational age were observed to be mentally retarded. The relative risk for exposure during the 8-15 week period is at least 4 times greater than that for exposure at 16-25 weeks after conception.

Dose-Response Models: The dose response for severe mental retardation has been examined in depth by Otake, Yoshimaru, and Schull (Ot87). Their results are shown in Figure 6-2. Within the critical gestational age period of 8-15 weeks, the prevalence of severe mental retardation can be linearly related to the absorbed dose received by the fetus. There is a highly significant increase in the occurrence of severe mental retardation with dose in Hiroshima and in the combined data from both cities. This increase is strongest in the children irradiated at 8-15 weeks after conception but a suggestive increase is also seen at 16-25 weeks after conception. In the data for both cities, the variation in frequency of occurrence with dose, when exposure occurred 8-15 weeks after conception, can be accounted for by a linear model, although there is some suggestion of a nonlinear component in the dose-response function for both the 8-15 and the 16-25 week periods (Figure 6-2).

Maximum likelihood analyses based on a simple linear model were made to estimate a possible threshold dose and its 95% confidence intervals (Ot87). When all cases were considered, the estimated lower bound of the threshold for the most sensitive period of 8-15 weeks after conception was zero. However, exclussion of cases with a possible nonradiation related etiology yields a threshold with a lower bound of 0.12 Gy for ungrouped data and 0.23 Gy when the data is stratified by dose interval. Both of the estimated thresholds, 0.39 and 0.46 Gy, respectively, are significantly different from zero. Further investigation, using an exponential linear model, found an estimated lower bound for a threshold of 0.09 Gy for

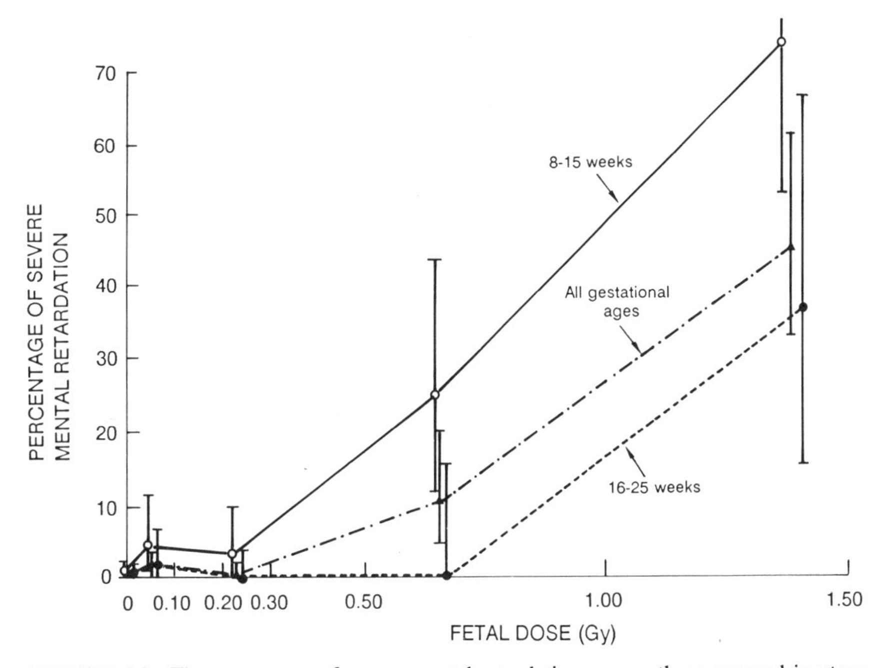

FIGURE 6-2 The percentage of severe mental retardation among those exposed in utero by dose and gestational age in Hiroshima and Nagasaki. The vertical lines indicate 90% confidence intervals (Ot87).

the grouped data and 0.15 Gy for the individual data for those exposed in the 8-15 week period. Similarly, a threshold was also indicated for the 16-25 week-period, with a lower bound of 0.21 Gy, based on a linear model with either the individual or the grouped data, and 0.22-0.25 Gy with the exponential linear model. However, the case for a threshold is not clear; linear regressions using a threshold predict a larger response than was actually observed at large doses (W. J. Schull, personal communication).

In summary, analysis of the epidemiologic data has identified the maximal sensitivity of the human brain to occur between 8 and 15 weeks of gestational development. During this period, the dose-effect relationship resulting from the new DS86 dosimetry system indicates a frequency of severe mental retardation of 43% at 1 Gy and suggests that a threshold for the effect may exist in the range 0.2 to 0.4 Gy (Ot87, IC88).

Uncertainties: A number of uncertainties are associated with these risk estimates. These include the limited number of cases, the appropriateness of the comparison group, errors in the estimation of the absorbed doses and the calculated prenatal ages at exposure, variation in the severity of mental retardation, and other confounding factors in the postbombing period, including malnutrition and disease (Sc86a).

Discussion: Significant harmful effects of radiation on the developing brain of children exposed in utero during the atomic bombings of Hiroshima and Nagasaki were observed only for those exposed during the periods 8-15 and 16-25 weeks after conception. During the period at 8-15 weeks, the period of maximum sensitivity, the dose-response relationship appeared to be different from that at subsequent gestational ages, indicating that radiation effects on cerebral growth and development vary with gestational age at exposure. This period of maximum radiation sensitivity is the time of the most rapid cell proliferation and migration of immature neurons from the ventricular and subventricular proliferative layers to the cerebral cortex (Do73, Ra75, Ra78). Radiation exposure during this period may be inferred to induce neuronal abnormalities and misarrangement of neurons, as well as decreasing the number of normal neurons. This inference appears to be supported by nuclear magnetic resonance images of the brains of severely mentally retarded children, in which abnormal collections of neurons in areas of disturbed brain architecture have been demonstrated (W. J. Schull, personal communication).

The data for 8-15 weeks after conception, based on the DS86 doses, fit either a linear or linear exponential dose-response relationship without a threshold. Otake et al. have pointed out that estimating a threshold for this effect is difficult and may depend on the clinical criteria for severe mental retardation. If exposure to radiation moves the distribution of intelligence downward in proportion to dose, as described below, the number of individuals with levels of intellectual function below the diagnostic threshold must necessarily increase as the dose increases (Ot87). Clinical selection of an arbitrary level for severe mental retardation dichotomizes the distribution of intelligence levels and could lead to an apparent threshold for this effect.

At 16-25 weeks after conception, differentiation accelerates, synaptogenesis that begins at about week 8 increases, and the functional cytoarchitecture of the brain takes place. During this period radiation may impair synaptogenesis, producing a functional deficit in brain connections. The response seen among the atomic-bomb survivors, irradiated during the period 16-25 weeks after conception, suggests that the evidence for a threshold is stronger during this period than during the 8-15 week interval.

No evidence of a radiation-related increase in mental retardation has been observed in survivors exposed earlier than 8 weeks after conception or later than 26 weeks after conception. The absence of an effect prior to the eighth week suggests that either the cells that were killed or inactivated at this stage of development are more readily replaced than those that were damaged later, or that the embryo fails to develop further. The final weeks of gestation are largely a time of continued cytoarchitectural and cellular differentiation and synaptogenesis, and the basic neuronal structure of the cerebrum is nearing completion at this time. Since differentiated cells are generally less radiosensitive than undifferentiated ones, measurable damage may require much higher doses and, given the small number of atomic-bomb survivors at these doses, may be more difficult to detect (Ot87).

Nonradiation-related explanations for the observed effects on the embryonic and fetal central nervous system that could affect these findings include: (1) genetic variation, (2) nutritional deprivation, (3) bacterial and viral infections during pregnancy, and (4) embryonic or fetal hypoxemia. It is possible that one or more of these factors could have confounded the observations. It is commonly presumed that radiation-related damage to the developing brain results largely, if not solely, from neuronal death. This assumption rests in part on the relatively large proportion of the mentally retarded who have small heads. There is a need, therefore, to determine what role, if any, these other possible causes of a relatively small brain may play in the radiation-related risk of mental retardation.

Intelligence Test Scores

Intelligence test (Koga) scores of individuals of 10-11 years of age who were exposed prenatally to the Hiroshima and Nagasaki atomic bombs have been analyzed, using estimates of the uterine absorbed dose based on the DS86 system of dosimetry (Sc88). As indicated in Figure 6-3, no radiation-related effect on intelligence is evident among survivors who were exposed in utero during the first seven weeks after conception or during week 26 or later. In contrast, children exposed at 8-15 weeks after conception and,

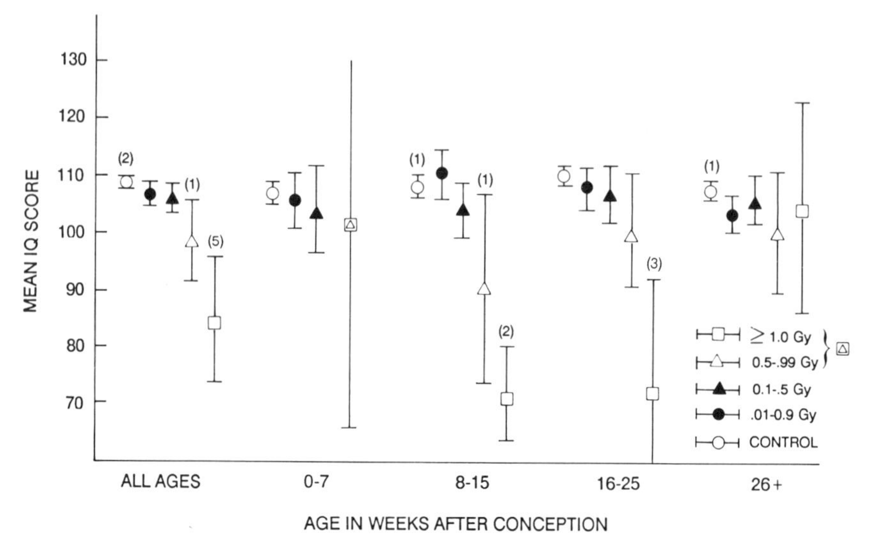

FIGURE 6-3 Mean IQ scores and 95% confidence limits by gestational age in weeks and fetal dose. The numbers in parentheses are severely retarded cases, IQ $\leq$ 64 (Sc86a).

to a lesser extent, those exposed at 16-25 weeks after conception show a progressive shift downward in individual scores with increasing exposure. Within the group exposed 8-15 weeks after conception, a linear model fits the regression of intelligence scores on dose somewhat better than linear-quadratic models. The diminution in intelligence score under the linear model is 21-29 points at 1 Gy and is somewhat greater (24-33 points) at 1 Gy when controls who received less than 0.01 Gy are excluded from the analysis (Sc88).

School Performance

In a study of the school performance of prenatally exposed atomic-bomb survivors, the DS86 sample included 929 children. As judged by a simple regression of school performance as a function of fetal dose, there is a highly significant decrease in school achievement in children exposed 8-15 weeks and 16-25 weeks after conception (Figure 6-4) (Ot88). This trend is strongest in the earlier school years. In the groups exposed within 0-7 weeks, or $\geq$26 weeks after conception, there is no evidence of a radiation-related effect on scholastic performance. These results parallel those obtained for prenatally exposed atomic-bomb survivors with regard to achievement on standard intelligence tests in childhood as discussed above (Sc88).

Summary—Japanese Results: The DS86 in utero sample consisted

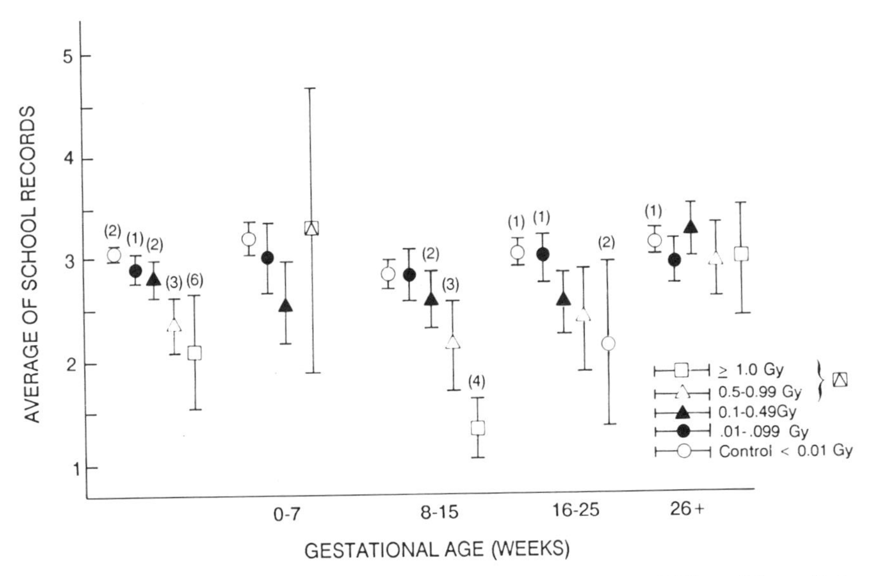

FIGURE 6-4 Average school subject score in the first grade with 95% confidence limits by gestational age and fetal dose (Ot88).

of almost 1,600 atomic-bomb survivors, including 30 individuals who were severely mentally retarded. A variety of dose-response models with and without a threshold have been fitted to the individual, as well as grouped, dose data. The highest risk of radiation damage to the embryonic and fetal brain occurred in individuals irradiated 8-15 weeks after conception. The frequency of severe mental retardation in the 8-15 week-old fetus is described by a simple linear, nonthreshold model. The risk at 1 Gy is about 43% with the DS86 dosimetry systems under a simple linear model, and about 48% when a linear exponential model is used. There is some indication of a threshold for severe mental retardation, but this is difficult to assess because there is a continuous diminution of intelligence with increasing dose. Using a 95% confidence interval, the grouped dose data suggest a lower bound on the threshold dose of about 0.1 Gy, whereas regressions using individual doses yield a lower bound of about 0.2 Gy. However, linear regressions which include thresholds are not consistent with the observations at doses greater than 1 Gy. When individual doses are used, damage to the fetus exposed at 16-25 weeks after conception seems to fit a linear-quadratic or quadratic regression and suggests a lower bound of about 0.2 Gy on a possible threshold dose.

Within the group exposed 8-15 weeks after conception, the regression of the intelligence test (Koga) score on absorbed dose is linear; the range of the decrease in intelligence test score is between 21 and 29 points at 1

Gy. Similarly, damage to the fetal brain at 8-15 weeks after conception is linearly related to fetal absorbed dose, as judged by a simple regression of school performance scores on dose.

Other Epidemiologic Studies

New York Tinea Capitis Study: Albert et al. (Al66) reported that children in New York, treated for tinea capitis by x irradiation, had a higher incidence of treated psychiatric disorders than those treated with chemotherapy. Shore et al. (Sh76) and Omran et al. (Om78) confirmed these observations in this series of 2,215 patients with tinea capitis and demonstrated a higher frequency of mild, nontreated forms of behavioral maladjustment and mental disease in the irradiated population.

Israel Tinea Capitis Study: Ron et al. (Ro82) evaluated several measures of mental and brain function in 10,842 Israeli children treated for tinea capitis by x-ray therapy (mean brain dose, 1.3 Gy) and two nonirradiated, tinea capitis-free comparison groups were used. While not all measures were statistically significant, there was a consistent trend for the irradiated children to exhibit subsequent behavioral impairment more often than those in the comparison group. The irradiated children had lower examination scores on scholastic aptitude, intelligence quotient, and psychological tests; completed fewer school grades; had increased admissions to mental hospitals for certain neuropsychiatric diseases; and had a slightly higher frequency of mental retardation.

Childhood Leukemia Patients: Meadows et al. (Me81) also reported lower intelligence quotient scores and disturbances in cognitive functions in children with acute lymphocytic leukemia who were treated with radiation to the brain.

Summary

The consequences of irradiation of the mammalian embryo and fetus during the period of major organogenesis may include teratogenic effects on various organs. In humans, mental retardation is the best documented of the developmental abnormalities following radiation exposure. In the Japanese atomic-bomb survivors who were irradiated in utero, the prevalence of radiation-related mental retardation was highest in those irradiated between 8 and 15 weeks after conception, decreased in those irradiated between 16 and 25 weeks, and was negligible or absent in those irradiated before 8 weeks or later than 25 weeks. In those irradiated between weeks 8 and 15, the prevalence of mental retardation appeared to increase with dose in a manner consistent with a linear, nonthreshold response, although the data do not exclude a threshold in the range of 0.2-0.4 Gy.

CATARACT OF THE EYE LENS

Radiation-induced opacification of the lens of the eye, or cataract formation, has been observed to result from a dose of radiation to the lens in excess of 0.6-1.5 Gy, depending on the dose rate and the linear energy transfer (LET) of the radiation, as well as on the sensitivity of the method used to examine the lens (ICRP84). The threshold for ophthalmologically detectable opacities in atomic-bomb survivors has been estimated to range, using T65 dosimetry, from 0.6 to 1.5 Gy (Ot82), whereas the threshold in persons treated with x rays to the eye has been observed to range from about 2 Gy when the dose was received in a single exposure to more than 5 Gy when the dose was received in multiple exposures over a period of weeks (Me72, ICRP84). The threshold for neutrons appears to be lower; that is, in patients treated with 7.5 MeV neutrons in multiple exposures over a period of 1 month, the threshold for a vision-impairing cataract was estimated to approximate 3-5 Gy (Ro76). By the same token, long-continued occupational exposure to 0.7-1 Gy of mixed neutron-gamma radiation has been observed to cause cataracts (Ha53, Lv74), whereas similar occupational exposure to comparable doses of x rays or gamma rays has not (ICRP84).

Although it is clear from the foregoing that detectable injury of the lens can result from a dose of as low as 1 Gy, depending on the dose rate and LET of the radiation, the threshold for a vision-impairing cataract under conditions of highly fractionated or protracted exposure is thought to be no less than 8 Sv (ICRP84). This dose exceeds the amount of radiation that can be accumulated by the lens through occupational exposure to irradiation under normal working conditions and greatly exceeds that which is likely to be accumulated by a member of the general population through other types of exposure.

LIFE SHORTENING

In laboratory mammals exposed to whole-body radiation, life expectancy decreases with increasing dose. From early experiments with rats and mice, the life-shortening effect of irradiation was interpreted as a manifestation of accelerated aging (Ru39, He44, Br52, Al57, Ca57). When analyzed in relation to the cause of death, however, the effect was not observed to be the same for all age-related diseases (Up60) but to result principally from an accelerated onset of neoplasia (Wa75).

Mortality from diseases other than cancer has not been consistently or significantly increased by irradiation in human populations (Be78, UN82), with the possible exception of an early cohort of U.S. radiologists (Wa56, Wa66, Se58, Se65, Ma75a, Ma75b) in whom the confounding influence of other risk factors cannot be excluded. The bulk of the epidemiologic data appear to be consistent, therefore, with the data from laboratory animals (UN82). Although the data do not support the view that radiation causes a nonspecific acceleration of the aging process, the life-shortening effects of a given dose in different species are similar when analyzed in terms of the upward displacement of the age-specific death rate for the species (Sa66, Sa70).

In the earlier literature, the mean survival time of animals exposed to low-level, whole-body radiation was reported, in a few instances, to exceed that of the controls. This phenomenon has since been interpreted by some observers as evidence for the existence of a beneficial, or hormetic, effect of small doses of radiation (Lu82, Hi83). In each such experiment, however, the survival of the nonirradiated controls was compromised by mortality from intercurrent infection. Even if such an effect of low-level irradiation were reproducible, which is uncertain, its biological significance and its relevance to human populations living under contemporary conditions of nutrition and sanitation are questionable (Sa62, UN82).

Relatively low doses of ionizing radiation can produce certain other types of effects which might be interpreted as beneficial (Sa87). For example, experimental studies have demonstrated prolongation of the life span in arthropods and single-celled organisms under certain conditions. Again however, the various types of molecular and cellular changes in biological systems (e.g., alterations in cell proliferation kinetics, changes in cell life cycle, induction of sterility, and other adaptive mechanisms) through which radiation may produce the observed effects are of doubtful relevance to the risks of radiation-induced mutagenic and carcinogenic effects in human populations.

FERTILITY AND STERILITY

General Considerations

Depending on their degree of maturation and differentiation, the germinal cells of the mammalian testis and ovary are highly radiosensitive (Fa72, Ha87).

The seminiferous epithelium of the testis maintains a steady state of spermatogenesis throughout reproductive life, which involves the active proliferation and differentiation of spermatogonial stem cells. Through this process, the stem cells sequentially give rise to type A and type B spermatogonia spermatocytes, spermatids, and, ultimately, the functional end cells, spermatozoa. In contrast, the female is born with a full complement of maturing oocytes that no longer undergo cell division. On the contrary, the number of oocytes in the ovary decreases throughout adult life through physiological attrition and, to a much lesser extent, ovulation.

Radiation damage to the reproductive cells of the mammalian testis or ovary can impair fertility and fecundity. If the dose is high enough, sterility may result; however, impairment of fertility requires a dose large enough to damage or deplete most of the reproductive cells. If the number or proportion of cells that are damaged remains sufficiently small, fertility is not impaired. Thus, the effect is dose-dependent, with a threshold which varies among species and individuals of differing susceptibility (ICRP84, Up87).

Testis

The germ cells of the human testis may be highly radiosensitive, depending on their degree of maturation (Fa72, Ha87). Type A spermatogonia appear to represent the most sensitive cell stage; later stages of spermiogenesis are highly radioresistant. Sufficient numbers of type A spermatogonia are killed by 0.15 Gy of acute x-radiation to interrupt spermatozoa production, leading to temporary infertility. After an x ray dose in excess of 3-5 Gy, whether delivered acutely or fractionated over a few days or weeks, permanent sterility may result (UN82). An x ray dose of 1.2-1.7 mGy/day has been observed to be tolerated indefinitely by dogs, without detectable effects on their sperm production (Ca68, Fe78, Fe79). Under continuous gamma-radiation exposure to 18 mGy/day, the testis of the mouse has been observed to maintain spermatogenesis, similarly, albeit at reduced levels, for as long as 16 weeks (Fa72).

Ovary

In the human ovary, mature oocytes represent the most sensitive germ cell stage, being killed in sufficient numbers by an acute exposure to 0.65-1.5 Gy to impair fertility temporarily. In contrast, a dose of 6-20 Gy may be tolerated by the ovaries if it is fractionated over a period of weeks (Lu72, Lu76). The threshold for permanent sterilization of the human ovary decreases with increasing age (UN82, ICRP84, Up87).

Conclusions

The estimated threshold dose equivalent for induction of temporary sterility in the adult human testis is 0.15 Sv; for permanent sterility it is 3.5 Sv when received as a single exposure. The corresponding threshold dose equivalent for permanent sterility in the adult ovary is 2.5-6.0 Sv received in a single exposure and 6.0 Sv when received in highly fractionated or protracted exposures (ICRP84).

REFERENCES

Al57 Alexander, P. 1957. Accelerated aging: Long term effect of exposure to ionizing radiations. Gerontologia 1:174-193.

Al66 Albert, R. E., A. R. Omran, E. W. Brauer et al. 1966. Follow-up study of patients treated by x-ray for tinea capitas. Am. J. Public Health 56:2114-2120.

Be78 Beebe, G. W., C. E. Land, and H. Kato. 1978. The hypothesis of radiation-accelerated aging and the mortality of Japanese A-bomb victims. Pp. 3-37 in Late Effects of Ionizing Radiation. Vienna: International Atomic Energy Agency.

Bi88 Bithell, J. F., and C. A. Stiller. 1988. A new calculation of the carcinogenic risk of obstetric x-raying. Stat. Med. 7:857-864.

Bl73 Blot, W. J., and R. W. Miller. 1973. Mental retardation following in utero exposure to the atomic bombs of Hiroshima and Nagasaki. Radiology 106:617-619.

Bl75 Blot, W. J. 1975. Review of thirty years study of Hiroshima and Nagasaki atomic bomb survivors. II. Biological effect. C. Growth and development following prenatal and children exposure to atomic radiation. J. Radiat. Res. 16(Suppl):82-88.

Br52 Brues, A. M., and G. A. Sacher. 1952. Analysis of mammalian radiation injury and lethality. Pp. 441-465 in Symposium on Radiobiology, J. J. Nickson, ed. New York: John Wiley.

Br87 Brent, R. L., D. A. Beckman, and R. P. Jensh. 1987. Relative radiosensitivity of fetal tissue. Adv. Radiat. Biol. 12:239-256.

Ca57 Casarett, G. W. 1957. Acceleration of Aging by Ionizing Radiation. UR-492.

Ca68 Casarett, G. W., and H. A. Eddy. 1968. Fractionation of dose in radiation-induced male sterility. Pp. 14.1-14.10 in Dose Rate in Mammalian Radiation Biology, D. G. Brown, R. G. Cragle, and T. R. Noonon, eds. USAEC CONF-680410.

Co84 Covelli, V., V. Di Majo, B. Bassani, S. Rebessi, M. Coppola, and G. Silini. 1984. Influence of age on life shortening and tumor induction after x-ray and neutron irradiation. Radiat. Res. 100:348-364.

Do73 Dobbing, J., and J. Sands. 1973. Quantitative growth and development of human brain. Arch. Dis. Child 48:757-767.

Fa72 Fabrikant, J. I. 1972. Radiobiology. Chicago: Year Book Medical.

Fa72b Fabrikant, J. I. 1920. Cell population kinetics in the eminiferous epithelian under continuous low dose rate radiation. Pp. 805-814 in Advances in Radiation Research, Biology and Medicine, Vol. II, J. F. Duplan and A. Chapiro, eds. New York: Gordon and Breach.

Fe78 Fedorova, N. L., and B. A. Markelov. 1978. Functional activity of dog's testicles at chronic and combined gamma radiation in the course of three years. Kosmicheskara Biologua Aviakomicheskaia Meditsina 12:42-46.

Fe79 Fedorova, N. L., and B. A. Markelov. 1979. Dog's spermatogenesis after interuption of three year's chronic gamma-irradiation. Rabiobiologiva 12:42-46.

Fu58 Furchtgott, E., and M. Echols. 1958. Activity and emotionality in pre- and neonatally x-irradiated rats. J. Comp. Physiol. Psychol. 51:541-545.

Fu75 Furchtgott, E. 1975. Ionizing radiation and the nervous system. In Biology of Brain Disfunction, G. E. Gaull, ed. New York: Plenum Press.

Ha53 Ham, W. T., Jr. 1953. Radiation cataract. Arch. Ophthalmol. 50:618-643.

Ha85 Harvey, E. B., J. D. Boice, Jr., M. Honeyman, and J. T. Fannery. 1985. Prenatal x-ray exposure and childhood cancer in twins. N. Engl. J. Med. 312:541-545.

Ha87 Hall, E. 1987. Radiobiology for the radiologist. New York: Harper & Row.

He44 Henshaw, P. S. 1944. Experimental roentgen injury. IV. Effects of repeated small doses of x-rays on the blood picture, tissue morphology and life span in mice. J. Natl. Cancer Inst. 4:513-522.

Hi83 Hickey, R. J., E. J. Bowers, and R. C. Clelland. 1983. Radiation hormesis, public health, and public policy: A commentary. Health Phys. 44:207-209.

Hs76 Hsu, T. H., and J. I. Fabrikant. 1976. Spermatogonial cell renewal under continuous irradiation at 1.8 and 4.5 rads per day. Pp. 157-168 in Biological and Environmental Effects of Low-Level Irradiation, Vol. 1. Report IAEA-SM-202/214. Vienna: International Atomic Energy Agency.

ICRP84 International Commission on Radiological Protection (ICRP). 1984. Non-stochastic Effects of Ionizing Radiation. ICRP Publication 41. Oxford: Pergamon.

ICRP86 International Commission on Radiological Protection. 1986. Developmental Effects of Irradiation on the Brain of the Embryo and Fetus. ICRP Publication 49. Oxford: Pergamon.

ICRP88 International Commission on Radiological Protection. In press. Statement from the 1987 Como Meeting of the ICRP. Oxford: Pergamon.

Ja70 Jablon, S., and H. Kato. 1970. Childhood cancer in relation to prenatal exposure to a-bomb radiation. ABC TR 26-70. Lancet ii:1000-1003.

Ka88 Kato, H., Y. Yoshimoto, and W. J. Schull. 1988. Risk of Cancer among in Utero Children Exposed to A-Bomb Radiation. RERF Technical Report. In press.

Le62 Levinson, B. 1962. Effects of neonatal irradiation on learning in rats. In Response of the Nervous System to Ionizing Radiation, T. J. Haley and R. S. Snider, eds. New York: Academic Press.

Lu72 Lushbaugh, C. C., and R. C. Ricks. 1972. Some cytokinetic and histopathologic considerations of irradiated male and female gonadal tissues. Pp. 228-248 in Frontiers of Radiation Therapy and Oncology, Vol. 6, J. M. Vaeth, ed. Basel: Karger.

Lu76 Lushbaugh, C. C., and G. W. Casarett. 1976. The effects of gonadal irradiation in clinical radiation therapy: A review. Cancer 37:1111-1120.

Lu82 Luckey, T. D. 1982. Physiological benefits from low levels of ionizing radiation. Health Phys. 43:771-789.

Lv74 Lvovskaya, E. N. 1974. The state of eye in persons occupied in roentgen-radiological facilities of Moscow. Proceedings of NIJGT i PZ:209-214.

Ma62 MacMahon, B. 1962. Prenatal x-ray exposure and childhood cancer. J. Natl. Cancer Inst. 28:1173-1191.

Ma66 Manosevitz, M., and J. R. Rostkowski. 1966. The effects of neonatal irradiation on postnatal activity and elimination. Radiat. Res. 28:701-707.

Ma75a Matanoski, G. M., R. Seltser, P. E. Sartwell, et al. 1975. The current mortality rates of radiologists and other physician specialists: Death rate from all causes and from cancer. Am. J. Epidemiol. 101:188-198.

Ma75b Matanoski, G. M., R. Seltser, P. E. Sartwell, et al. 1975. The current mortality rates of radiologists and other physician specialists: Specific causes of death. Am. J. Epidemiol. 101:199-210.

Ma82 Martinez, P. F. A. 1982. Neuroanatomy. Development and Structure of the Central Nervous System. Philadelphia: W. B. Saunders.

Me72 Merriam, G. R., A. Schechter, and E. F. Focht. 1972. The effects of ionizing radiation on the eye. Front. Radiat. Ther. Oncol. 6:346-385.

Me81 Meadows, A. T., J. Gordan, D. J. Massari, P. Littman, J. Fergusson, and K. Moss. 1981. Declines in IQ scores and cognative disfunctions in children with acute lymphocytic leukemia treated with cranial irradiation. Lancet ii:1015-1018.

Mi76 Miller, R. W., and J. H. Mulvihill. 1976. Small head size after atomic irradiation. Teratology 14:335-338.

Mi86 Miller, R. W., and J. D. Boice, Jr. 1986. Radiogenic cancer after prenatal or childhood exposure. In Radiation Carcinogenesis, A. Upton et al., eds. New York: Elsevier.

Mo74 Mole, R. H. 1974. Antenatal irradiation and childhood cancer: Causation or coincidence? Br. J. Cancer 30:199-208.

Mo84 Monson, R. R., and B. MacMahon. 1984. Prenatal x-ray exposure and cancer in children. In Radiation Carcinogenesis: Epidemiology and Biological Significance, J. D. Boice, Jr., and J. F. Fraumeni. Jr., eds. New York: Raven Press.

NRC72 National Research Council Advisory Committee on the Biological Effects of Ionizing Radiations. 1972. The Effects on Populations of Exposure to Low Levels of Ionizing Radiations (BEIR I). Washington, D.C.: National Academy of Sciences.

NRC80 National Research Council Committee on the Biological Effects of Ionizing Radiation. 1980. The Effects on Populations of Exposure to Low Levels of Ionizing Radiation (BEIR III). Washington, D.C.: National Academy of Sciences.

Om78 Omran, A. R., R. E. Shore, R. A. Markoff et al. 1978. Follow-up study of patients treated by x-ray epilation for tinea capitas: Psychiatric and psychomotor evaluation. Am. J. Public Health 68:561-567.

Ot82 Otake, M., and W. J. Schull. 1982. The relationship of gamma and neutron radiation to posterior lenticular opacities among atomic bomb survivors in Hiroshima and Nagasaki. Radiat. Res. 92:574-595.

Ot83 Otake, M., and W. J. Schull. 1983. In Utero Exposure to A-Bomb Radiation and Mental Retardation. A Reassessment. RERF Technical Report No. 1-83.

Ot86 Otake, M., and W. J. Schull. 1986. Analysis and interpretation on deficits of the central nervous system observed in the in utero exposed survivors of Hiroshima and Nagasaki. Jpn. J. Appl. Stat. 15:163-180.

Ot87 Otake, M., H. Yoshimaru, and W. J. Schull. 1987. Severe Mental Retardation among the Prenatally Exposed Survivors of the Atomic Bombing of Hiroshima and Nagasaki: A Comparison of the Old and New Dosimetry Systems. RERF Technical Report 16-87.

Ot88 Otake, M., W. J. Schull, Y. Fujikoshi, and H. Yoshimaru. 1988. Effect on School Performance of Prenatal Exposure to Ionizing Radiation in Hiroshima: A Comparison of the T65DR and DS86 Dosimetry Systems. RERF Technical Report 2-88. In press.

Pr87 Preston, D. L., and D. A. Pierce. 1987. The Effect of Changes in Dosimetry on Cancer Mortality Risk Estimates in the Atomic Bomb Survivors. RERF Technical Report 9-87.

Ra75 Rakic, P. 1975. Cell migration and neuronal ectopias in the brain. Pp. 95-129 in Morphogenesis and Malformation of the Face and Brain, D. Bergsma, ed. New York: Alan R. Liss.

Ra78 Rakic, P. 1978. Neuronal migration and contact guidance in the primate telencephalon. Postgrad. Med. J. 54(Suppl. 1):Z5-40.

Ro76 Roth, J., M. Brown, M. Catterall et al. 1976. Effects of fast neutrons on the eye. Br. J. Ophthalmol. 60:236-244.

Ro82 Ron, E., B. Modan, S. Flora, I. Harkedar, and R. Gureurt. 1982. Mental function following scalp irradiation during childhood. Am. J. Epidemiol. 116:149-160.

Ro87 Roesch, W. C. 1987. Reassessment of Atomic Bomb Radiation Dosimetry in Hiroshima and Nagasaki: Final Report. Hiroshima: Radiation Effects Research Foundation.

Ru39 Russ, S., and G. M. Scott. Biological effects of gamma-irradiation (Series II). Br. J. Radiol. 12:440-441.

Ru78 Rubin, P., and G. W. Casarett. 1978. Clinical radiation pathology, Vols. I and II. Philadelphia: W. B. Saunders.

Sa62 Sacher, G. A., and E. Trucco. 1962. A theory of the improved performance and survival produced by small doses of radiation and other poisons. Pp. 244-251 in Biological Aspects of Aging, N. W. Shock, ed. New York: Columbia University Press.

Sa66 Sacher, G. A. 1966. The Gompertz transformation in the study of the injury-mortality relationship: application to late radiation effects and aging. Pp. 411-441 in Radiation and Aging, P. J. Lindop and G. A. Sacher, eds. London: Taylor and Francis.

Sa70 Sacher, G. A., D. Grahn, R. J. M. Fry et al. 1970. Epidemiological and cellular effects of chronic radiation exposure: A search for relationship. Pp. 13-38 in First European Symposium on Late Effects of Radiation, P. Metalli, ed. Rome: Comitato Nazionale Energia Nucleare.

Sa87 Sagan, L. A. 1987. What is hormesis and why haven't we heard about it before. Health Phys. 52:521-525.

Sc86a Schull, W. J., and M. Otake. 1986. Effects on Intelligence of Prenatal Exposure to Ionizing Radiation. RERF Technical Report 7-86.

Sc86b Schull, W. J., and M. Otake. 1986. Neurological deficit and in utero exposure to the atomic bombing of Hiroshima and Nagasaki: A reassessment and new directions. Pp. 399-419 in Radiation Risks to the Developing Nervous System, H. Kriegel et al., eds. New York: Gustav Fischer Verlag.

Sc88 Schull, W. J., M. Otake, and H. Yoshimaru. 1988. Effect on Intelligence Test Score of Prenatal Exposure to Ionizing Radiation in Hiroshima and Nagasaki: A Comparison of the Old and New Dosimetry Systems. Revised Technical Report 3-88. In preparation.

Se58 Seltser, R., and P. E. Sartwell. 1958. Ionizing radiation and longevity of physicians. J. Am. Med. Assoc. 166:585-587.

Se65 Seltser, R., and P. E. Sartwell. 1965. The influence of occupational exposure to radiation on the mortality of American radiologists and other medical specialists. Am. J. Epidemiol. 81:2-22.

Sh76 Shore, R. E., R. E. Albert, and B. S. Pasternack. 1976. Follow-up study of patients treated by x-ray for tinea capitas: Resurvey of post treatment illness and mortality experience. Arch. Environ. Health 1:17-24.

Sh88 Shimbun, C. 1988. Threshold dose for severe mental retardation of in utero exposed children determined by RERF study. Personal communication, 14 January.

St56 Stewart, A., J. Webb, D. Giles, and D. Hewitt. 1956. Malignant disease in childhood and diagnostic irradiation in utero. Lancet ii:447-448.

St58 Stewart, A., J. Webb, and D. Hewitt. 1958. A survey of childhood malignancies. Br. Med. J. 1:1495-1508.

UN77 United Nations Scientific Committee on the Effects of Atomic Radiation (UNSCEAR). 1977. Sources and Effects of Ionizing Radiation. Report E.77.IX.1. New York: United Nations.

UN82 United Nations Scientific Committee on the Effects of Atomic Radiation (UNSCEAR). 1982. Ionizing Radiation: Sources and Biological Effects. Report E.82.IX.8. New York: United Nations.

UN86 United Nations Scientific Committee on the Effects of Atomic Radiation (UNSCEAR). 1986. Genetic and Somatic Effects of Ionizing Radiation. Report E.86.IX.9. New York: United Nations.

Up60 Upton, A. C., A. W. Kimball, J. Furth, K. W. Christenberry, and W. H. Benedict. 1960. Some delayed effects of atom-bomb radiations in mice. Cancer Res. 20(No. 8, Part II):1-93.

Up87 Upton, A. R. 1987. Cancer induction and non-stochastic effects. Br. J. Radiol. 60:1-16.

Wa56 Warren, S. 1956. Longevity and causes of death from irradiation in physicians. J. Am. Med. Assoc. 162:464-468.

Wa66 Warren, S., and O. M. Lombard. 1966. New data on the effects of ionizing radiation on radiologists. Arch. Environ. Health 13:415-421.

Wa75 Walburg, H. E. 1975. Radiation-induced life-shortening and premature aging. Adv. Radiat. Biol. 7:145-179.

Wo67 Wood, J. W., K. G. Johnson, Y. Omori, S. Kawamoto, and R. J. Keehn. 1967. Mental retardation in children exposed in utero, Hiroshima and Nagasaki. Am. J. Public Health 57:1381-1390.

Ya62 Yamazaki, J. N., L. E. Bennett, and C. D. Clemente. 1962. Behavioral and histological effects of head irradiation in new born rats. In Response of the Nervous System to Ionizing Radiation, T. J. Haley and R. S. Snider, eds. New York: Academic Press.

Yo88 Yoshimoto, Y., H. Kato, and W. J. Schull. 1988. Risk of Cancer among in Utero Children Exposed to A-Bomb Radiation: 1950-84. RERF Technical Report 4-88. Hiroshima: Radiation Effects Research Foundation.

7

Low Dose Epidemiologic Studies

INTRODUCTION

As pointed out in Chapter 1, studies of the imputed effects of irradiation at low doses and low dose rates fulfill an important function even though they do not provide sufficient information for calculating numerical estimates of radiation risks. They are the only means available now for determining that risk estimates based on data accumulated at higher doses and higher dose rates do not underestimate the effects of low-level radiation on human health. As also discussed in Chapter 1, there is good reason to postulate, on the basis of animal studies, that the carcinogenic effectiveness of low-LET radiations is reduced at low dose rates, although the available human data do not suffice to confirm this hypothesis.

In its review of low dose studies reported since the BEIR III report (NRC80), this Committee considered populations exposed to radiation from a number of different sources: diagnostic radiography, fallout from nuclear weapons testing, nuclear installations, radiation in the work place, and high levels of natural background radiation. Studies of prenatal exposures to diagnostic x rays are discussed in Chapter 6.

DIAGNOSTIC RADIOGRAPHY: ADULT-ONSET MYELOID LEUKEMIA

A case-control study of patients with chronic myelogenous leukemia (CML) (Pr88) found that during the 3-20 years prior to their diagnosis, more cases than controls had x-ray examinations of the back, gastrointestinal (GI) tract, and kidneys; and cases more often had GI tract and radiographs of the back taken on multiple occasions. A total of 5 cases and 0 controls had GI tract series done on four or more separate occasions, and 11 cases and 1 control had back x rays done on five or more occasions. The odds ratio for exposure to 0-0.99, 1.00-9.99, 10.00-19.99, and $\geq$ 20.0 Gray in the 3-20 years prior to diagnosis were 1.0, 1.4, 1.7, and 2.4, respectively (*p* for the highest exposure category, $p < 0.05$). The association was strongest for the period 6-10 years prior to diagnosis, and the effect of radiation exposure during this period remained significant after consideration of other risk factors in a logistic regression analysis. It was estimated that 23% of cases were attributable to exposure to diagnostic x rays during the period 3-20 years prior to the date of diagnosis of the case (17% during the 6-10 years prior to diagnosis).

These recent findings support the association of adult-onset myelogenous leukemia (ML) with certain types of radiographic examinations and with multiple such examinations. The findings are similar to those of the case-control study in New Zealand which found that risk of ML increased with the frequency of x-ray examination of the back and GI tract (Gu64). In the earlier British and tri-state leukemia studies, it was also noted that patients with ML were more likely than controls to have had multiple radiographic examinations (St62, Gi72).

A study that was without positive risk findings involved a smaller number of patients (63 patients with ML, including some children) and used nonleukemia patients as controls. The controls were matched to the cases by having visited the same clinic at two distinct times (the year when the patient was diagnosed with ML and the year when the patient first visited the clinic before diagnosis of ML) (Li80). This algorithm for control selection may have introduced a serious bias, since controls selected from among repeat clinical patients are likely to have received more medical attention (including more x-ray examinations) than the general population.

Summary

The issue as to how much adult-onset ML is attributable to diagnostic radiography is still unresolved. Questions that have been raised include: (1) whether the excess radiography may have been for preleukemic conditions; (2) whether the association between ML and radiography was due to confounding by the conditions for which x rays were taken; (3) whether there were possible sources of bias (selection, recall, etc.); (4) host susceptibility variables; and 5) dosimetry. Two studies that have attempted to evaluate questions 1 and 2 have found little to suggest that much of the observed association was attributable to these sorts of confounding (St62, Pr88); only in the period immediately preceding diagnosis did patients with ML have more x rays because of infections or vague illnesses, and the strongest association of ML with radiography was seen not during this period but during the previous period. The positive studies found that the reasons for trunk x rays were distributed similarly in cases and controls, but that for any given reason prompting relatively high bone marrow doses, cases had more repeat exams.

Potential bias is always a concern in case-control studies. Another concern and a major limitation of all case-control studies of ML associated with diagnostic radiography is that the dosimetry is uncertain. Doses for a typical examination are, therefore, usually assigned if dose estimates are made at all. A recent dosimetry survey of diagnostic radiographic procedures performed in the United Kingdom shows that the range of doses administered for each type of examination is wide (Sh86).

FALLOUT FROM NUCLEAR WEAPONS TESTING

In the late 1970s, several studies reported excess cancer, primarily leukemia, among persons who were exposed to fallout from nuclear weapons tests. These included residents of Utah and neighboring states downwind of the Nevada Test Site (NTS) and veterans who had participated in the tests. Estimates of the doses to most organs in both groups was reported to be sufficiently low (less than 50 milliGray for all tests combined) so that no detectable increase in risk would have been predicted on the basis of cancer risk estimates derived from high-dose studies. A possible exception was the dose to the thyroid, which exceeded 500 mGy in some individuals (studies of thyroid tumors are reviewed in Chapter 5).

Cancer Among Residents Downwind from NTS

A survey of death rates from excess cases of childhood leukemia in Utah from 1944 to 1975 was reported in 1979 (Ly79). Based on preliminary data on fallout patterns, the state was divided into two parts; counties with above average and supposedly below average levels of exposure. The time periods considered were chosen so that there were two "unexposed" cohorts (deaths occurring before 1951 or in children born after 1958) and one "exposed" cohort (those under age 15 at any time from 1951 to 1958). In the "low-exposure" counties, all three cohorts had mortality rates that were comparable to the rates for the U.S. population as a whole. In the "high-exposure" counties, the "unexposed" cohorts had rates that were lower than the rates for the United States as a whole, while the "exposed" cohort had rates slightly higher than U.S. rates and about 2.4 times higher than the rates for the "unexposed" cohorts. Land (La79) subsequently pointed out that death rates for all other childhood cancers in this study showed the opposite pattern; that is, a lower rate for the "exposed" cohort in the "high exposure" area. This suggested that the apparent increase in leukemia rates might have been an artifact of diagnostic error. Land et al. (La84) later reexamined the association, using mortality data from the National Center for Health Statistics for 1950 to 1978, and found that while leukemia death rates were about 50% higher in the "exposed" than in the "unexposed" cohorts, they were not significantly different at a 90% confidence level. Moreover, compared to the "unexposed cohort," rates for eastern Oregon, Iowa, and the total United States were also higher by about the same amount. Land et al. concluded that there was no pattern of excess leukemia mortality that supported a causal association with fallout exposure and that the excess reported reflected an anomalously low rate in southern Utah during the period 1944 to 1949. Both studies suffer from the fact that comparisons are based on aggregate groups and may not reflect any associations among individuals. Futhermore, Beck and Krey (Be83) have since shown that the levels of fallout were not, in fact, higher in the "high exposure" counties than in the "low exposure" counties, contrary to the original supposition. A case-control study is currently in progress to examine the association between leukemia and individual estimates of doses, which includes 1,179 patients with leukemia and 5,380 people who died from other causes among Mormon residents of Utah from 1952 to 1981.

In 1984, Johnson (Jo84) reported on results of a retrospective cohort study of cancers in Mormon families who were listed in both the 1951 and 1961 telephone directories for towns in southwestern Utah and neighboring parts of Nevada and Arizona. Self-reports of cancer and other diseases among those that could be located in 1981 were obtained by volunteers. A total of 288 cases of cancer were reported in this group, compared with 179 cases of cancer expected on the basis of rates for all Utah Mormons. The major excesses (observed/expected) were for leukemia (31/7.0), thyroid cancer (20/3.1), breast cancer (35/23.0), melanoma (12/4.5), bone cancer (8/0.7), and brain tumors (9/3.9). A subgroup of 239 persons who reported acute effects from fallout exposure showed even higher rates of cancer (33 cases of cancer at all sites observed compared with 7.1 expected cases). The cancers reported were not medically confirmed and were likely to have been overreported; Lyon and Schuman (Ly84) point out that the female:male ratio was about 70% higher in this study than nationally, suggesting overreporting of female cases, and that only 126 deaths from all causes were reported, whereas at least 192 deaths from cancer would have been expected.

Johnson's reliance on data gathered by volunteers appears to be a weak point in his study. Machado et al. (Ma87) analyzed cancer rates from the National Center for Health Statistics for the three counties of southwestern Utah covered by the survey over the periods 1955-1980 for leukemia and 1964-1980 for other cancers, and found no excesses of either single or grouped sites, with the exception of leukemia (62/42.8 for people of all ages, 9/3.2 for those from 0-14 years old).

Cancer Among Participants in Nuclear Weapons Tests

U. S. Weapons Tests

In 1980, Caldwell et al. (Ca80) reported that among the 3,224 participants of the nuclear test explosion Smoky, nine cases of leukemia occurred through 1977, compared with 3.5 expected cases. In a later report (Ca83), the number of cases of leukemia increased to 10/4.0 and data were provided on cancer at other sites through 1979. The total number of observed cases of cancer was 112, compared with 117.5 expected; there was a significant increase only in leukemia incidence and mortality. In 1984, four cases of polycythemia vera were observed, compared with 0.2 expected (Ca84). Robinette et al. (Ro85) expanded the study to include a cohort of 46,186 participants in one or more of five test series at the NTS or the Pacific Proving Ground (PPG). The excess cases of leukemia among the participants of the Smoky test were confirmed, but only 46 deaths from leukemia were observed in the participants of the other PPG tests, compared with 52.4 expected deaths. No one series showed a significant excess of leukemia, and there was also no consistent excess for any other cancer site.

British Weapons Test

Darby et al. (Da88) described a cohort study of 22,347 British participants in nuclear weapons tests and related experimental programs in Australia and the Pacific Ocean and 22,325 matched controls. For all causes of death RR = 1.01; for all cancers RR = 0.96. Leukemias and multiple myeloma occurred significantly more often in participants than controls; 22 versus 6 cases and 6 versus 0 cases, respectively. However, for participants at both test sites, the death rates were only slightly higher in participants than expected, based on national rates (SMR = 113 and 111 respectively), while the death rates were much lower than expected in the controls (SMR = 32 and 0, respectively). There was no association with the type or degree of radiation exposure.

Canadian Studies

Raman et al. (Ra87) carried out a cohort study of 954 Canadian military personnel who had been involved in clean-up operations after nuclear reactor accidents at Chalk River Nuclear Laboratories or who had observed nuclear weapons blasts in the United States or Australia; two matched controls were selected from military records for each exposed subject. No differences in cause-specific mortality between cases and controls and no trends by degree of exposure were found; the study size was small, and only very large differences would have been detectable.

Leukemia from Global Fallout

Archer (Ar87) compared time trends in global fallout across the United States with trends in leukemia rates. Fallout activity was estimated from measurements of beta emissions, airborne particulates, precipitation, and ^{131}I in milk. Fallout appeared to peak in 1957 and 1962. Death rates for acute and myeloid leukemia in children aged 5-9 years rose to an initial peak in 1962 and a secondary peak in 1968; no such pattern was observed for other types of leukemia. Leukemia death rates (for all ages and all cell types) peaked in the decade 1960-1969 and were consistently highest in states with high ^{90}Sr levels in the diet, milk, and bones (based on surveys by the Public Health Services from 1957 to 1970) and lowest in states with low ^{90}Sr levels. The excess of myeloid and acute leukemia deaths was estimated to be about 6.5 per 10^4 PYGy (based on an estimated average cumulative dose of 4 mGy).

Darby and Doll (Da87) reviewed data on childhood leukemia incidence rates and fallout exposures in England and Wales, Norway, and Denmark. Fallout exposures rose rapidly between 1962 and 1965 and declined slowly thereafter. In England and Wales there was about a 10% increase in incidence rates up to 1979, possibly attributable to improvements in diagnosis, whereas incidence rates in Norway and Denmark declined slightly after 1960 during the period of highest population exposure from fallout. The data were thus interpreted to provide no convincing evidence of an increase in incidence that could be attributed to fallout.

Summary

There are several possible explanations for the cancer excesses that have been reported in the studies cited. The possibility that they may represent chance variations may explain the excess cases of leukemia associated with the Smoky nuclear test, although that test was unusual in ways that are discussed below. Chance may also explain the differences in results from the three studies of leukemia in Utah residents that are based on reported death rates. Although there appeared to be a small excess in southwestern Utah, the causality of fallout exposure cannot be assessed from these studies; the case-control study in progress may help resolve this uncertainty.

On the other hand, associations may be real and reflect an underestimation either of the doses or of the risk per unit dose. This may be the case for the Smoky nuclear test, which was the highest-yield tower detonation at the NTS. Fallout was particularly heavy, 10 to 20 times greater than at other detonations in this test series (Ha81). The leukemias occurred most frequently in two groups: those near the hypocenter and those ferried in by helicopters within hours of the test. Whether these doses could have been large enough to explain the excess is uncertain.

Although there was a wide variation in individual doses among participants at nuclear tests (Ro85), the collective dose could not have been underestimated sufficiently to explain the excess if the risk coefficients derived from high-dose studies were correct and not underestimated. The most likely explanation is that the observed excess cases of leukemia are random overestimates of the risk coefficients. In view of the uncertainty in both sets of estimates, the discrepancy may be small; Archer's estimate of the risk coefficient for leukemia based on his data on global fallout is only slightly higher than that based on data for the atomic-bomb survivors (Ar87).

CANCER AMONG INDIVIDUALS NEAR NUCLEAR INSTALLATIONS

Nuclear Reactor Accidents

It is still too early to assess whether any cancer excess will occur following the Three Mile Island or Chernobyl nuclear reactor accidents. The collective dose equivalent resulting from the radioactivity released in the Three Mile Island accident was so low that the estimated number of excess cancer cases to be expected, if any were to occur, would be negligible and undetectable (Fa81). For the Chernobyl accident, preliminary estimates suggest that up to 10,000 excess cancer deaths could occur over the next 70 years among the 75 million Soviet citizens exposed to the radioactivity released during the accident, against a background of 9.5 million cases of cancer that would occur spontaneously; hence the excess would not be detectable. However, among the 116,000 people evacuated from immediate high-exposure areas in the Ukraine and Byelorussia, there might be a detectable increase in the cases of leukemia and solid cancer (An88, No86).

Leukemia Among Individuals Near British Nuclear Reprocessing Plants

In the district near the Sellafield nuclear reprocessing plant in northern England, 6 leukemia deaths in children aged 0-24 years occurred from 1968 to 1974, compared with 1.4 expected cases (Ga84), and 19 incident cases occurred (10.5 expected) (IAG84). Follow-up studies of two cohorts, one of children born to women resident in the Seascale Civil Parish during 1950-1983 (Ga87a) and one of children born elsewhere but attending schools in

Seascale were performed (Ga87b). There were five deaths from leukemia (0.53 expected) in the former cohort; no deaths were found in the latter cohort.

Within 12.5 km of Dounreay, a nuclear reprocessing plant in northern Scotland, five cases of leukemia occurred (0.5 expected) in children of the same age from 1979 to 1984 (He86). Darby and Doll (Da87) reviewed the data on radiation exposures in Dounreay and concluded that the excess cases were not explainable by radioactive discharges from that nuclear installation. Recently, the influx of a large number of new workers with a concomitant increase in viral infections has been proposed as a causative factor for childhood leukemia in Dounreay and Sellafield (Ki88).

Cancer Among Individuals Near Other Nuclear Installations

Roman et al. (Ro87) described a cluster of 29 cases of leukemia (14.4 expected) in children aged 0-4 years living within 10 km of one or more nuclear facilities in southern England. Hole and Gillis (Ho86) reported a cluster of 31 cases of leukemia (24.3 expected) in children aged 0-14 years living in regions adjacent to four nuclear facilities in western Scotland. These reports are difficult to interpret, owing to the bias due to first observing an apparent cluster and then defining the population at risk and the time period of risk. To avoid this bias, Baron (Ba84) examined cancer mortality in individuals living near 14 nuclear and 5 nonnuclear facilities in England and Wales and found no overall pattern of increasing cancer SMRs in individuals living around the nuclear facilities. A more comprehensive survey of cancer incidence and mortality near nuclear installations for the period 1959-1980 is reported by the United Kingdom Office of Population Censuses and Surveys (Co87a, Fo87). The investigators found significant overall excesses of cancer mortality due to lymphoid leukemia and brain cancer in children and due to liver cancer, lung cancer, Hodgkin's disease, all lymphomas, unspecified brain and central nervous system tumors, and all malignancies in adults; however, the mortality rates in the control areas were lower than expected, and there has not been a general increase in cancer rates in individuals living in the vicinity of nuclear installations. Moreover, there were no consistent, positive, or statistically significant trends in cancer rates with distance from the nuclear installations. Beral (Be87) noted that the incidence of leukemia and all cancers in children were significantly elevated in all exposed areas combined (excluding Sellafield) compared with those in control areas, whereas mortality was not. Cook-Mozaffari (Co87) confirmed the observation, but suggested that the differences may be due to a variation in case registration, possibly owing to social class differences.

Recently, Openshaw et al. have demonstrated how cancer clusters can be identified objectively using a Geographic Analysis Machine (Op88). This methodology was applied to mortality data for acute lymphoblastic leukemia in children living in the Northern and Northwestern regions of England. Again, Seascale, near Sellafield, in Cumbria was identified as an area having unusually high mortality. Although this type of analysis requires a large computational effort, it appears to free studies of cancer clusters from bias due to the selection of an arbitrary risk area and the effects of arbitrary administrative boundries.

Clapp et al. (Cl87) reported excess cases of leukemia and other hematologic malignancies in five Massachusetts towns located near a nuclear reactor. There were 13 excess cases of myelogenous leukemia in males (5.2 expected); the excess cases were mainly in adults, and the possible confounding effect of occupational factors was not considered.

No excess cases of cancer have been found around either the Rocky Flats nuclear reprocessing plant in Colorado (Cr87) or the San Onofre nuclear power plant in California (En83).

Summary

It is difficult to assess the significance of the reports of excess cancer cases near nuclear installations in Great Britain; it appears highly unlikely that all were caused by chance, although the anecdotal nature of some of the observations makes testing of significance impossible. Available radiation dosimetry information also makes it seem unlikely that the excesses or clusters could be explained by the very low radiation exposures. While there has not been a general increase in cancer rates in individuals living in the vicinity of nuclear installations (Co87a,b, Fo87), there does appear to be an excess in the number of cases of childhood leukemia, particularly in individuals living around installations before 1955 and among children born in the region. Whether the excesses will be found to be balanced by a comparable number of deficiences around other nuclear installations, or whether they will prove to occur more consistently than not, are questions calling for further study.

EPIDEMIOLOGIC STUDIES OF WORKERS EXPOSED TO LOW DOSE, LOW-LET RADIATION

A number of epidemiologic studies of individuals exposed occupationally to low levels of low-LET radiation have been reported. Although, because of limited size and exposure, such studies cannot contribute directly to the estimation of stable radiation risk estimates, they are of use for assessing whether such estimates are substantially in error. Occupational studies have several noteworthy advantages and disadvantages.

Occupational exposures are generally monitored, but there may still remain considerable uncertainty about exposures measured in early years (see for example In87). Also there may be multiple exposures both to external sources and internal emitters of radiation, and to many potentially carcinogenic chemicals, which may make any specific radiation effect difficult to isolate. Occupational cohorts are usually well defined and their individual members well identified, which facilitates follow-up, but the healthy worker effect—the tendency for working populations to have lower rates of mortality than those of the general population, primarily because of selection factors—means that comparisons between an occupational cohort and the general population can be difficult to interpret.

Epidemiologic Studies of Workers

Table 7-1 summarizes the available details on those occupational studies that have been published to date. In these studies, workers were monitored for their exposure to low-LET ionizing radiation. The power of such studies to detect a significant increase in risk depends on the number of observed deaths from the cause of interest. Several of these studies have yet to accumulate a sufficient number of deaths to reach any sensible conclusions relating to individual types of cancer. The most consistent result, observed to date from the studies shown in Table 7-1, is that the risk estimates for all types of cancer combined and for all types of leukemia combined are consistent with the risk estimates provided in the present report, since no studies have reported results which differ significantly from the null. In terms of individual cancers, a significant and dose-related effect has been observed for multiple myelomas in the Hanford study (Gi89) and in the British Nuclear Fuels study (Sm86); in the latter case, data from individuals who received the dose 15 years before death are excluded. A significant excess of prostate cancer has been observed in the United Kingdom Atomic Energy Authority study (Be85), but this excess seems to be associated, in part, with exposures to multiple forms of radiation, including tritium and other internal nuclides. Excesses of prostate cancer were also seen in the British Nuclear Fuels and the Oak Ridge National Laboratory studies (Ch85), but these excesses were not significant and were not dose related. In addition to multiple myelomas and prostate cancer, dose-response effects as a result of exposure to external gamma radiation have been reported for bladder cancer and all lymphatic and hematopoietic cancers by the British Nuclear Fuels study (doses received in the 15 years prior to death are excluded), and for lung cancer by the Oak Ridge Y-12 Plant study (Ch88). In the latter study, part of the dose to the lung was due to alpha radiation.

TABLE 7-1 Some Epidemiologic Studies of Workers Monitored for External Gamma Radiation

Study	Type of Operation	Years of Employment	Last Year of Follow-up	No. of Individuals in Study	No. of Radiation Workers[a]	Dose[b] (mSv)	Total No. of Deaths
United Kingdom Atomic Energy Authority, UK (Be85)	Reactor research and development	1946–1979	1979	39,546 (29,173 males, 10,373 females)	20,382 (18,759 males, 1,623 females)	32.4	3,373
British Nuclear Fuels Limited (Windscale Plant, Sellafield Plant), UK (Sm86)	Plutonium production, fuel reprocessing, waste treatment, fast reactor fuel fabrication	1946–1975	1983	14,000 (11,402 males, 2,598 females)	10,157	124.0	2,277
Hanford Site, Washington State, USA (Gi89)	Reactor research and development	1944–1978	1981[c]	44,100 (31,500 males, 12,600 females)	36,235	43.6	7,249[d]
Oak Ridge National Laboratory, Tennessee, USA (Ch85)	Reactor research and development, plutonium production, chemical processing, and separation of isotopes	1943–1972	1977	8,375 males	7,778	17.3	966

Table 7-1 *Continued*

Study	Type of Operation	Years of Employment	Last Year of Follow-up	No. of Individuals in Study	No. of Radiation Workers[a]	Dose[b] (mSv)	Total No. of Deaths
Oak Ridge Y-12 Plant, Tennessee, USA (Ch88)	Uranium enrichment, weapon fabrication, isotope research	1947–1974	1979	6,781 males	5,278	9.6	862
Rocky Flats Nuclear Weapons Plant, USA (Wi87)	Plutonium weapons fabrication	1952–1979	1979	5,413 males		41.3	409
Atomic Energy of Canada Limited, Canada (Ho87)	Reactor research and development	1950–1981	1981	13,570 (10,278 males, 3,292 females)	7,685 (6,626 males, 1,239 females)	46.8 males; 3.86 females	946
Ontario Hydro, Canada (An86)	Power reactor operation	1970–1985	1985	23,997 males	5,039		2,860

[a]No. of individuals reported or estimated to be monitored.
[b]Mean whole body gamma dose per radiation worker.
[c]Deaths occurring in the State of Washington in the years 1982–1985 were also evaluated.
[d]Observed deaths from 1945–1981.

Summary

The studies have provided no evidence to date that risk estimates for leukemias and other types of cancer combined are in error, based on extrapolation from high-dose studies. For individual cancer sites, only for multiple myelomas and prostate cancer is there any suggestion that associations were seen in more than one study. In interpreting the latter associations, however, the potential biases discussed in Chapter 1 must be borne in mind. In particular, the problem of multiple comparisons and the tendency for both researchers and editors to focus on positive as opposed to null results. It must also be pointed out that the absence of any associations in a number of studies essentially offers no meaningful evidence, because of the very small numbers of observed deaths. Continued monitoring of these and other occupational cohorts in the future is highly desirable. When possible, standardization and pooling of study results should improve the interpretation and the overall significance of these studies. To date the evidence does not contradict or imply the possible inaccuracy of risk estimates derived from high-dose studies.

HIGH NATURAL BACKGROUND RADIATION

There are regions in the world where outdoor terrestrial background gamma radiation levels appreciably exceed the normal range (about 0.2-0.6 mGy per year). Such regions exist in Brazil, India, People's Republic of China, Italy, France, Iran, Madagascar, and Nigeria (UN82). Because the total dose rate of low-LET natural background radiation is low, and the lifetime dose of such radiation accumulated by any one person is small (<0.1 Gy), it is difficult to determine whether there are any variations in disease rates associated with changes in natural background radiation levels and, if so, whether such variations are consistent with the health effects estimated by extrapolation from the observed effects of high-dose and high dose-rate exposures.

A cautious approach is warranted in the interpretation of geographically based mortality surveys. Although "beneficial" effects of radiation have been alleged on the basis of reduced mortality in high background areas in the United States (Hi81), analyses that include an adjustment for altitude indicate no "beneficial" effects (We86). While mortality rates for both cancer and cardiovascular disease are lower in areas of the United States having high levels of natural radiation, such areas are found primarily in high altitude locations. This apparently "beneficial" effect of radiation may, in fact, be an example of confounding, since conditions of reduced oxygen pressure stimulate a wide array of physiological adaptations, which could themselves be protective (Fr75).

Recently, childhood cancers have been analyzed in relation to natural radiation levels in England (Kn88), and although reported associations were observed, their interpretation is complicated by the general problems of correlational analyses (see Chapter 1, Epidemiological Principles).

Guarapari, Brazil

This village of approximately 12,000 inhabitants is located in an area where local soil contains monazite sands, which is the source of gamma and alpha radiation received by the townspeople. The radioactivity in monazite comes primarily from thorium. The average annual absorbed dose to an inhabitant of this area, based on lithium fluoride dosimetry, is about 6.4 mSv (640 mrem), which is roughly 6 times the global average background radiation dose level (excluding radon progeny in the lung) (Ba75). Studies of the health of this population are limited, but a cytogenetic study of 200 individuals, in comparison with a control group from a similar village, reported an increase in the total number of chromosome aberrations (Ba75).

Kerala, India

The population living along the Kerala Coast of India is exposed to about 4 times the normal level of natural background radiation (excluding radon progeny in the lung). Because of the presence of monazite in the soil (thorium concentration, 8.0-10.5%, by weight), the average absorbed dose rate for the 70,000 people living in the region has been estimated to be about 3.8 mGy/yr (380 mrad/yr) (Go71). The incidence of both Down syndrome and chromosome aberrations has been reported to be increased in this population (Ko76).

Yanjiang County, Guangdong Province, People's Republic of China

The most extensive observations on the health effects of high natural background radiation have been those made on the mortality experience of the population in Guangdong Province, People's Republic of China. In this area, which contains monazite with high levels of thorium, uranium and radium, individuals are exposed to about 3-4 mSv (300-400 mrem) of gamma radiation per year. The population of this region has been studied extensively for both genetic and carcinogenic effects (We86, Ta86). A sample of 70,000 individuals in this area and a geographically adjacent control area, receiving a normal background of radiation of 1 mSv/year (100 mrem/year), were followed for the period 1970-1985, with approximately 1 million person-years of follow-up in each area.

On analysis, site-specific, age-adjusted cancer mortality rates did not differ between the high natural background area and the control area. For total cancer mortality, the observed cancer rate was higher in the normal background area, although the difference was not statistically significant. Known risk factors affecting cancer mortality rates were generally comparable in the two areas, although there were some cultural and educational differences. Chromosome aberrations and a higher reactivity of T lymphocytes were found in individuals in the high natural background area. There were no differences for a large number of hereditary diseases or congenital defects in children. The prevalence of Down syndrome was greater in the high-background region, but this was discounted because the residents of the control area had a lower prevalence of Down syndrome than those of surrounding counties, who had rates similar to those living in the high natural background area.

Summary

In areas of high natural background radiation, an increased frequency of chromosome aberrations has been noted repeatedly. The increases are consistent with those seen in radiation workers and in persons exposed at high dose levels, although the magnitudes of the increases are somewhat larger than predicted. No increase in the frequency of cancer has been documented in populations residing in areas of high natural background radiation.

REFERENCES

An86 Anderson, T. W. 1986. Ontario Hydro Mortality 1970-1985. Ontario Hydro Report, Canada.

An88 Anspaugh, L. R., R. J. Catlin, and M. Goldman. 1988. The global impact of the Chernobyl reactor accident. Science 242:1513-1519.

Ar87 Archer, V. E. 1987. Association of nuclear fallout with leukemia in the United States. Arch. Environ. Health 42:263-271.

Ay84 Ayesh, R., J. R. Idle, J. C. Ritchie, M. J. Crothers, and M. R. Hetzel. 1984. Metabolic oxidation phenotypes as markers for susceptibility to lung cancer. Nature 312:169-170.

Ba75 Barcinski, M. A., M. D. C. A. Abreu, J. C. C. De Almeida, J. M. Naya, L. G. Fonseca, and L. E. Castro. 1975. Cytogenetic investigation in a Brazilian population living in an area of high natural radioactivity. Am. J. Hum. Genet. 27:802-806.

Ba84 Baron, J. A. 1984. Cancer mortality in small areas around nuclear facilities in England and Wales. Br. J. Cancer 50:815-829.

Be83 Beck, H. L., and P. W. Krey. 1983. Radiation exposure in Utah from Nevada nuclear tests. Science 220:18-24.

Be85 Beral, V., H. Inskip, P. Fraser et al. 1985. Mortality of employees of the United Kingdom Atomic Energy Authority, 1946-1979. Br. Med. J. 291:440-447.

Be87 Beral, V. 1987. Cancer near nuclear installations (letter). Lancet i:556.

Ca80 Caldwell, G. S., D. B. Kelley, and C. W. Heath. 1980. Leukemia among participants in military maneuvers at a nuclear bomb test: A preliminary report. J. Am. Med. Assoc. 244:1575-1578.

Ca83 Caldwell G. G., D. Kelley, M. Zack, H. Falk, and C. W. Heath. 1983. Mortality and cancer frequency among military nuclear test (Smoky) participants, 1957 through 1979. J. Am. Med. Assoc. 250(5):620-624.

Ca84 Caldwell, G. C., D. B. Kelley, C. W. Heath, Jr., and M. Zack. 1984. Polycythemia vera among participants of a nuclear weapons test. J. Am. Med. Assoc. 252:662-664.

Ch85 Checkoway, H., R. M. Mathew, C. M. Shy, et al. 1985. Radiation, work experience, and cause specific mortality among workers at an energy research laboratory. Br. J. Ind. Med. 42:525-533.

Ch88 Checkoway H., N. Pearce, D. J. Crawford-Brown, et al. 1988. Radiation doses and cause specific mortality among workers at a nuclear materials fabrication plant. Am. J. Epidemiol. 127:255-266.

Cl87 Clapp, R. W., S. Cobb, C. K. Chan, and B. Walker, Jr. 1987. Leukaemia near Massachusetts nuclear power plant. Lancet ii:1324-1325.

Co87a Cook-Mozaffari, P., F. L. Ashwood, T. Vincent, et al. 1987. Cancer incidence and mortality in the vicinity of nuclear installations. England and Wales, 1950-1980. Studies on Medical and Population Subjects, No. 51. London: Her Majesty's Stationery Office.

Co87b Cook-Mozaffari, P. 1987. Cancer near nuclear installations (letter). Lancet i:855-856.

Cr87 Crump, K. S., T. H. Ng, and R. G. Cuddihy. 1987. Cancer incidence patterns in the Denver metropolitan area in relation to the Rocky Flats plant. Am. J. Epidemiol. 126(1):127-135.

Da87 Darby, S. C., and R. Doll. 1987. Fallout, radiation doses near Dounreay, and childhood leukemia. Br. Med. J. 294:603-607.

Da88 Darby, S. C., G. M. Kendall, T. P. Fell, et al. 1988. A summary of mortality and incidence of cancer in men from the United Kingdom who participated in the United Kingdom's atmospheric nuclear weapon tests and experimental programs. Br. Med. J. 296:332-338.

En83 Enstrom, J. E. 1983. Cancer mortality patterns around the San Onofre nuclear power plant, 1960-1978. Am. J. Public Health 73(1):83-92.

Ev86 Evans, J. S., J. E. Wennberg, and B. J. McNeil. 1986. The influence of diagnostic radiography on the incidence of breast cancer and leukemia. N. Engl. J. Med. 315:810-815.

Fa81 Fabrikant, J. I. 1981. Health effects of the nuclear accident at Three Mile Island. Health Phys. 40:151-161.

Fo87 Forman, D., P. Cook-Mozaffari, S. Darby et al. 1987. Cancer near nuclear installations. Nature 329:499-505.

Fr75 Fraisancho, A. R. 1975. Functional adaptation to high altitude hypoxia. Science 187:313-319.

FSG88 Fallout Study Group. 1988. A case-control study of leukemia in Utah and its relationship to fallout from nuclear weapons testing. Unpublished manuscript.

Ga84 Gardner, M. J., and P. D. Winter. 1984. Mortality in Cumberland during 1959-78 with reference to cancer in young people around Windscale (letter). Lancet i:216-217.

Ga87a Gardner, M. J., A. J. Hall, S. Downes, and J. D. Terrell. 1987. Follow up study of children born to mothers resident in Seascale, West Cumbria (birth cohort). Br. Med. J. 295:822-827.

Ga87b Gardner, M. J., A. J. Hall, S. Downes, and J. D. Terrell. 1987. Follow up study of children born elsewhere but attending schools in Seascale, West Cumbria (schools cohort). Br. Med. J. 295:819-822.

Gi72 Gibson, R., S. Graham, A. Lilienfeld, L. Schuman, J. E. Dowd, and M. L. Levin. 1972. Irradiation in the epidemiology of leukemia among adults. J. Natl. Cancer Inst. 48:301-311.

Gi89 Gilbert, E. S., J. G. R. Petersen, and J. A. Buchanan. 1989. Mortality of workers at the Hanford site: 1945-1981. Health Phys. (in press).

Go71 Gopal-Ayengar, A. R., K. Sundaram, K. B. Mistry, C. M. Sunta, K. S. V. Nambi, S. P. Kathuria, A. S. Basu, and M. David. 1971. Evaluation of the long-term Effects of high background radiation on selected population groups on the Kerala Coast. Pp. 31-51 in Proceedings of the 4th International Conference on Peaceful Uses of Atomic Energy, Vol. 11.

Gu64 Gunz, F., and H. Atkinson. 1964. Medical radiation and leukemia: A retrospective survey. Br. Med. J. 1:389.

Ha81 Harris, P. S., C. Lowery, A. G. Nelson, S. Obermiller, W. J. Ozeroff, and E. Weary. 1981. Shot Smoky, a test of the Plumbob Series, 31 August, 1957. Technical Report DNA 6004F. Washington, D.C.: Defense Nuclear Agency.

He86 Heasman, M. A., I. W. Kemp, J. D. Urquhart, and R. Black. 1986. Childhood leukaemia in northern Scotland (letter). Lancet i:266.

Hi81 Hickey, R. J., E. J. Bowers, D. E. Spence, B. S. Zemel, A. B. Clelland, and R. C. Clelland. 1981. Low level ionizing radiation and human mortality: Multiregional epidemiological studies. Health Phys. 40:625-641.

Hi87 Hickey, R. J., E. J. Bowers, D. E. Spence, B. S. Zeuill, A. B. Cleland, and R. C Cleland. 1987. Low level ionizing radiation and human mortality: multiregional epidemiological studies. Health Phys. 40:625-641.

Ho86 Hole, D. J., and C. R. Gillis. 1986. Childhood leukemia in the west of Scotland. Lancet 2:525-525.

Ho87 Howe, G. R., J. L. Weeks, A. B. Miller, et al. 1987. A follow-up study of radiation workers employed by Atomic Energy of Canada Limited: Mortality, 1950-1981.

IAG84 Independent Advisory Group. 1984. Investigation of the possible increased incidence of cancer in West Cumbria. London: Her Majesty's Stationery Office.

In86 Interlaboratory Task Group on Health and Environmental Aspects of the Soviet Nuclear Accident. June 1987. Health and Environmental Consequences of the Chernobyl Nuclear Power Plant Accident. DOE/ER-0332. Springfield, Va.: National Technical Information Service.

In87 Inskip, H., V. Beral, P. Fraser, et al. 1987. Further assessment of the effects of occupational radiation exposure in the United Kingdom Atomic Energy Authority mortality study. Br. J. Ind. Med. 44:149-160.

Jo84 Johnson, C. J. 1984. Cancer incidence in an area of radioactive fallout downwind from the Nevada test site. J. Am. Med. Assoc. 251:230-136.

Ki88 Kinlen, L. 1988. Evidence for an infective cause of childhood leukemia: comparison of a Scottish new town with nuclear reprocessing sites in Britain. Lancet ii:1323-1326.

Kn88 Knox, E. G., A. M. Stewart, E. A. Gilman, and G. W. Kneale. 1988. Background radiation and childhood cancer. J. Radiol. Prot. 8(1):9-18.

Ko76 Kochupillan, N., I. C. Verma, M. S. Grewal, and V. Ramalingaswami. 1976. Down's syndrome and related abnormalities in an area of high background radiation in coastal Kerala. Nature 262:60-61.

La79 Land, C. E. 1979. The hazards of fallout or of epidemiologic research. N. Engl. J. Med. 300:431-432.

La84 Land, C. E., F. W. McKay, and S. G. Machado. 1984. Childhood leukemia and fallout from the Nevada nuclear tests. Science 223:139-144.

Li80 Linos, A., J. Gray, A. Orvis, et al. 1980. Low dose radiation and leukemia. N. Engl. J. Med. 302:1101.

Ly79 Lyon, J. K., M. R. Klauber, J. W. Gardner, and K. S. Udall. 1979. Childhood leukemias associated with fallout from nuclear testing. N. Engl. J. Med. 300:397-402.

Ly84 Lyon, J. L., and K. L. Schuman. 1984. Radioactive fallout and cancer (letter). J. Am. Med. Assoc. 252(14):1854-1855.

Ma87 Machado, S. G., C. E. Land, and F. W. McKay. 1987. Cancer mortality and radioactive fallout in southwestern Utah. Am. J. Epidemiol. 125:44-61.

NRC80 National Research Council, Committee on the Biological Effects of Ionizing Radiations. 1980. The Effect on Populations of Exposure to Low Levels of Ionizing Radiation (BEIR III). Washington, D.C.: National Academy Press. 524 pp.

No86 Norman C., and D. Dickson. 1986. The aftermath of Chernobyl. Science 223:1141-1143.

Op88 Openshaw, S., M. Charlton, A. W. Craft, and J. M. Birch. 1988. Investigation of leukemia clusters by use of a geographical analysis machine. Lancet i:272-273.

Pi83 Pickle, L. W., L. M. Brown, and W. J. Blot. 1983. Information available from surrogate respondents in case-control interview studies. Am. J. Epidemiol. 118:99.

Pr88 Preston-Martin, S., D. C. Thomas, M. C. Yu, and B. E. Henderson. Submitted. Diagnostic radiography as a risk factor for chronic myeloid leukemia.

Ra87 Raman, S., G. S. Dulberg, R. A. Spasoff, and T. Scott. 1987. Mortality among Canadian military personnel exposed to low dose radiation. Can. Med. Assoc. J. 136:1051-1056.

Ro84 Rowley, J. B. 1984. Biological implications of consistent chromosome rearrangements in leukemia and lymphoma. Cancer Res. 44:3159-3168.

Ro85 Robinette, C. D., S. Jablon, and T. L. Preston. 1985. Studies of participants in nuclear tests. Washington, D.C.: National Research Council.

Ro87 Roman, E., V. Beral, L. Carpenter, et al. 1987. Childhood leukaemia in the West Berkshire and Basingstoke and North Hampshire District Health Authorities in relation to nuclear establishments in the vicinity. Br. Med. J. 294:597-602.

Sh86 Shrimpton, P. C., B. F. Wall, D. G. Jones, E. S. Fisher, M. C. Hillier, G. M. Kendall, and R. M. Harrison. 1986. Doses to patients from routine diagnostic x-ray examinations in England. Br. J. Radiol. 59:749-758.

Sm86 Smith, P. G., and A. J. Douglas. 1986. Mortality of workers at the Sellafield plant of British Nuclear Fuels. Br. Med. J. 293:845-854.

St62 Stewart, A., W. Pennypacker, and R. Barber. 1962. Adult leukemia and diagnostic x-rays. Br. Med. J. 2:882.

St73a Stewart, A. 1973. An epidemiologist takes a look at radiation risks. DHEW Publ. No. [FDA] 738024, BRH/DBE 73-2. Washington, D.C.: U.S. Department of Health, Education, and Welfare.

St73b Stewart, A. 1973. The carcinogenic effects of low level radiation. A re-appraisal of epidemiologic methods and observations. Health Phys. 24:223-240.

Ta86 Tao, Z., and L. Wei. 1986. An epidemiological investigation of mutational diseases in the high background radiation area of Yangjiang, China. J. Radiat. Res. 27:141-150.

UN82 United Nations Scientific Committee on the Effects of Atomic Radiations (UNSCEAR). 1982. Ionizing Radiation: Souces and Biological Effects. United Nations publication. E.82.IX.8 06300p. New York: United Nations, 773 pp.

We86 Wei, L. X., Y. R. Zha, Z. F. Tao, et al. 1986. Recent advances of health survey in high background areas in Yangjiang, China. Pp. 1-17 in Proceedings of the International Symposium on Biological Effects of Low Level Radiation.

Wi87 Wilkinson, G.S., G. L. Tietjen, L. D. Wiggs, W. A. Galke, J. F. Acquavella, M. Reyes, G. L. Voelz, and R. J. Waxweiler. 1987. Mortality among plutonium and other radiation workers at a plutonium weapons facility. Am. J. Epidemiol. 125:231-250.

Glossary

Absolute risk. An expression of excess risk based on the assumption that the excess risk from radiation exposure *adds* to the underlying (baseline) risk by an increment dependent on dose but independent of the underlying natural risk.

Absorbed dose. The mean energy imparted by ionizing radiation to an irradiated medium per unit mass. Units: gray (Gy), rad.

Activity. The mean number of decays per unit time of a radioactive nuclide. Units: becquerel (Bq), curie (Ci).

Additive interaction model (AIM). The assumption that the total risk from exposures to radiation and to another risk factor is equal to the sum of the excess risks from the two taken separately.

Adenosarcoma. A mixed tumor which consists of a substance like embryonic connective tissue together with glandular elements.

Alpha particle. Two neutrons and two protons bound as a single particle that is emitted from the nucleus of certain radioactive isotopes in the process of decay or disintegration.

Aneuploid. Having numbers of chromosomes not equal to exact multiples of the haploid number. Down syndrome is an example.

Ankylosing spondylitis. Arthritis of the spine.

Ataxia telangiectasia (AT). An inherited disorder associated with an increased risk of cancer, lymphoma in particular, and characterized by immunologic, chromosomal, and DNA defects.

Background radiation. The amount of radiation to which a member of the population is exposed from natural sources, such as terrestrial radiation due to naturally occurring radionuclides in the soil, cosmic radiation originating in outer space, and naturally occurring radionuclides deposited in the human body.

Baseline rate. The cancer experience observed in a population in the absence of the specific agent being studied; the baseline rate might, however, include cancers from a number of other causes, such as smoking, background radiation, etc.

Becquerel (Bq). SI unit of activity. (see Units)

BEIR III. Refers to the third National Research Council's Committee on *B*iological *E*ffects of *I*onizing *R*adiation, as well as to the report published by this committee in 1980.

Beta particle. A charged particle emitted from the nucleus of certain unstable atomic nuclei (radioactive elements), having the charge and mass of an electron.

Cancer. A malignant tumor of potentially unlimited growth, capable of invading surrounding tissue or spreading to other parts of the body by metastasis.

Carcinogen. An agent that may cause cancer. Ionizing radiations are physical carcinogens; there are also chemical and biologic carcinogens and biologic carcinogens may be external (e.g., viruses) or internal (genetic defects).

Carcinoma. A malignant tumor (cancer) of epithelial origin.

Case-control study. An epidemiological study in which people with disease and a similarly composed group of people without disease are compared in terms of exposures to a putative causative agent.

Cell culture. The growing of cells in vitro (a glass container) in such a manner that the cells are no longer organized into tissues.

Chromosonal nondisjunction. Either a gain or a loss of chromosomes that occurs when cell division leading to either egg or sperm production goes awry. This results in aneupoidy.

Cohort study. Or follow-up study; an epidemiological study in which groups of people are identified with respect to the presence or absence of exposure to a disease-causing agent and the outcomes in terms of disease rates are compared.

Competing risks. Other causes of death which affect the value of the risk being studied. Persons dying from other causes are not at risk of dying from the factor in question.

Confidence interval. A measure of the reliability of a risk estimate. A 90% confidence interval means that 9 times out of 10 the estimated risk would be within the specified interval.

Curie. (Ci). A unit of activity equal to 3.7×10^{10} disintegrations/s. (see Units)

DNA. Deoxyribonucleic acid; the genetic material of cells.

Dominant mutation. The mutation is dominant if it produces its effect in the presence of an equivalent normal gene from the other parent.

Dose. See absorbed dose.

Dose-distribution factor. A factor which accounts for modification of the dose effectiveness in cases in which the radionuclide distribution and the resultant dose are nonuniform.

Dose-effect (dose-response) model. A mathematical formulation of the way the effect (or biological response) depends on dose.

Dose equivalent. A quantity that expresses, for the purposes of radiation protection and control, an assumed equal biological effectiveness of a given absorbed dose on a common scale for all kinds of ionizing radiation. SI unit is the Sievert. (see Units)

Dose rate. The quantity of absorbed dose delivered per unit time.

Dose Rate Effectiveness Factor (DREF). A factor by which the effect caused by a specific dose of radiation changes at low as compared to high dose rates.

Doubling dose. The amount of radiation needed to double the natural incidence of a genetic or somatic anomaly.

Electron volt. (eV). A unit of energy = 1.6×10^{-12} ergs–1.6×10^{-19} J; 1 eV is equivalent to the energy gained by an electron in passing through a potential difference of 1 V; 1 keV–1,000 eV; 1 MeV–1,000,000 eV.

Epidemiology. The study of the determinants of the frequency of disease in man. The two main types of epidemiological studies of chronic disease are cohort (or follow-up) studies and case control (or retrospective) studies.

Etiology. The science or description of cause(s) of disease.

Euploid. Having uniform exact multiples of the haploid number of chromosomes.

Fallout. Radioactive debris from a nuclear detonation or other source, usually deposited from air-borne particulates.

Fluoroscopy. A method of visualizing internal structures by directing x rays through an object (e.g., part of the body) onto a fluorescent screen.

Fractionation. The delivery of a given total dose of radiation as several smaller doses, separated by intervals of time.

Gamma radiation. Also gamma rays; short wavelength electromagnetic radiation of nuclear origin, similar to x rays but usually of higher energy (100 keV to 9 MeV).

Geometric mean. The geometric mean of a set of positive numbers is the exponential of the arithmetic mean of their logarithms. The geometric mean of a lognormal distribution is the exponential of the mean of the associated normal distribution.

Geometric standard deviation (GSD). The geometric standard deviation of a lognormal distribution is the exponential of the standard deviation of the associated normal distribution.

Gray (Gy). SI unit of absorbed dose. (see Units)

Half-life, biologic. Time required for the body to eliminate half of an administered dose of any substance by regular processes of elimination; it is approximately the same for both stable and radioactive isotopes of a particular element.

Half-life, radioactive. Time required for a radioactive substance to lose 50% of its activity by decay.

ICDA. The *I*nternational *C*lassification of *D*iseases *A*dapted for use in the U.S. The ICD is periodically revised by the World Health Organization; the 8th ICDA is adapted from the 8th ICD and was issued in 1972.

Incidence. Or incidence rate; the rate of occurrence of a disease within a specified period of time, often expressed as number of cases per 100,000 individuals per year.

In utero. In the womb, i.e., before birth.

In vitro. (Literally, in glass), in culture or in the test-tube (as opposed to in vivo, in the living individual).

In vivo. In the living organism.

Ionizing radiation. Radiation sufficiently energetic to dislodge electrons from an atom. Ionizing radiation includes x and gamma radiation, electrons (beta radiation), alpha particles (helium nuclei), and heavier charged atomic nuclei. Neutrons ionize indirectly by colliding with atomic nuclei.

Isotopes. Nuclides that have the same number of protons in their nuclei, and hence the same atomic number, but that differ in the number of neutrons, and therefore in the mass number; chemical properties of isotopes of a particular element are almost identical.

Kerma. *K*enetic *E*nergy *R*eleased in *Ma*terial. A unit of exposure, expressed in rad, that represents the kinctic energy transferred to charged particles per unit mass of irradiated medium when indirectly ionizing (uncharged) particles, such as photons or neutrons, traverse the medium. If all of the kinetic energy is absorbed "locally," the kerma is equal to the absorbed dose.

Latent period. The period of time between exposure and expression of the disease. After exposure to a dose of radiation, there is a delay in several years (the minimum latent period) before any cancers are seen.

Life-span study (LSS). Life-span study of the Japanese atomic-bomb survivors; the sample consists of 120,000 persons, of whom 82,000 were exposed to the bombs, mostly at low doses.

Life table. A table showing the number of persons who, of a given number born or living at a specified age, live to attain successive higher ages, together with the numbers who die in each age interval.

Linear energy transfer (LET). Average amount of energy lost per unit track length.

Low LET. Radiation characteristic of light charged particles such as electrons produced by x rays and gamma rays where the distance between ionizing events is large on the scale of a cellular nucleus.

High LET. Radiation characteristic of heavy charged particles such as protons and alpha particles where the distance between ionizing events is small on the scale of a cellular nucleus.

Linear (L) model. Also, linear dose-effect relationship; expresses the effect (e.g., mutation or cancer) as a direct (linear) function of dose.

Linear-quadratic (LQ) model. Also, linear-quadratic dose-effect relationship; expresses the effect (e.g., mutation or cancer) as partly directly proportional to the dose (linear term) and partly proportional to the square of the dose (quadratic term). The linear term will predominate at lower doses, the quadratic term at higher doses.

Lymphosarcoma. A sarcoma of the lymphoid tissue. This does not include Hodgkin's disease.

Monosomy. The absence of one chromosome from the complement of an otherwise diploid cell.

Monte Carlo Calculation. The evaluation of a probability distribution by means of random sampling.

Mortality (rate). The rate to which people die from a disease, e.g., a specific type of cancer, often expressed as number of deaths per 100,000 per year.

Multiplicative interaction model (MIM). The assumption that the relative risk (the relative excess risk plus one) resulting from the exposure to two risk factors is the product of the relative risks from the two factors taken separately.

Neoplasms. Any new and abnormal growth, such as a tumor; neoplastic disease refers to any disease that forms tumors, whether malignant or benign.

Neutron. Uncharged subatomic particle capable of producing ionization in matter by collision with charged particles.

Nonstochastic. Describes effects whose severity is a function of dose; for these, a threshold may occur; some nonstochastic somatic effects are cataract induction, nonmalignant damage to skin, hematological deficiencies, and impairment of fertility.

Nuclide. A species of atom characterized by the constitution of its nucleus, which is specified by its atomic mass and atomic number (Z), or by its number of protons (Z), number of neutrons (N), and energy content.

Oncogenes. Genes which carry the potential for cancer.

Person-gray. Unit of population exposure obtained by summing individual dose-equipment values for all people in the exposed population. Thus, the number of person-grays contributed by 1 person exposed to 1 Gy is equal to that contributed by 100,000 people each exposed to 10 μGy.

Person-years-at-risk (PYAR). The number of persons exposed times the number of years after exposure minus some lag period during which the dose is assumed to be unexpressed (minimum latent period).

Prevalence. The number of cases of a disease in existence at a given time per unit population, usually 100,000 persons.

Probability of causation. A number that expresses the probability that a given cancer, in a specific tissue, has been caused by a previous exposure to a carcinogenic agent, such as radiation.

Progeny. The decay products resulting after a series of radioactive decay. Progeny can also be radioactive, and the chain continues until a stable nuclide is formed.

Projection model. A mathematical model that simultaneously described the excess cancer risk at different levels of some factor such as dose, time after exposure, or baseline level of risk, in terms of a parametric function of that factor. It becomes a projection model when data in a particular range of observations is used to assign values to the parameters in order to estimate (or project) excess risk for factor values outside that range.

Promoter. An agent which is not by itself carcinogenic, but which can amplify the effect of a true carcinogen by increasing the probability of late-stage cellular changes needed to complete the carcinogenic process.

Protraction. The spreading out of a radiation dose over time by continuous delivery at a lower dose rate.

Quadratic-dose model. A model which assumes that the excess risk is proportional to the square of the dose.

Quality factor. (Q). An LET dependent factor by which absorbed doses are multiplied to obtain (for radiation-protection purposes) a quantity which corresponds more closely to the degree of biological effect produced by x or low-energy gamma rays. Dose in Gy × Q = Dose equivalent in Sv.

Rad. A unit of absorbed dose. Replaced by the gray in SI units. (see Units)

Radioactivity. The property of some nuclides of spontaneously emitting particles or gamma radiation, emitting x radiation after orbital electron capture, or undergoing spontaneous fission.

Artificial radioactivity. Man-made radioactivity produced by fission, fusion, particle bombardment, or electromagnetic irradiation.

Natural radioactivity. The property of radioactivity exhibited by more than 50 naturally occurring radionuclides.

Radiogenic. Caused by radiation.

Radioisotopes. A radioactive atomic species of an element with the same atomic number and usually identical chemical properties.

Radionuclide. A radioactive species of an atom characterized by the constitution of its nucleus.

Radiosensitivity. Relative susceptibility of cells, tissues, organs, and organisms to the injurious action of radiation; radiosensitivity and its antonym, radioresistance, are used in a comparative sense rather than an absolute one.

Recessive gene disorder. This requires that a pair of genes, one from each parent, be present in order for the disease to be manifest. An example is cystic fibrosis.

Relative biological effectiveness (RBE). Biological potency of one radiation as compared with another to produce the same biological endpoint. It is numerically equal to the inverse of the ratio of absorbed doses of the two radiations required to produce equal biological effect. The reference radiation is often 200-kV x rays.

Relative risk. An expression of excess risk relative to the underlying (baseline) risk; if the excess equals the baseline risk the relative risk is 2.

Rem. (*r*ad *e*quivalent, *m*an); unit of dose equivalent. The dose equivalent in "rem" is numerically equal to the absorbed dose in "rad" multiplied by the "quality factor" (see Quality factor), the distribution factor and any other necessary modifying factor.

RERF. *R*adiation *E*ffects *R*esearch *F*oundation; a binationally funded Japanese foundation chartered by the Japanese Welfare Ministry under an agreement between the the U.S.A. and Japan. The RERF is the successor to the ABCC (Atomic Bomb Casualty Commission).

Risk coefficient. The increase in the annual incidence or mortality rate per unit dose: (1) absolute risk coefficient is the observed minus the expected number of cases per person year at risk for a unit dose; (2) the relative-risk coefficient is the fractional increase in the baseline incidence or mortality rate for a unit dose.

Risk estimate. The number of cases (or deaths) that are projected to occur in a specified exposed population per unit dose for a specified exposure regime and expression period: number of cases per person-gray or, for radon, the number of cases per person cumulative working-level month.

Rem. A unit of dose equivalent. Replaced by the sievert. (see Units)

Sarcoma. A malignant growth arising in tissue of mesodermal origin (connective tissue, bone, cartilage or striated muscle).

Sex-linked mutation (or X-linked). A mutation associated with the X chromosome. It will usually only manifest its effect in males (who have only a single X chromosome).

SI units. The International System of Units as defined by the General Conference of Weights and Measures in 1960. These units are generally based on the meter/kilogram/second units, with special quantities for radiation including the becquerel, gray, and sievert.

Sievert. The SI unit of radiation dose equivalent. It is equal to dose in grays times a quality factor times other modifying factors, for example, a distribution factor; 1 sievert (Sv) equals 100 rem.

Specific activity. Total activity of a given nuclide per gram of a compound, element, or radioactive nuclide.

Specific energy. The actual energy per unit mass deposited per unit volume in a given event. This is a stochastic quantity as opposed to the average value over a large number of instances (i.e., the absorbed dose).

Spline. A curve of predetermined shape; a spline with 1 knot has a single inflection point and thus two different segments.

Squamous cell carcinoma. A cancer composed of cells that are scaly or platelike.

Standard mortality ratio (SMR). Standard mortality ratio is the ratio of the disease or accident mortality rate in a certain specific population compared with that in a standard population. The ratio is based on 100 for the standard so that an SMR of 200 means that the test population has twice the mortality from that particular cause of death.

Stochastic. Random events leading to effects whose probability of occurrence in an exposed population (rather than severity in an affected individual) is a direct function of dose; these effects are commonly regarded as having no threshold; hereditary effects are regarded as being stochastic; some somatic effects, especially carcinogenesis, are regarded as being stochastic.

Target theory (hit theory). A theory explaining some biological effects of radiation on the basis that ionization, which occurs in a discrete volume (the target) within the cell, directly causes a lesion that later results in a physiological response to the damage at that location; one, two, or more hits (ionizing events within the target) may be necessary to elicit the response.

Threshold hypothesis. The assumption that no radiation injury occurs below a specified dose.

Time-since-exposure (TSE) model. A model in which the risk is not constant but varies with the time after exposure.

Transformed cells. Tissue culture cells changed in vitro from growing in an orderly pattern and exhibiting contact inhibition to growing in a pattern more like that of cancer cells, due to the loss of contact inhibition.

Transolocation. A chromosome aberration resulting from chromosome breakage and subsequent structural rearrangement of the parts between the same or different chromosomes.

Trisomy. The presence of an additional (third) chromosome of one type in an otherwise diploid cell.

Tumorigenicity. Ability of cells to proliferate into tumors when inoculated into a specified host organism under specified conditions.

Units[a]	Conversion Factors
Becquerel (SI)	1 disintegration/s = 2.7×10^{-11} Ci
Curie	3.7×10^{10} disintegrations/s = 3.7×10^{10} Bq
Gray (SI)	1 J/kg – 100 rad
Rad	100 erg/g – 0.01 Gy
Rem	0.01 Sievert
Sievert (SI)	100 rem

[a] International Units are designated SI.

UNSCEAR. *U*nited *N*ations *S*cientific *C*ommittee on the *E*ffects of *A*tomic *R*adiation publishes periodic reports on sources and effects of ionizing radiation.

x radiation. Also x rays; penetrating electromagnetic radiation, usually produced by bombarding a metallic target with fast electrons in a high vacuum.

Xeroderma pigmentosum (XP). An inherited disease in which skin cells are highly susceptible to sun-induced cancer; XP cells have a defect in DNA repair after ultraviolet irradiation which apparently accounts for the propensity for this neoplasm.

Index

B

E

H

N

Q

R

S

T

U

V

W

X

Y